Ship Hydrostatics and Stability

Second Edition

Ship Hydrostatics and Stability

Second Edition

Adrian B. Biran
Technion, Faculty of Mechanical Engineering

Rubén López-Pulido
Former Representative of Spain to the IMO

With contributions by
Javier de Juana Gamo

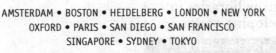

AMSTERDAM • BOSTON • HEIDELBERG • LONDON • NEW YORK
OXFORD • PARIS • SAN DIEGO • SAN FRANCISCO
SINGAPORE • SYDNEY • TOKYO
Butterworth-Heinemann is an imprint of Elsevier

Butterworth-Heinemann is an imprint of Elsevier
The Boulevard, Langford Lane, Kidlington, Oxford OX5 1GB, UK
225 Wyman Street, Waltham, MA 02451, USA

First edition 2003
Second edition 2014

Notice
No responsibility is assumed by the publisher for any injury and/or damage to persons or property as a matter of products liability, negligence or otherwise, or from any use or operation of
any methods, products, instructions or ideas contained in the material herein. Because of rapid
advances in the medical sciences, in particular, independent verification of diagnoses and drug
dosages should be made

British Library Cataloguing in Publication Data
A catalogue record for this book is available from the British Library

Library of Congress Cataloging-in-Publication Data
A catalog record for this book is available from the Library of Congress

ISBN–13: 978-0-08-098287-8

For information on all Butterworth-Heinemann publications
visit our web site at books.elsevier.com

Printed and bound in the US
13 14 15 16 17   10 9 8 7 6 5 4 3 2 1

Working together
to grow libraries in
developing countries

www.elsevier.com • www.bookaid.org

Dedication

To my wife Suzi.

Adrian

To the Neapolitan professor Giulio Russo-Krauss (1940–2010), as he was and always will be for me a 'Scuola di Vita', and to my son, Ventura (2011), as I wish I could be in the future the same for him.

Rubén

Contents

Biography

Adrian Biran

Adrian Biran received a *Cum laude* Diploma of Engineering, in the field of ship engineering, from the Bucharest Polytechnic Institute, and Master of Science and Doctor of Science degrees from the TECHNION—Israel Institute of Technology, with theses related to software for Marine Engineering and Naval Architecture. After graduating he worked as Design Engineer, Chief of Department and Project Leader at IPRONAV—the Institute for Ship Design, Bucharest. In continuation he managed the design office of the Bucharest Studios and worked as Project Leader at IPA—the Institute for Automation Design, Bucharest. In Israel he worked as Senior Engineer at the Israel Shipyards, Haifa, and as Research Fellow and Research Engineer at the Technion R&D Foundation. In parallel with the above activities Adrian Biran served as project instructor at the Military Technical Academy, Bucharest, and at the Beersheba University. He has been for many years an adjunct teacher at the TECHNION—Israel University of Technology, since 1995 as Adjunct Associate Professor. Adrian Biran is the author of several technical papers on various subjects including Naval Architecture. He wrote in Romanian a book of popular science about ships. Together with Moshe Breiner, Adrian wrote a book on *MATLAB for Engineers* that was published in three English, three German, two French, and two Greek editions. In 2003 he published with Butterworth-Heinemann, a book on *Ship Hydrostatics and Stability* that was immediately translated into Turkish. In 2005 the book was reprinted with contributions by Rubén López Pulido. A chapter on stability regulations was reproduced in the Maritime Engineering Reference Book edited by Anthony Molland. In 2011 he published *What every Engineer should know about MATLAB and Simulink*.

Rubén López-Pulido

Rubén López-Pulido is a Member of the Royal Institution of Naval Architects (RINA), UK, and the Spanish Association of Naval Architects (AINE). He is former Maritime Attaché of the Embassy of Spain in London and Representative of Spain to the International Maritime Organization (IMO) of UN. He was Naval Architect and Software Engineer at SENER working as developer of Naval Architecture Modules of FORAN System, a CAD/CAM/CAE software for ship design and shipbuilding. He was Hydrodynamicist and Stability of Ships

Researcher at the CEHINAV (ETSIN-UPM) Madrid Ship Model Basin. In 2010 he received the Spanish National Best Outstanding Career Award for naval architects under 35 years of age, to recognize outstanding professional service and contributions to the maritime and shipping business, naval architecture, and marine engineering community. He holds an MSc in Maritime Technology (ETSIN-UPM), an MPhil in Moral and Political Philosophy (UNED), and an MA in International Security and Strategic Studies (UGR). He is an executive postgraduate of the London School of Economics and Political Science (LSE) and the IESE Business School. He belongs since 2009 to the civil servant's Senior Executive Service of the Spanish Government, and since 2012 he is Head of the Emergency Coordination and Crisis Management Unit for Transport and Infrastructures of the Spanish Government.

Preface to the Second Edition

Six years have passed since the issue of the enlarged reprint of the book. New intact and damage stability regulations have been adopted in the meantime, mainly by IMO, but also by the German Navy. While in the past the regulations were prescriptive and based on deterministic models, the new orientation turns towards goal-based and risk-assessment approaches. New ship forms increased the vulnerability to parametric roll and the occurrence of large roll angles and loss of containers have been frequently reported. Extensive research is carried on for a better understanding of this phenomenon, as well as of not-yet fully understood capsizing modes, such as dead ship condition, pure loss of stability and broaching-to. One aim of the research is to develop so-called second-generation criteria of stability. As it is recognized now that stability depends not only on the design of ships, but also on their loading and operation, as well as on environmental conditions, another aim of the research, and of IMO, is to issue guidance documentation for ship masters. The old deterministic approach to damage calculations has been replaced in large part by the probabilistic approach. Nevertheless, as the old mariners' saying states, 'There is always stormy weather ahead,' the accident of the cruise liner *Costa Concordia*, in 2012, will trigger new changes as it has unveiled new challenges for Naval Architects and experts in maritime regulations.

All these developments made a new edition necessary. We updated the sections that have become obsolete and inserted the highlights of the recent regulations and research results. In doing so we are taking advantage of the fact that our contributor Javier de Juana has been attending the IMO meetings during years, and one of the authors has been Permanent Delegate as Spanish representative to the IMO. In addition, we corrected errors and added a few exercises and explanations that proved useful during the lectures delivered by one of the authors at the Technion.

We are pleased to thank those who helped us in our endeavour. While translating the previous edition into Turkish, Professor Hüseyin Yilmaz reported several errors. Thomas Wardecki and Andreas Rinke provided details on the German-Navy regulations presently in force. We thank Miguel Palomares and Lorenzo Mayol of IMO for their help, and Luis Pérez-Rojas, Leonardo Fernández-Jambrina, Antonio Rodriguez, Jesús Valle, Antonio Souto, and Jorge Vicario for providing important insight for some chapters and the cover figure, a modification of the hull forms of the DTMB combatant 5415. We acknowledge the courtesy of Luis García Bernáldez and Verónica Alonso of Sener who allowed us to describe some of the main features of the FORAN computer system. We thank The Mathworks for their continuing support and permit to use their marvelous and powerful software throughout the book. This second edition was made possible only by the dedicated work of Hayley Gray, Charlie Kent and Susan Li, all of Elsevier, UK.

Finally, the authors want to thank their wives, Suzi and Noelia, for their patience, understanding and forgiveness for the time stolen from that due to their families.

Adrian Biran and Rubén López-Pulido, 2013

Preface to the First Reprint

Using the book in two consecutive academic years we discovered several typos and errors. They are corrected in this reprint and the author thanks those students, and especially Eyal Lahav, who have read the book with attention and transmitted their comments.

Several Naval Architects involved in university education or in maritime legislation sent comments and suggestions. Bertram Volker corrected orthographical errors in German terms. Lawrence Doctors recommended to insert a theorem regarding wall-sided floating bodies with negative initial metacentric height. Dan Livneh drew the attention of the author to the new approach of classification societies to the parametric roll of container ships. The most extensive contributions are due to Rubén López-Pulido who corrected a few examples, transmitted updated information on IMO, volunteered to add the Spanish translations of important technical terms and prepared a Spanish index for the end of the book. All these contributions are implemented in this reprint.

Additional software was included on the companion sites of this book. Short descriptions appear in an Appendix at the end of the book.[1]

The author acknowledges the continuous support of The Mathworks, and personally that of Courtney Esposito who provided the latest updates of MATLAB.

Finally, the author thanks Jonathan Simpson and Miranda Turner for their encouragement to update the book and for their editorial help.

Adrian Biran, 2005

---

[1]In this edition not included in the book, but on the website of the book.

Preface

This book is based on a course of Ship Hydrostatics delivered during a quarter of a century at the Faculty of Mechanical Engineering of the Technion–Israel Institute of Technology. The book reflects the author's own experience in design and R&D and incorporates improvements based on feedback received from students.

The book is addressed in the first place to undergraduate students for whom it is a first course in Naval Architecture or Ocean Engineering. Many sections can be also read by technicians and ship officers. Selected sections can be used as reference text by practising Naval Architects.

Naval Architecture is an age-old field of human activity and as such it is much affected by tradition. This background is part of the beauty of the profession. The book is based on this tradition but, at the same time, the author tried to write a modern text that considers more recent developments, among them the theory of *parametric resonance*, also known as *Mathieu effect*, the use of personal computers, and new regulations for intact and damage stability.

The Mathieu effect is believed to be the cause of many marine disasters. German researchers were the first to study this hypothesis. Unfortunately, in the first years of their research they published their results in German only. The German Federal Navy—*Bundesmarine*—elaborated stability regulations that allow for the Mathieu effect. These regulations were subsequently adopted by a few additional navies. Proposals have been made to consider the effect of waves for merchant vessels too.

Very powerful personal computers are available today; their utility is enhanced by many versatile, user-friendly software packages. PC programmes for hydrostatic calculations are commercially available and their prices vary from several hundred dollars, for the simplest, to many thousands for the more powerful. Programmes for particular tasks can be written by a user familiar with a good software package. To show how to do it, this book is illustrated with a few examples calculated in Excel and with many examples written in MATLAB. MATLAB is an increasingly popular, comprehensive computing environment characterized by an interactive mode of work, many built-in functions, immediate graphing facilities and easy programming paradigms. Readers who have access to MATLAB, even to the Students' Edition, can readily use those examples. Readers who do not work in MATLAB can convert the examples to other programming languages.

Several new stability regulations are briefly reviewed in this book. Students and practising Naval Architects will certainly welcome the description of such rules and examples of how to apply them.

About this book

Theoretical developments require an understanding of basic calculus and analytic geometry. A few sections employ basic vector calculus, differential geometry or ordinary differential equations. Students able to read them will gain more insight into matters explained in the book. Other readers can skip those sections without impairing their understanding of practical calculations and regulations described in the text.

Chapter 1 introduces the reader to basic terminology and to the subject of hull definition. The definitions follow new ISO and ISO-based standards. Translations into French, German and Italian are provided for the most important terms.

The basic concepts of hydrostatics of floating bodies are described in Chapter 2; they include the conditions of equilibrium and initial stability. By the end of this chapter the reader knows that hydrostatic calculations require many integrations. Methods for performing such integrations in Naval Architecture are developed in Chapter 3.

Chapter 4 shows how to apply the procedures of numerical integration to the calculation of actual hydrostatic properties. Other matters covered in the same chapter are a few simple checks of the resulting plots, and an analysis of how the properties change when a given hull is subjected to a particular class of transformations, namely the properties of affine hulls.

Chapter 5 discusses the statical stability at large angles of heel and the curve of statical stability.

Simple models for assessing the ship stability in the presence of various heeling moments are developed in Chapter 5. Both static and dynamic effects are considered, as well as the influence of factors and situations that negatively affect stability. Examples of the latter are displaced loads, hanging loads, free liquid surfaces, shifting loads, and grounding and docking. Three subjects closely related to practical stability calculations are described in Chapter 7: weight and trim calculations and the inclining experiment.

Ships and other floating structures are approved for use only if they comply with pertinent regulations. Regulations applicable to merchant ships, ships of the US Navy and UK Navy, and small sail or motor craft are summarily described in Chapter 8.

The phenomenon of parametric resonance, or Mathieu effect, is briefly described in Chapter 9. The chapter includes a simple criterion of distinguishing between stable and unstable solutions and examples of simple simulations in MATLAB.

Ships of the German Federal Navy are designed according to criteria that take into account the Mathieu effect: they are introduced in Chapter 10.

Chapters 8 and 10 deal with intact ships. Ships and some other floating structures are also required to survive after a limited amount of flooding. Chapter 11 shows how to achieve this goal by subdividing the hull by means of watertight bulkheads. There are two methods of calculating the ship condition after damage, namely the method of lost buoyancy and the method of added weight. The difference between the two methods is explained by means of a simple example. The chapter also contains short descriptions of several regulations for merchant and for naval ships.

Chapters 8, 10 and 11 inform the reader about the existence of requirements issued by bodies that approve the design and the use of ships and other floating bodies, and show how simple models developed in previous chapters are applied in engineering calculations. Not all the details of those regulations are included in this book, neither all regulations issued all over the world. If the reader has to perform calculations that must be submitted for approval, it is highly recommended to find out which are the relevant regulations and to consult the complete, most recent edition of them.

Chapter 12 goes beyond the traditional scope of Ship Hydrostatics and provides a bridge towards more advanced and realistic models. The theory of linear waves is briefly introduced and it is shown how real seas can be described by the superposition of linear waves and by the concept of spectrum. Floating bodies move in six degrees of freedom and the spectrum of those motions is related to the sea spectrum. Another subject introduced in this chapter is that of tank stabilizers, a case in which surfaces of free liquids can help in reducing the roll amplitude.

Chapter 13 is about the use of modern computers in hull definition, hydrostatic calculations and simulations of motions. The chapter introduces the basic concepts of Computer Graphics and illustrates their application to hull definition by means of the MultiSurf and SurfaceWorks packages. A roll simulation in SIMULINK, a toolbox of MATLAB, exemplifies the possibilities of modern simulation software.

Using this book

Boldface words indicate a key term used for the first time in the text, for instance **length between perpendiculars**. Italics are used to emphasize, for example *equilibrium of moments*. Vectors are written with a line over their name: $\overline{KB}$, $\overline{GM}$. Listings of MATLAB programmes, functions and file names are written in typewriter characters, for instance `mathisim.m`.

Basic ideas are exemplified on simple geometric forms for which analytic solutions can be readily found. After mastering these ideas the students should practise on real ship data provided in examples and exercises, at the end of each chapter. The data of an existing vessel, called *Lido 9*, are used throughout the book to illustrate the main concepts. Data of a few other real-world vessels are given in additional examples and exercises.

I am closing this preface by paying a tribute to the memory of those who taught me the profession, Dinu Ilie and Nicolae Pârâianu, and of my colleague in teaching, Pinchas Milch.

Acknowledgments

The first acknowledgements should certainly go to the many students who took the course from which emerged this book. Their reactions helped in identifying the topics that need more explanations. Naming a few of those students would imply the risk of being unfair to others.

Many numerical examples were calculated with the aid of the programme system ARCHIMEDES. The TECHNION obtained this software by the courtesy of Heinrich Söding, then Professor at the Technical University of Hannover, now at the Technical University of Hamburg. Included with the programme source there was a set of test data that describe a vessel identified as *Ship No*. 83074. Some examples in this book are based on that data.

Sol Bodner, coordinator of the Ship Engineering Program of the Technion, provided essential support for the course of Ship Hydrostatics. Itzhak Shaham and Jack Yanai contributed to the success of the programme.

Paul Münch provided data of actual vessels and *Lido Kineret, Ltd* and the *Özdeniz Group, Inc.* allowed us to use them in numerical examples. Eliezer Kantorowitz read initial drafts of the book proposal. Yeshayahu Hershkowitz, of Lloyd's Register, and Arnon Nitzan, then student in the last graduate year, read the final draft and returned helpful comments. Reinhard Siegel, of AeroHydro, provided the drawing on which the cover of the book is based, and helped in the application of MultiSurf and SurfaceWorks.

Richard Barker drew the attention of the author to the first uses of the term Naval Architecture. Our common love for the history of the profession enabled a pleasant and interesting dialogue.

Naomi Fernandes of MathWorks, Baruch Pekelman, their agent in Israel, and his assistants enabled the author to use the latest MATLAB developments.

The author thanks Addison-Wesley Longman, especially Karen Mosman and Pauline Gillet, for permission to use material from the book *MATLAB for Engineers* written by him and Moshe Breiner.

And finally the author thanks the editors of Elsevier Science, Rebecca Hamersley, Rebecca Rue, Sallyann Deans and Nishma Shah for their cooperation and continuous help. It was the task of Nishma Shah to bring the project into production.

Adrian Biran, 2003

Definitions, Principal Dimensions

1.1 Introduction

The subjects treated in this book are the basis of the profession called **Naval Architecture**. The term *Naval Architecture* comes from the titles of books published in the 17th century. For a long time the oldest such book we were aware of was Joseph Fursttenbach's *Architectura Navalis* published in Frankfurt in 1629. The bibliographical data of a beautiful reproduction are included in the references listed at the end of this book. Close to 1965 an older Portuguese manuscript was rediscovered in Madrid, in the Library of the Royal Academy of History. The work is due to João Baptista Lavanha and is known as *Livro Primeiro da Architectura Naval*, that is "First book on Naval Architecture." The traditional dating of the manuscript is 1614. The following is a quotation from a translation due to Richard Barker:

> *"Architecture consists in building, which is the permanent construction of any thing. This is done either for defence or for religion, and utility, or for navigation. And from this partition is born the division of Architecture into three parts, which are Military, Civil and Naval Architecture.*
>
> *...*
>
> *And Naval Architecture is that which with certain rules teaches the building of ships, in which one can navigate well and conveniently."*

The term may be still older. Thomas Digges (English, 1546–1595) published in 1579 an *Arithmeticall Militarie Treatise*, named *Stratioticos* in which he promised to write a book on "Architecture Nautical." He did not do so. Both the British Royal Institution of Naval Architects—RINA—and the American Society of Naval Architects and Marine Engineers—SNAME—opened their web sites for public debates on a modern definition of Naval Architecture. Out of the many proposals appearing there, that provided by A. Blyth, FRINA, looked to us both concise and comprehensive:

> *"Naval Architecture is that branch of engineering which embraces all aspects of design, research, developments, construction, trials, and effectiveness of all forms of man-made vehicles which operate either in or below the surface of any body of water."*

If Naval Architecture is a branch of engineering, what is engineering? In the New Encyclopedia Britannica (1989) we find:

> *"Engineering is the professional art of applying science to the optimum conversion of the resources of nature to the uses of mankind. Engineering has been defined by the Engineers Council for Professional Development, in the United States, as the creative application of 'scientific principles to design or develop structures, machines ...' "*

This book deals with the scientific principles of Hydrostatics and Stability. These subjects are treated in other languages in books bearing titles such as *Ship theory* (for example Doyère, 1927; Godino, 1956; Mirokhin et al., 1989) or *Ship statics* (for example Hervieu, 1985; Godino, 1956). Further scientific principles to be learned by the Naval Architect include Hydrodynamics, Strength, Motions on Waves, and more. The "art of applying" these principles belongs to courses in ship design.

1.2 Marine Terminology

Like any other field of engineering, Naval Architecture has its own vocabulary composed of *technical terms*. While a word may have several meanings in common language, when used as a technical term, in a given field of technology, it has one meaning only. This enables unambigous communication within the profession, hence the importance of clear definitions.

The technical vocabulary of a people with long maritime tradition has pecularities of origins and usage. As a first important example in English let us consider the word **ship**; it is of Germanic origin. Indeed, to this day the equivalent Danish word is *skib*, the Dutch, *schep*, the German, *Schiff* (pronounce "shif"), the Norwegian *skip* (pronounce "ship"), and the Swedish, *skepp*. For mariners and Naval Architects a ship has a soul; when speaking about a ship they use the pronoun "she."

Another interesting term is **starboard**; it means the right-hand side of a ship when looking forward. This term has nothing to do with stars. Pictures of Viking vessels (see especially the Bayeux Tapestry) show that they had a steering board (paddle) on their right-hand side. In Norwegian a "steering board" is called "styri bord." In old English the Nordic term became "steorbord" to be later distorted to the present-day "starboard." The correct term should have been "steeringboard." German uses the exact translation of this term, "Steuerbord."

The left-hand side of a vessel was called *larboard*. Hendrickson (1997) traces this term to "lureboard," from the Anglo-Saxon word "laere" that meant empty, because the steerman stood on the other side. The term became "lade-board" and "larboard" because the ship could be loaded from this side only. Larboard sounded too much like starboard and could be confounded with this. Therefore, more than 200 years ago the term was changed to **port**. In fact, a ship with a steering board on the right-hand side can approach to port only with her left-hand side.

1.3 The Principal Dimensions of a Ship

In this chapter, we introduce the principal dimensions of a ship, as defined in the international standard ISO 7462 (1985). The terminology in this document was adopted by some national standards, for example the German standard DIN 81209-1. We extract from the latter publication the symbols to be used in drawings and equations, and the symbols recommended for use in computer programmes. Basically, the notation agrees with that used by **SNAME** and with the *ITTC Dictionary of Ship Hydrodynamics* (RINA, 1978). Much of this notation has been used for a long time in English-speaking countries.

Beyond this chapter, many definitions and symbols appearing in this book are derived from the above-mentioned sources. Different symbols have been in use in continental Europe, in countries with a long maritime tradition. Hervieu (1985), for example, opposes the introduction of Anglo-Saxon notation and justifies his attitude in the Introduction of his book. If we stick in this book to a certain notation, it is not only because the book is published in the UK, but also because English is presently recognized as the world's *lingua franca* and the notation is adopted in more and more national standards. As to spelling, we use the British one. For example, in this book we write "centre," rather than "center" as in the American spelling, "draught" and not "draft," and "moulded" instead of "molded."

To enable the reader to consult technical literature using other symbols, we shall mention the most important of them. For ship dimensions we do this in Table 1.1, where we shall give also translations into French and German of the most important terms, following mainly ISO 7462 and DIN 81209-1. In addition, Italian terms will be inserted and they conform to Italian technical literature, for example Costaguta (1981), and Spanish terms that conform, for

Table 1.1   Principal ship dimensions and related terminology

English Term	Symbol	Computer Notation	Translations
After (aft) perpendicular	*AP*		Fr perpendiculaire arrière, G Hinteres Lot, I perpendiclare addietro, S perpendicular de popa
Baseline	*BL*		Fr ligne de base, G Basis, I linea base, S línea base
Bow			Fr proue, l'avant, G Bug, I prora, prua, S proa
Breadth	*B*	B	Fr largeur, G Breite, I larghezza, S manga
Camber			Fr bouge, G Balkenbucht, I bolzone, S brusca
Centreline plane		CL	F plan longitudinal de symétrie, G Mittschiffsebene, I piano di simmetria, piano diametrale, S plano de crujía, crujía, plano diametral, plano de simetría
Depth	*D*	DEP	Fr creux, G Seitenhöhe, I altezza di costruzione, puntale, S puntal
Depth, moulded			Fr creux sur quille, G Seitenhöhe, I altezza di costruzione, S puntal de trazado (puntale)
Design waterline	*DWL*	DWL	Fr flottaison normale, G Konstruktionswasserlinie (*KWL*) I linea d'acqua del piano di costruzione, S línea de agua de diseño
Draught	*T*	T	Fr tirant d'eau, G Tiefgang, I immersione, S calado
Draught, aft	T_A	TA	Fr tirant d'eau arrière, G hinterer Tiefgang, I immersione a poppa, S calado a popa
Draught, amidships	T_M	TM	Fr tirant d'eau milieu, G mittleres Tiefgang, I immersione media, S calado en la maestra
Draught, extreme			Fr profondeur de carène hors tout, G größter Tiefgang, I pescaggio, S calado máximo
Draught, forward	T_F	TF	Fr tirant d'eau avant, G vorderer Tiefgang, I immersione a prora, S calado a proa
Draught, moulded			Fr profondeur de carène hors membres, S calado de trazado
Forward perpendicular	*FP*		Fr perpendiculaire avant, G vorderes Lot, I perpendicolare avanti, S perpendicular de proa

(continued)

Table 1.1 Continued.

English Term	Symbol	Computer Notation	Translations
Freeboard	f	FREP	Fr franc-bord, G Freibord, I franco-bordo, S francobordo
Heel angle	ϕ_s	HEELANG	Fr bande, gîte, G Krängungswinkel, I angolo di inclinazione trasversale, S ángulo de escora, escora
Length between perpendiculars	L_{pp}	LPP	Fr longueur entre perpendiculaires, G Länge zwischen den Loten, I lunghezza tra le perpendicolari S eslora entre perpendiculares
Length of waterline	L_{WL}	LWL	Fr longueur à la flottaison, G Wasserlinienlänge, I lunghezza di galleggiamento, S eslora en la flotación
Length overall	L_{OA}		Fr longueur hors tout, G Länge über alles, I lunghezza fuori tutto, S eslora máxima
Length overall submerged	L_{OS}		Fr longueur hors tout immergé, G Länge über alles unter Wasser. I lunghezza massima opera viva, S eslora máxima de la obra viva
Lines plan			Fr plan de formes, G Linienriß, I piano di costruzione, piano delle linee, S plano de formas
Load waterline	DWL	DWL	Fr ligne de flottaison en charge, G Konstruktionswasserlinie, I linea d'acqua a pieno carico, S línea de agua a máxima carga
Midships			Fr couple milieu, G Haupspant, I sezione maestra, S sección maestra
Moulded			Fr hors membres, G auf Spanten, I fuori ossatura, S fuera de miembros, de trazado
Port		P	Fr bâbord, G Backbord, I sinistra, S babor
Sheer			Fr tonture, G Decksprung, I insellatura, S arrufo
Starboard		S	Fr tribord, G Steuerbord, I dritta, S estribor
Station			Fr couple, G Spante, I ordinata, S cuaderna (del plano de formas)
Stern, poop			Fr arrière, poupe, G Hinterschiff, I poppa, S popa
Trim			Fr assiette, G Trimm, I differenza d'immersione, S assieto, trimado, diferencia de immersión
Waterline	WL	WL	Fr ligne d'eau, G Wasserlinie, I linea d'acqua, S linea de agua

example, to Aláez Zazurca (2004). The translations will be marked by "Fr" for French, "G" for German, "I" for Italian, and "S" for Spanish. Almost all ship hulls are symmetric with respect with a longitudinal plane (plane *xz* in Figure 1.6). In other words, ships present a "port-to-starboard" symmetry. The definitions take this fact into account. Those definitions are explained in Figures 1.1–1.4.

The outer surface of a steel or aluminum ship is usually not smooth because not all plates have the same thickness. Therefore, it is convenient to define the hull surface of such a ship on the inner surface of the plating. This is the **Moulded surface** of the hull. Dimensions measured to this surface are qualified as **Moulded**. By contrast, dimensions measured to the outer surface of the hull or of an appendage are qualified as **extreme**. The moulded surface is used in the first stages of ship design, before designing the plating, and in test-basin studies.

Figure 1.1   Length dimensions

Figure 1.2   How to measure the length between perpendiculars

Figure 1.3   The case of a keel not parallel to the load line

Figure 1.4   Breadth, depth, draught, and camber

The **baseline**, shortly BL, is a line lying in the longitudinal plane of symmetry and parallel to the designed summer load waterline (see next paragraph for a definition). It appears as a horizontal in the lateral view of the hull surface. The baseline is used as the longitudinal axis, that is the x-axis of the system of coordinates in which hull points are defined. Therefore, it is recommended to place this line so that it passes through the lowest point of the hull surface. Then, all z-coordinates will be positive.

Before defining the dimensions of a ship we must choose a reference waterline. ISO 7462 recommends that this **load waterline** be the **designed summer load line**, that is the waterline up to which the ship can be loaded, in sea water, during summer when waves are lower than in winter. The qualifier "designed" means that this line was established in some design stage. In later design stages, or during operation, the load line may change. It would be very inconvenient to update this reference and change dimensions and coordinates; therefore, the "designed" datum line is kept even if no more exact. A notation older than ISO 7462 is DWL, an abbreviation for "Design Waterline."

The **after perpendicular**, or **aft perpendicular**, noted AP, is a line drawn perpendicularly to the load line through the after side of the rudder post or through the axis of the rudder stock. The latter case is shown in Figures 1.1 and 1.3. For naval vessels, and today for some merchant ships, it is usual to place the AP at the intersection of the aftermost part of the moulded surface and the load line, as shown in Figure 1.2. The **forward perpendicular**, FP, is drawn perpendicularly to the load line through the intersection of the foreside of the stem with the load waterline. Mind the slight lack of consistency: while all moulded dimensions are measured to the moulded surface, the FP is drawn on the outer side of the stern. The distance between the after and the forward perpendicular, measured parallel to the load line, is called **length between perpendiculars** and its notation is L_{pp}. An older notation was *LBP*. We call **length overall**, L_{OA}, the length between the ship extremities. The **length overall submerged**, L_{OS}, is the maximum length of the submerged hull measured parallel to the designed load line.

We call **station** a point on the baseline, and the transverse section of the hull surface passing through that point. The station placed at half L_{pp} is called **midships**. It is usual to note the midship section by means of the symbol shown in Figure 1.5(a). In German literature we usually find the simplified form shown in Figure 1.5(b).

The **moulded depth**, D, is the height above baseline of the intersection of the underside of the deck plate with the ship side (see Figure 1.4). When there are several decks, it is necessary to specify to which one refers the depth.

The **moulded draught**, T, is the vertical distance between the top of the keel to the designed summer load line, usually measured in the midships plane (see Figure 1.4). There may be appendages protruding below the keel, for example the sonar dome of a warship. Then, it is necessary to define an **extreme draught** that is the distance between the lowest point of the hull or of an appendage and the designed load line.

Certain ships are designed with a keel that is not parallel to the load line. Some tugs and fishing vessels display this feature. To define the draughts associated with such a situation let us refer to Figure 1.3. We draw an auxiliary line that extends the keel afterward and forwards. The distance between the intersection of this auxiliary line with the aft perpendicular and the load line is called **aft draught** and is noted with T_A. Similarly, the distance between the load line and the intersection of the auxiliary line with the forward perpendicular is called **forward draught** and is noted with T_F. Then, the draught measured in the midship section is known as

(a)   **(b)**

Figure 1.5   (a) Midships symbol in English literature and (b) midships symbol in German literature

midships draught and its symbol is T_M. The difference between depth and draught is called **freeboard**; in DIN 81209-1 it is noted by f.

The breadth of the waterplane in the midships section is called **moulded breadth** or **moulded beam** and we note it by B. The maximum breadth may occur in another section; for fast ships usually aft of midships. Also, the deck breadth may be larger than the moulded breadth.

The **moulded volume of displacement** is the volume enclosed between the submerged, moulded hull, and the horizontal waterplane defined by a given draught. This volume is noted by ∇, a symbol known in English-language literature as *del*, and in European literature as *nabla*. In English we must use two words, "submerged hull," to identify the part of the hull below the waterline. Romance languages use for the same notion only one word derived from the Latin "carina." Thus, in French it is "carène," while in Catalan, Italian, Portuguese, Romanian, and Spanish it is called "carena."

In many ships the deck has a transverse curvature that facilitates the drainage of water. The vertical distance between the lowest and the highest points of the deck, in a given transverse section, is called **camber** (see Figure 1.4). According to ISO 7460 the camber is measured in mm, while all other ship dimensions are given in m. A common practice is to fix the camber amidships as 1/50 of the breadth in that section and to fair the deck toward its extremities (for the term "fair" see Section 1.4.3). In most ships, the intersection of the deck surface and the plane of symmetry is a curved line with the concavity upwards. Usually, that line is tangent to a horizontal passing at a height equal to the ship depth, D, in the midship section, and runs upwards toward the ship extremities; it is higher at the bow. This longitudinal curvature is called **sheer** and is illustrated in Figure 1.1. The deck sheer helps in preventing the entrance of waves and is taken into account when establishing the load line in accordance with international conventions.

1.4 The Definition of the Hull Surface

1.4.1 Coordinate Systems

The DIN 81209-1 standard recommends the system of coordinates shown in Figure 1.6. The x-axis runs along the ship and is positive forwards, the y-axis is transversal and positive to port, and the z-axis is vertical and positive upwards. The origin of coordinates lies at the intersection of the centreline plane with the transversal plane that contains the aft perpendicular. The international standards ISO 7460 and 7463 recommend the same positive senses as DIN 81209-1 but do not specify a definite origin. Other systems of coordinates are possible. For example, a system defined as above, but having its origin in the midship section, has some advantages in the display of certain hydrostatic data. Computer programmes written in the USA use a system of coordinates with the origin of coordinates in the plane of the

Figure 1.6   System of coordinates recommended by DIN 81209-1

forward perpendicular, FP, the x-axis positive afterwards, the y-axis positive to starboard, and the z-axis positive upwards. For dynamic applications taking the origin in the centre of gravity simplifies the equations. However, it should be clear that to each loading condition corresponds one centre of gravity, while a point like the intersection of the aft perpendicular with the baseline is independent of the ship loading. The system of coordinates used for the hull surface can be also employed for the location of weights. By its very nature, the system in which the hull is defined is fixed in the ship and moves with her. To define the various **floating conditions,** that is the positions that the vessel can assume, we use another system, fixed in space, that is defined in ISO 7463 as x_0, y_0, z_0. Let this system initially coincide with the system x, y, z. A vertical translation of the system x, y, z with respect to the space-fixed system x_0, y_0, z_0 produces a draught change.

If the ship-fixed z-axis is vertical, we say that the ship floats in an *upright condition.* A rotation of the ship-fixed system around an axis parallel to the x-axis is called **heel** (Figure 1.7) if it is temporary, and **list** if it is permanent. The heel can be produced by lateral wind, by the centrifugal force developed in turning, or by the temporary, transverse displacement of weights. The list can result from incorrect loading or from flooding. If the transverse inclination is the result of ship motions, it is time-varying and we call it **roll**.

When the ship-fixed x-axis is parallel to the space-fixed x_0-axis, we say that the ship floats on **even keel**. A static inclination of the ship-fixed system around an axis parallel to the ship-fixed y-axis is called **trim**. If the inclination is dynamic, that is a function of time resulting from ship motions, it is called **pitch**. A graphic explanation of the term trim is given in Figure 1.7. The trim is measured as the difference between the forward and the aft draught. Thus, the trim is positive if the ship is **trimmed by the head**. As defined here the trim is measured in metres.

Figure 1.7 Heel and trim

1.4.2 Graphic Description

In most cases the hull surface cannot be defined by simple analytical equations. To cope with the problem, Naval Architects have drawn lines obtained by cutting the hull surface with sets of parallel planes. Readers may find an analogy with the definition of the earth surface in topography by *contour lines*. Each contour line connects points of constant height above sea level. Similarly, we represent the hull surface by means of lines of constant x, constant y, and constant z. Thus, cutting the hull surface by planes parallel to the yOz plane we obtain the **transverse** sections noted in Figure 1.8 as $St0$ to $St10$, that is **Station 0, Station 1, ... Station 10**. Cutting the same hull by horizontal planes (planes parallel to the base plane xOy), we obtain the **waterlines** marked in Figure 1.9 as $WL0$ to $WL5$. Finally, by cutting the same hull with longitudinal planes parallel to the xOz plane, we draw the **buttocks** shown in Figure 1.10. The most important buttock is the line $y = 0$ known as **centreline**; for almost all ship hulls it is a plane of symmetry.

Figure 1.8 Stations

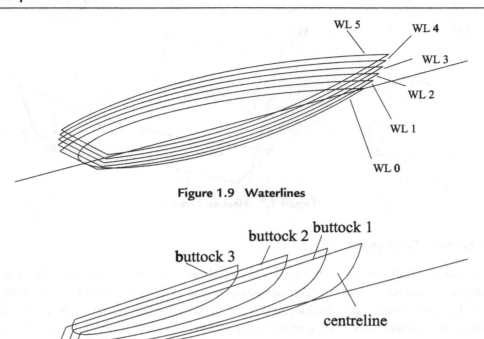

Figure 1.9   Waterlines

Figure 1.10   Buttocks

Stations, waterlines, and buttocks are drawn together in the **lines drawing**. Figure 1.11 shows one of the possible arrangements, probably the most common one. As stations and waterlines are symmetric for almost all ships, it is sufficient to draw only a half of each one. Let us take a look to the right of our drawing; we see the set of stations represented together in the **body plan**. The left half of the body plan contains Stations 0–4, that is the stations of the **afterbody**, while the right half is composed of Stations 5–10, that is the **forebody**. The set of buttocks, known as **sheer plan**, is placed at the left of the body plan. Beneath is the set of waterlines. Looking with more attention to the lines drawing we find out that each line appears as curved in one projection, and as straight lines in the other two. For example, stations appear as curved lines in the body plan, as straight lines in the sheer and in the waterlines plans. For readers familiar with the terminology of descriptive geometry the straight lines are also the *traces* of the cutting planes.

The station segments having the highest curvature are those in the **bilge** region, that is between the bottom and the ship side. Often no buttock or waterlines cuts them. To check what happens there it is usual to draw one or more additional lines by cutting the hull surface with one or more planes perpendicular to the transversal planes of stations and making an angle with the horizontal. A good practice is to incline the plane so that it will be

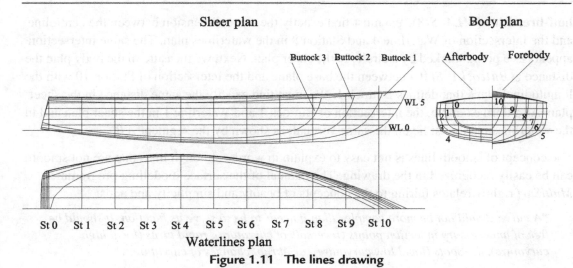

Sheer plan

Body plan

Buttock 3 Buttock 2 Buttock 1   Afterbody Forebody

WL 5

WL 0

St 0 St 1 St 2 St 3 St 4 St 5 St 6 St 7 St 8 St 9 St 10

Waterlines plan

Figure 1.11 The lines drawing

approximately normal to the station lines in the region of highest curvature. The intersection of such a plane with the hull surface is appropriately called **diagonal**.

Figure 1.11 was produced by modifying under MultiSurf (a product of AeroHydro, Inc.) a model provided with that software.

1.4.3 Fairing

The curves appearing in the lines drawing must fulfil two kinds of conditions: they must be coordinated and they must be "smooth," except where functionality requires for abrupt changes. Lines that fulfil these conditions are said to be **fair**. We are going to be more specific. In the preceding section we have used three projections to define the ship hull. From descriptive geometry we may know that two projections are sufficient to define a point in three-dimensional space. It follows that the three projections in the lines drawing must be coordinated, otherwise one of them may be false. Let us explain this idea by means of Figure 1.12. In the body plan, at the intersection of Station 8 with Waterline 4, we measure that

WL 4

St 8

$y(WL4, St8)$

$z(Buttock1, St10)$

$y(WL4, St8)$

Figure 1.12 Fairing

half-breadth $y(WL4, St8)$. We must find exactly the same dimension between the centreline and the intersection of Waterline 4 and Station 8 in the waterlines plan. The same intersection appears as a point, marked by a circle, in the sheer plan. Next, we measure in the body plan the distance $z(Buttock1, St10)$ between the base plane and the intersection of Station 10 with the longitudinal plane that defines Buttock 1. We must find exactly the same distance in the sheer plan. As a third example, the intersection of Buttock 1 and Waterline 1 in the sheer plan and in the waterlines plan must lie on the same vertical, as shown by the segment AB.

The concept of smooth lines is not easy to explain in words, although lines that are not smooth can be easily recognized in the drawing. The manual of the surface modelling programme *MultiSurf* rightly relates fairing to the concepts of beauty and simplicity and adds:

> *"A curve should not be more complex than it needs to be to serve its function. It should be free of unnecessary inflection points (reversals of curvature), rapid turns (local high curvature), flat spots (local low curvature), or abrupt changes of curvature..."*

With other words, a "curve should be pleasing to the eye" as one famous Naval Architect was fond of saying. For a formal definition of the concept of **curvature** see Chapter 13, Computer methods.

The fairing process cannot be satisfactorily completed in the lines drawing. Let us suppose that the lines are drawn at the scale 1:200. A good, young eye can identify errors of 0.1 mm. At the ship scale this becomes an error of 20 mm that cannot be accepted. Therefore, for many years it was usual to redraw the lines at the scale 1:1 in the **moulding loft** and the fairing process was completed there.

Some time after 1950, both in East Germany (the former DDR) and in Sweden, an optical method was introduced. The lines were drawn in the design office at the scale 1:20, under a magnifying glass. The drawing was photographed on glass plates and brought to a projector situated above the workshop. From there the drawing was projected on plates so that it appeared at the 1:1 scale to enable cutting by optically guided, automatic burners.

The development of hardware and software in the second half of the 20th century allowed the introduction of computer-fairing methods. Historical highlights can be found in Kuo (1971). When the hull surface is defined by algebraic curves, as explained in Chapter 13, the lines are smooth by construction. Recent computer programmes include tools that help in completing the fairing process and checking it, mainly the calculation of curvatures and **rendering**. A rendered view is one in which the hull surface appears in perspective, shaded, and lighted so that surface smoothness can be summarily checked. For more details see Chapter 13.

1.4.4 Table of Offsets

In shipyard practice it has been usual to derive from the lines plan a digital description of the hull known as **table of offsets**. Today, programmes used to design hull surface produce

Table 1.2 Table of offsets

	St	0	1	2	3	4	5	6	7	8	9	10
	x	0.000	0.893	1.786	2.678	3.571	4.464	5.357	6.249	7.142	8.035	8.928
WL	z						Half-Breadths					
0	0.360		0.900	1.189	1.325	1.377	1.335	1.219	1.024	0.749	0.389	
1	0.512	0.894	1.167	1.341	1.440	1.463	1.417	1.300	1.109	0.842	0.496	0.067
2	0.665	1.014	1.240	1.397	1.482	1.501	1.455	1.340	1.156	0.898	0.564	0.149
3	0.817	1.055	1.270	1.414	1.495	1.514	1.470	1.361	1.184	0.936	0.614	0.214
4	0.969	1.070	1.273	1.412	1.491	1.511	1.471	1.369	1.201	0.962	0.648	0.257
5	1.122	1.069	1.260	1.395	1.474	1.496	1.461	1.363	1.201	0.972	0.671	0.295

automatically this document. An example is shown in Table 1.2. The numbers correspond to Figure 1.11. The table of offsets contains **half-breadths** measured at the stations and on the waterlines appearing in the lines plan. The result is a table with two entries in which the offsets (half-breadths) are grouped into columns, each column corresponding to a station, and in rows, each row corresponding to a waterline. Table 1.2 was produced in MultiSurf.

1.5 Coefficients of Form

In ship design it is often necessary to classify the hulls and to find relationships between forms and their properties, especially the hydrodynamic properties. The **coefficients of form** are the most important means of achieving this. By their definition, the coefficients of form are non-dimensional numbers.

The **block coefficient** is the ratio of the moulded displacement volume, ∇ (see Figure 1.13), to the volume of the parallelepiped (rectangular block) with the dimensions L, B, and T:

$$C_B = \frac{\nabla}{LBT} \tag{1.1}$$

In Figure 1.14 we see that C_B indicates how much of the enclosing parallelepiped is filled by the submerged hull.

Figure 1.13 The submerged hull

Figure 1.14 The definition of the block coefficient, C_B

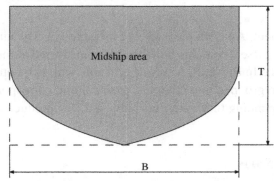

Figure 1.15 The definition of the midship-section coefficient, C_M

The **midship coefficient**, C_M, is defined as the ratio of the midship-section area, A_M, to the product of the breadth and the draught, BT:

$$C_M = \frac{A_M}{BT} \tag{1.2}$$

Figure 1.15 enables a graphical interpretation of C_M.

The **prismatic coefficient**, C_P, is the ratio of the moulded displacement volume, ∇, to the product of the midship-section area, A_M, and the length, L:

$$C_P = \frac{\nabla}{A_M L} = \frac{C_B LBT}{C_M BTL} = \frac{C_B}{C_M} \tag{1.3}$$

In Figure 1.16 we can see that C_P is an indicator of how much of a cylinder with constant section A_M and length L is filled by the submerged hull. Let us note the **waterplane area** by A_W. Then, we define the **waterplane-area coefficient** by

$$C_{WL} = \frac{A_W}{LB} \tag{1.4}$$

A graphic interpretation of the waterplane coefficient can be deduced from Figure 1.17.

Figure 1.16 The definition of the prismatic coefficient, C_P

Figure 1.17 The definition of the waterplane coefficient, C_{WL}4

Figure 1.18 The definition of the vertical prismatic coefficient, C_{VP}.

The **vertical prismatic coefficient** is calculated as

$$C_{VP} = \frac{\nabla}{A_W T} \tag{1.5}$$

For a geometric interpretation see Figure 1.18.

Other coefficients are defined as ratios of dimensions, for instance L/B, known as **length-breadth ratio**, and B/T known as "B over T." The **length coefficient of Froude**, or **length-displacement ratio** is

$$ⓂÌ = \frac{L}{\nabla^{1/3}} \tag{1.6}$$

and, similarly, the **volumetric coefficient**, ∇/L^3.

Table 1.3   Coefficients of form and related terminology

English Term	Symbol	Computer Notation	Translations European Symbol
Block coefficient	C_B	CB	Fr coefficient de bloc, δ, G Blockkoeffizient, I coeficiente di finezza (bloc), S coeficiente de bloque
Coefficient of form			Fr coefficient de remplissage, G Völligkeitsgrad, I coefficiente di carena, S coeficiente de forma
Displacement	Δ		Fr déplacement, G Verdrängung, I dislocamento, S desplazamiento
Displacement mass	Δ	DISPM	Fr déplacement, masse, G Verdrängungsmasse
Displacement volume	∇	DISPV	Fr volume de la carène, G Verdrängungs Volumen, I volume di carena, S volumen de carena
Midship coefficient	C_M	CMS	Fr coefficient de remplissage au maître couple, β, G Völligkeitsgrad der Hauptspantfläche, I coefficiente della sezione maestra, S coeficiente de la (sección) maestra
Midship-section area	A_M		Fr aire du couple milieu, G Spantfläche, I area della sezione maestra, S área de la (sección) maestra
Prismatic coefficient	C_P	CPL	Fr coefficient prismatique, ϕ, G Schärfegrad, I coefficiente prismatico o longitudinale, S coeficiente prismático longitudinal
Vertical prismatic coefficient	C_{VP}	CVP	Fr coefficient de remplissage vertical, ψ, I coeficiente di finezza prismatico verticale, S coeficiente prismático vertical
Waterplane area	A_W	AW	Fr aire de la surface de flottaison, G Wasserlinienfläche, I area del galleggiamento, S área de la flotación
Waterplane-area coefficient	C_{WL}		Fr coefficient de remplissage de la flottaison, α, G Völligkeitsgrad der Wasserlinienfläche, I coefficiente del piano di galleggiamento, S coeficiente de afinamiento de la flotación

Table 1.3 shows the symbols, the computer notations, the translations of the terms related to the coefficients of form, and the symbols that have been used in continental Europe.

We leave to an exercise the proof that the vertical prismatic coefficient, C_{VP}, depends on other coefficients.

1.6 Summary

The material treated in this book belongs to the field of Naval Architecture. The terminology is specific to this branch of engineering and is based on a long maritime tradition. The terms and symbols introduced in the book comply with recent international and corresponding national standards. So do the definitions of the main dimensions of a ship. Familiarity with the terminology and the corresponding symbols enables good communication between specialists all over the world and correct understanding and application of international conventions and regulations.

In general, the hull surface defies a simple mathematical definition. Therefore, the usual way of defining this surface is by means of curves obtained by cutting the surface with sets of planes parallel to the planes of coordinates. Let the x-axis run along the ship, the y-axis be transversal, and the z-axis, vertical. The sections of constant x are called **stations**, those of constant z, **waterlines**, and the contours of constant y, **buttocks**. The three sets must be coordinated and the curves be **fair**, a concept related to simplicity, curvature, and beauty.

Sections, waterlines, and buttocks are represented together in the **lines plan**. Line plans are drawn at a reducing scale; therefore, an accurate fairing process cannot be carried out on the drawing board. In the past it was usual to redraw the lines on the **moulding loft**, at the 1:1 scale. In the second half of the 20th century the introduction of digital computers and the progress of software, especially computer graphics, made possible new methods that will be briefly discussed in Chapter 13.

In early ship design it is necessary to choose an appropriate hull form and estimate its hydrodynamic properties. These tasks are facilitated by characterizing and classifying the ship forms by means of non-dimensional coefficients of form and ratios of dimensions. The most important coefficient of form is the block coefficient defined as the ratio of the **displacement volume** (volume of the submerged hull) to the product of ship length, breadth, and draught. An example of ratio of dimensions is the length-breadth ratio.

1.7 Examples

Example 1.1 (Coefficients of a fishing vessel). In INSEAN (1962) we find the test data of a fishing vessel hull called *C.484* and whose principal characteristics are:

$$
\begin{aligned}
L_{WL} &= 14.251 \text{ m} \\
B &= 4.52 \text{ m} \\
T_M &= 1.908 \text{ m} \\
\nabla &= 58.536 \text{ m}^3 \\
A_M &= 6.855 \text{ m}^2 \\
A_W &= 47.595 \text{ m}^2
\end{aligned}
$$

We calculate the coefficients of form as follows:

$$C_B = \frac{\nabla}{L_{pp}BT_M} = \frac{58.536}{14.251 \times 4.52 \times 1.908} = 0.476$$

$$C_{WL} = \frac{A_W}{L_{WL}B} = \frac{47.595}{14.251 \times 4.52} = 0.739$$

$$C_M = \frac{A_M}{BT} = \frac{6.855}{4.52 \times 1.908} = 0.795$$

$$C_P = \frac{\nabla}{A_M L_{WL}} = \frac{58.536}{6.855 \times 14.251} = 0.599$$

and we can verify that

$$C_P = \frac{C_B}{C_M} = \frac{0.476}{0.795} = 0.599$$

1.8 Exercises

Exercise 1.1 (Vertical prismatic coefficients). Find the relationship between the vertical prismatic coefficient, C_{VP}, the waterplane-area coefficient, C_{WL}, and the block coefficient, C_B.

Exercise 1.2 (Coefficients of *Ship 83074*). Table 1.4 contains data belonging to the hull we called *Ship 83074*. The length between perpendiculars, L_{pp}, is 205.74 m, and the breadth, B, 28.955 m. Complete the table and plot the coefficients of form against the draught, T. In Naval Architecture it is usual to measure the draught along the vertical axis, and other data—in our case the coefficients of form—along the horizontal axis (see Chapter 4).

Table 1.4 Coefficients of form of *Ship 83074*

Draught T m	Displacement Volume ∇ m^3	Waterplane Area A_{WL} m^2	C_B	C_{WL}	C_M	C_P
3	9029	3540.8	0.505	0.594	0.890	0.568
4	12632	3694.2			0.915	
5	16404	3805.2			0.931	
6	20257	3898.7			0.943	
7	24199	3988.6			0.951	
8	28270	4095.8			0.957	
9	32404	4240.4			0.962	

Table 1.5   Data of tanker hull C.786

L_{WL}	205.468	m
B	27.432	m
T_m	10.750	m
∇	46341	m^3
A_m	$0.220 \cdot \nabla^{2/3}$	m^2
A_{wl}	$3.648 \cdot \nabla^{2/3}$	m^2

Exercise 1.3 (Coefficients of hull C.786).   Table 1.5 contains data taken from INSEAN (1963) and referring to a tanker hull identified as C.786:

Calculate the coefficients of form and check that

$$\frac{C_B}{C_M} = C_P$$

CHAPTER 2

Basic Ship Hydrostatics

Chapter Outline

2.1 Introduction

This chapter deals with the **conditions of equilibrium** and **initial stability of floating bodies**. We begin with a derivation of **Archimedes' principle** and the definitions of the notions of **centre of buoyancy** and **displacement**. Archimedes' principle provides a particular formulation of the law of *equilibrium of forces* for floating bodies. The law of *equilibrium of moments* is formulated as **Stevin's law** and it expresses the relationship between the *centre of gravity* and the *centre of buoyancy* of the floating body. The study of *initial stability* is the study of the behaviour in the neighbourhood of the position of equilibrium. To derive the condition of initial stability we introduce Bouguer's concept of **metacentre**.

To each position of a floating body correspond one centre of buoyancy and one metacentre. Each position of the floating body is defined by three parameters, for instance the triple {*displacement, angle of heel, angle of trim*}; we call them the **parameters of the floating condition**. If we keep two parameters constant and let one vary, the centre of buoyancy travels along a curve and the metacentre along another. If only one parameter is kept constant and two vary, the centre of buoyancy and the metacentre generate two surfaces. In this chapter we shall briefly show what happens when the displacement is constant. The discussion of the case in which only one angle (that is, either heel or trim) varies leads to the concept of **metacentric evolute**.

The treatment of the above problems is based on the following assumptions:

1. the water is incompressible;
2. viscosity plays no role;
3. surface tension plays no role;
4. the water surface is plane;
5. the floating bodies are perfectly rigid.

The first assumption is practically exact in the range of water depths we are interested in. The second assumption is exact in static conditions (that is without motion) and a good approximation at the very slow rates of motion discussed in ship hydrostatics. In Chapter 12 we shall point out to the few cases in which viscosity should be considered. The third assumption is true for the sizes of floating bodies and the wave heights we are dealing with. The fourth assumption is never true, not even in the sheltered waters of a harbour. However, this hypothesis allows us to derive very useful, general results, and calculate essential properties of ships and other floating bodies. It is only in Chapter 9 that we shall leave the assumption of a plane water surface and see what happens in waves. In fact, the theory of ship hydrostatics was developed during 200 years under the hypothesis of a plane water surface and only in the middle of the 20th century it was recognized that this assumption cannot explain the capsizing of a few ships that were considered stable by that time. As to the fifth assumptions, it allows us to work with concentrated forces.

The results derived in this chapter are general in the sense that they do not assume particular body shapes. Thus, no symmetry must be assumed such as it usually exist in ships (port-to-starboard symmetry) and still less symmetry about two planes, as encountered, for instance, in Viking ships, some ferries, some offshore platforms and most buoys. The results hold the same for single-hull ships as for catamarans and trimarans. The only problem is that the treatment of the problems for general-form floating bodies requires "more" mathematics than the calculations for certain simple or symmetric solids. To make this chapter accessible to a larger audience, although we derive the results for body shapes without any form restrictions, we also exemplify them on parallelepipedic and other simply-defined floating body forms. Reading only those examples is sufficient to understand the ideas involved and the results obtained in this chapter. However, only the general derivations can provide the feeling of generality and a good insight into the problems discussed here.

2.2 Archimedes' Principle

2.2.1 A Body with Simple Geometrical Form

A body immersed in a fluid is subjected to an upwards force equal to the weight of the fluid displaced.

The above statement is known as **Archimedes' principle**. One legend has it that Archimedes (Greek, lived in Syracuse—Sicily—between 287 and 212 BC) discovered this law while taking a bath and that he was so happy that he ran naked in the streets shouting "I have found" (in Greek "*Heureka*," see entry "eureka" in Merriam-Webster, 1991). The legend may be nice, but it is most probably not true. What is certain is that Archimedes used his principle to assess the amount of gold in gold-silver alloys.

Archimedes' principle can be derived mathematically if we know another law of basic hydrostatics, namely Pascal's principle. Most textbooks contain only a brief, unconvincing proof based on intuitive considerations of equilibrium. A more elaborate proof is given here and we prefer it because only thus it is possible to decide whether Archimedes' principle applies or not in a given case. In Exercise 2.4 we show how a wrong conclusion can be reached if one ignores the physical model behind Archimedes' principle.

Let us consider a fluid whose specific gravity is γ. Then, at a depth z the **pressure** in the fluid equals γz. This is the weight of the fluid column of height z and unit-area cross-section. The pressure at a point is the same in all directions and this statement is known as *Pascal's principle*. The proof of this statement can be found in many textbooks on fluid mechanics, such as Douglas et al. (1979, p. 24).

In this subsection we calculate the *hydrostatic forces* acting on a body having a simple geometric form. The general derivation is contained in the next subsection. We consider a simple-form solid as shown in Figure 2.1; it is a parallelepipedic body whose horizontal, rectangular cross-section has the sides B and L. We consider the body immersed to the *draught T*. Let us call the top *face 1*, the bottom *face 2*, and number the vertical faces with 3–6. Figure 2.1b shows the diagrams of the liquid pressures acting on faces 4 and 6. To obtain the *absolute pressure* we must add the force due to the atmospheric pressure p_0. Those who like mathematics will say that the hydrostatic force on face 4 is the *integral* of the pressures on that face.

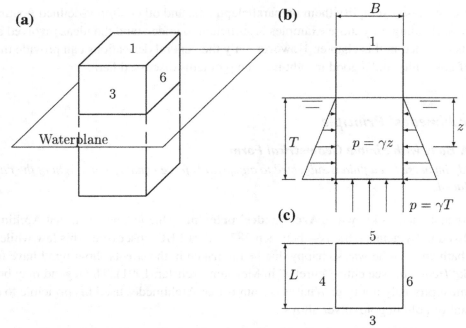

Figure 2.1   Hydrostatic forces on a body with simple geometrical form

Assuming that forces are positive in a rightwards direction, and adding the force due to the atmospheric pressure, we obtain

$$F_4 = L \int_0^T \gamma z \, dz + p_0 L T = \frac{1}{2} \gamma L T^2 + p_0 L T \qquad (2.1)$$

Similarly, the force on face 6 is

$$F_6 = -L \int_0^T \gamma z \, dz - p_0 L T = -\frac{1}{2} \gamma L T^2 - p_0 L T \qquad (2.2)$$

As the force on face 6 is equal and opposed to that on face 4 we conclude that the two forces cancel each other.

The reader who does not like integrals can reason in one of the following two ways:

1. The force per unit length of face 4, due to liquid pressure, equals the area of the triangle of pressures. As the pressure at depth T is γT, the area of the triangle equals

$$\frac{1}{2} T \times \gamma T = \frac{1}{2} \gamma T^2$$

Then, the force on the total length L of face 4 is

$$F_4 = L \times \frac{1}{2}\gamma T^2 + p_0 LT \tag{2.3}$$

Similarly, the force on face 6 is

$$F_6 = -L \times \frac{1}{2}\gamma T^2 - p_0 LT \tag{2.4}$$

The sum of the two forces F_4, F_6 is zero.

2. As the pressure varies linearly with depth, we calculate the force on unit length of face 4 as equal to the depth T times the mean pressure $\gamma T/2$. To get the force on the total length L of face 4 we multiply the above result by L and adding the force due to atmospheric pressure we obtain

$$F_4 = \frac{1}{2}\gamma LT^2 + p_0 LT$$

Proceeding in the same way we find that the force on face 6, F_6, is equal and opposed to the force on face 4. The sum of the two forces is zero. In continuation we find that the forces on faces 3 and 5 cancel one another. The only forces that remain are those on the bottom and on the top face, that is faces 2 and 1. The force on the top face is due only to atmospheric pressure and equals

$$F_1 = -p_0 LB \tag{2.5}$$

and the force on the bottom,

$$F_2 = p_0 LB + \gamma LBT \tag{2.6}$$

The resultant of F_1 and F_2 is an upwards force given by

$$F = F_2 + F_1 = \gamma LBT + p_0 LB - p_0 LT = \gamma LBT \tag{2.7}$$

The product LBT is actually the volume of the immersed body. Then, the force F given by Eq. (2.7) is the weight of the volume of liquid displaced by the immersed body. This verifies Archimedes' principle for the solid considered in this subsection.

We saw above that the atmospheric pressure does not play a role in the derivation of Archimedes' principle. Neither does it play any role in most other problems we are going to treat in this book; therefore, we shall ignore it in future.

Let us consider in Figure 2.2 a "zoom" of Figure 2.1. It is natural to see the resultant of the forces of pressure as applied at the point **P** situated in the centroid of face 2. The meaning of this sentence is that, for any coordinate planes, the moment of the force γLBT applied at the point **P** equals the integral of the moments of pressures. In the same figure, the point **B** is the *centre of volume* of the solid. If our solid would be made of a homogeneous material, the point **B** would be its *centre of gravity*. We see that **P** is situated exactly under **B**, but at double

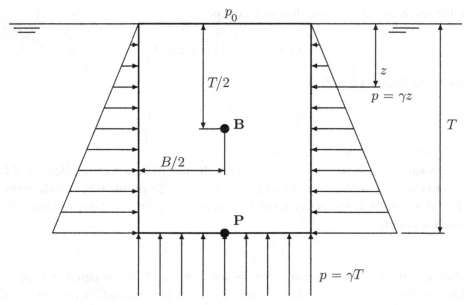

Figure 2.2   Zoom of Figure 2.1

draught. As a vector can be moved along its line of action, without changing its moments, it is commonly admitted that the force γLBT is applied in the point **B**. A frequent statement is: the force exercised by the liquid is applied in the centre of the displaced volume. The correct statement should be: "We can consider that the force exercised by the liquid is applied in the centre of the displaced volume." The force γLBT is called **buoyancy force**.

Above we have analysed the case of a solid that protrudes the surface of the liquid. Two other cases may occur; they are shown in Figure 2.3. We study again the same body as before. In Figure 2.3a the body is situated somewhere between the free surface and the bottom. Pressures are now higher; on the vertical faces their distribution follows a trapezoidal pattern. We can still show that the sum of the forces on faces 3–6 is zero. It remains to sum the forces on faces 1 and 2, that is on the top and the bottom of the solid. The result is

$$\gamma(z + H)LB - \gamma zLB = \gamma LBH \tag{2.8}$$

As γLBH is the weight of the liquid displaced by the submerged body, this is the same result as that obtained for the situation in Figures 2.1 and 2.2, that is Archimedes' principles holds in this case too.

In Figure 2.3b we consider the solid lying on the sea bottom (or lake, river, basin bottom) and assume that no liquid infiltrates under the body. Then no liquid pressure is exercised on face 2. The net hydrostatic force on the body is $\gamma z_1 LB$ and it is directed downwards. Archimedes' principle does not hold in this case. For equilibrium we must introduce a sea-bottom reaction, R, equal to the weight of the body plus the pressure force $\gamma z_1 LB$. The force necessary to lift the body from the bottom is equal to that reaction. However, immediately that the water can

Figure 2.3 Two positions of submergence

exercise its pressure on face 2, a buoyancy force is developed and the body seems lighter. It is as if when on the bottom the body is "sucked" with a force $\gamma z_1 L B$.

Figure 2.3b shows a particular case. Upwards hydrostatic forces can develop in different situations, for example:

- if the submerged body has such a shape that the liquid can enter **under** part of its surface. This is the case of most ships;
- the bottom is not compact and liquid pressures can act through it. This phenomenon is taken into account in the design of dams and breakwaters where it is called *uplift*.

In the two cases mentioned above the upwards force can be less than the weight of the displaced liquid. A designer should always assume the worst situation. Thus, to be on the safe side, when calculating the force necessary to bring a weight to the surface one should not count on the existence of the uplift. On the other hand, when calculating a deadweight—such as a concrete block—for an anchoring system, the existence of uplift forces should be taken into account because they can reduce the friction forces (between deadweight and bottom) that oppose horizontal pulls.

2.2.2 The General Case

In Figure 2.4 we consider a submerged body and a system of Cartesian coordinates, x, y, z, where z is measured vertically and downwards. The only condition we impose at this stage is

Figure 2.4   Archimedes' principle—vertical force

that no straight-line parallel to one of the coordinate axes pierces the body more than twice. We shall give later a hint on how to relax this condition, generalizing thus the conclusions to any body form. Let the surface of the body be $\mathcal{S}$, and let P be the horizontal plane that cuts in $\mathcal{S}$ the largest *contour*. The plane P divides the surface $\mathcal{S}$ into two surfaces, $\mathcal{S}_1$ situated above P, and $\mathcal{S}_2$ under P. We assume that $\mathcal{S}_1$ is defined by

$$z = f_1(x, y)$$

and $\mathcal{S}_2$ by

$$z = f_2(x, y)$$

The hydrostatic force on an element dA of $\mathcal{S}_1$ is pdA. This force is directed along the normal, $\mathbf{n}$, to $\mathcal{S}_1$ in the element of area. If the cosine of the angle between $\mathbf{n}$ and the vertical axis is $\cos(\mathbf{n}, z)$, the vertical component of the pressure force on dA equals $\gamma f_1(x, y) \cos(\mathbf{n}, z) dA$. As $\cos(\mathbf{n}, z) dA$ is the projection of dA on a horizontal plane, that is $dx dy$, we conclude that the vertical hydrostatic force on $\mathcal{S}_1$ is

$$\gamma \int \int_{\mathcal{S}_1} f_1(x, y) dx dy \qquad (2.9)$$

Let us consider now an element of $\mathcal{S}_2$ "opposed" to the one we considered on $\mathcal{S}_1$. We reason as above, taking care to change signs. We conclude that the hydrostatic force on $\mathcal{S}_2$ is

$$-\gamma \int \int_{\mathcal{S}_2} f_2(x, y) dx dy \qquad (2.10)$$

and the total force on $\mathcal{S}$,

$$F = \gamma \int \int_{\mathcal{S}} [f_1(x, y) - f_2(x, y)] dx dy \qquad (2.11)$$

The integral in Eq. (2.11) yields the volume of the submerged body. Thus, F equals the weight of the liquid displaced by the submerged body. It remains to show that the horizontal components of the resultant of hydrostatic pressures are equal to zero. We use Figure 2.5 to prove this for the component parallel to the x axis. The force component parallel to the x-axis acting on the element of area dA is

$$p \cos(\mathbf{n}, x) dA = \gamma z dy dz$$

Figure 2.5 Archimedes' principle—force parallel to the Ox-axis

On the other side of the surface, at the same depth z, there is an element of area such that the hydrostatic force on it equals

$$p\cos(\mathbf{n}, x)dA = -\gamma z dy dz$$

The sum of both forces is zero. As the whole surface S consists of such "opposed" pairs dA, the horizontal component in the x-direction is zero. By a similar reasoning we conclude that the horizontal component in the y-direction is zero too. This is also the result predicted by intuition. In fact, if the resultant of the horizontal components would not be zero we would obtain a "free" propulsion force.

This completes the proof of Archimedes' principle for a body shape subjected to the only restriction that no straight-line parallel to one of the coordinate axes intersects the body more than twice.

Could we relieve the above restriction and show that Archimedes' principle holds for any submerged body regardless of its shape? To do this we follow a reasoning similar to that employed sometimes in the derivation of Gauss' divergence theorem in vector analysis (see, for example, Borisenko and Tarapov, 1979). Figure 2.6a shows a body that does not fulfil the condition we imposed until now. In fact, in the right-hand part of the body a vertical line pierces four times the enclosing surface. The dashed line is the trace of the plane that divides the total volume of the body into two volumes, **1, 2**, such that each of them cannot be pierced more than twice by any line parallel to one of the coordinate axis.

Let us consider now the upper volume, **1**, in Figure 2.6b. Two forces act on this body:

   – the resultant of hydrostatic pressures, $\mathbf{P_1}$, on the external surface;
   – the force $\mathbf{R_{2,1}}$ exercised by the volume **2** on volume **1**.

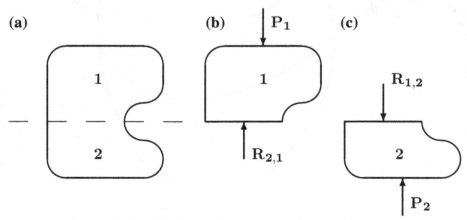

Figure 2.6 Extending Archimedes' principle

Similarly, let us consider the lower volume, **2**, in Figure 2.6c. Two forces act on this body:

– the resultant of hydrostatic pressures, $\mathbf{P}_2$, acting on the external surface;
– the force $\mathbf{R}_{1,2}$ exercised by the volume **1** on volume **2**.

As the forces $\mathbf{R}_{2,1}$ and $\mathbf{R}_{1,2}$ are equal and opposed, putting together the volumes **1** and **2** means that the sum of all forces acting on the total volume is $\mathbf{P}_1 + \mathbf{P}_2$, that is the force predicted by Archimedes' principle. Let us find the x- and y-coordinates of the point through which acts the buoyancy force. To do so we calculate the moments of this force about the xOz and yOz planes and divide them by the total force. The results are

$$x_P = \frac{\int\int_S x\gamma z[f_1(x, y) - f_2(x, y)]dxdy}{\int\int_S \gamma z[f_1(x, y) - f_2(x, y)]dxdy} \tag{2.12}$$

$$= \frac{\int\int_S xz[f_1(x, y) - f_2(x, y)]dxdy}{\int\int_S z[f_1(x, y) - f_2(x, y)]dxdy} \tag{2.13}$$

$$y_P = \frac{\int\int_S y\gamma z[f_1(x, y) - f_2(x, y)]dxdy}{\int\int_S \gamma z[f_1(x, y) - f_2(x, y)]dxdy} \tag{2.14}$$

$$= \frac{\int\int_S yz[f_1(x, y) - f_2(x, y)]dxdy}{\int\int_S z[f_1(x, y) - f_2(x, y)]dxdy} \tag{2.15}$$

These are simply the x- and y-coordinates of the centre of the submerged volume. We conclude that *the buoyancy force passes through the centre of the submerged volume, B (centre of the displaced volume of liquid).*

2.3 The Conditions of Equilibrium of a Floating Body

A body is said to be in **equilibrium** if it is not subjected to accelerations. Newton's second law shows that this happens if the sum of all forces acting on that body is zero and the sum of the moments of those forces is also zero. Two forces always act on a floating body: the weight of that body and the buoyancy force. In this section we show that the first condition for equilibrium, that is the one regarding the sum of forces, is expressed as Archimedes' principle. The second condition, regarding the sum of moments, is stated as *Stevin's law*.

Further forces can act on a floating body, for example those produced by wind, by centrifugal acceleration in turning or by towing. The influence of those forces is discussed in Chapter 6.

2.3.1 Forces

Let us assume that the bodies appearing in Figures 2.1 and 2.3a float freely. Then, the weight of each body and the hydrostatic forces acting on it are in equilibrium. Archimedes' principle can be reformulated as:

The weight of the volume of water displaced by a floating body is equal to the weight of that body.

The weight of the fluid displaced by a floating body is appropriately called **displacement**. We denote the displacement by the upper-case Greek letter *delta*, that is Δ. If the weight of the floating body is W, then we can express the equilibrium of forces acting on the floating body by

$$\Delta = W \tag{2.16}$$

For the volume of the displaced liquid we use the symbol ∇ defined in Chapter 1. In terms of the above symbols Archimedes' principle yields the equation

$$\gamma \nabla = W \tag{2.17}$$

If the floating body is a ship, we rewrite Eq. (2.17) as

$$\gamma C_B LBT = \sum_{i=1}^{n} W_i \tag{2.18}$$

where W_i is the weight of the ith item of ship weight. For example, W_1 can be the weight of the *ship hull*, W_2, of the *outfit*, W_3, of the *machinery*, and so on. The symbol C_B and the letters L, B, T have the meanings defined in Chapter 1.

In hydrostatic calculations Eq. (2.18) is often used to find the draught corresponding to a given displacement, or the displacement corresponding to a measured draught. In Ship Design Eq. (2.18) is used either as a **design equation** (see, for example, Manning, 1956), or as an **equality constraint** in design **optimization** problems (see, for example, Kupras, 1976).

Instead of the displacement weight we may work with the **displacement mass**, $\rho \nabla$, where ρ is the **density** of the surrounding water. Then, Eq. (2.18) can be rewritten as

$$\rho C_B LBT = \sum_{i=1}^{n} m_i \tag{2.19}$$

where m_i is the mass of the ith ship item. The DIN standards define, indeed, Δ as mass, and use Δ_F for displacement weight. The subscript "F" stands for "force." In the following chapters of this book we shall use the displacement mass rather than the displacement weight. The unit of mass is the *tonne* of 1000 kg, with the symbol 't'.

Table 2.1 Some foreign names for the point B

Language	Term	Meaning
French	Centre de carène	Centre of submerged hull
German	Formschwerpunkt	Centre of gravity of solid
Italian	Centro di carena	Centre of submerged hull
Portuguese	Centro do carena	Centre of submerged hull
Spanish	Centro de empuje	Centre of buoyancy force
	Centro de carena	Centre of submerged hull

To remember the meaning of the symbol Δ, let us think that the word "delta" begins with a "d," like the word "displacement" (we ignore the fact that in contemporary-Greek "delta" is actually read as "thelta"). As to the symbol ∇, it resembles "V," the initial letter of the word "volume."

The point **B** is called in English **centre of buoyancy**. There are languages in which the name of the point **B** recognizes the fact that **B** is not a centre of pressure. Table 2.1 gives a few examples. This is, of course, a matter of semantics. The line of action of the buoyancy force always passes through the point **B**.

2.3.2 Moments

In this subsection we discuss the second condition of equilibrium of a floating body: the sum of the moments of all forces acting on it must be zero. This condition is fulfilled in Figure 2.7a where the *centre of gravity*, G, and the centre of buoyancy, B, of the floating body are on the same **vertical** line. The weight of the body and the buoyancy force are equal—that is Δ—opposed, and act along the same line. The sum of their moments about any reference is zero.

Let us assume that the centre of gravity moves in the same plane, to a new position, G_1 (Figure 2.7b). The sum of the moments is no more zero; it causes a clockwise inclination of the body, by an angle ϕ. A volume submerges at right, another volume emerges at left. The

Figure 2.7 Stevin's Law, 1

result is that the centre of buoyancy moves to the right, to a new point that we mark by B_ϕ. The floating body will find a position of equilibrium when the two points G_1 and B_ϕ will be on the same vertical line. This situation is shown in Figure 2.7c.

There is a possibility of redrawing Figure 2.7 so that all situations are shown in one figure. To do this, instead of showing the body inclined clockwise by an angle ϕ, and keeping the waterline constant, we keep the position of the body constant and draw the waterline inclined counterclockwise by the angle ϕ. Thus, in Figure 2.8 the waterline corresponding to the initial position is $W_0 L_0$. The weight force, equal to Δ, acts through the initial centre of gravity, G_0; it is vertical, that is perpendicular to the waterline $W_0 L_0$. The buoyancy force, also equal to Δ, acts through the initial centre of buoyancy, B_0: it is vertical, that is perpendicular to the initial waterline.

We assume now that the centre of gravity moves to a new position, G_1. The floating body rotates in the same direction, by an angle ϕ, until it reaches a position of equilibrium in which the new waterline is $W_\phi L_\phi$. The new centre of buoyancy is B_ϕ. The line connecting G_1 and B_ϕ is vertical, that is perpendicular to the waterline $W_\phi L_\phi$. The weight and the buoyancy force act along this line.

Thus, in the case of a floating body, the second condition of equilibrium is satisfied if the centre of gravity and the centre of gravity are on the same vertical line. This condition is attributed to Simon Stevin (Simon of Bruges, Flanders, 1548–1620). Stevin is perhaps better known for other studies, among them one on decimal fractions that helped to establish the

Figure 2.8   Stevin's Law, 2

notation we use today, the discovery in 1586 of the law of composition of forces for perpendicular forces, and a demonstration of the impossibility of perpetual motion.

In Figures 2.7 and 2.8 we assumed that while the body rotates to a new position, no opening, such as a hatch, window, or vent, enters the water. If this assumption is not correct, the body can either reach equilibrium under more complex conditions (see Chapter 11), or sink.

2.4 A Definition of Stability

In the preceding section we learnt the conditions of equilibrium of a floating body. The question we ask in the next section is how to determine if a condition of equilibrium is stable or not. Before answering this question we must define the notion of **stability**. This concept is general; we are interested here in its application to floating bodies.

Let us consider a floating body in equilibrium and assume that some force or moment causes a **small change** in its position. Three situations can occur when that force or moment ceases to act:

1. The body returns to its initial position; we say that the condition of equilibrium is **stable**.
2. The position of the body continues to change. We say in this case that the equilibrium is **unstable**. In practical terms this can mean, for example, that the floating body capsizes.
3. The body remains in the displaced position until the smallest perturbation causes it to return to the initial position or to continue to move away from the initial position. Now we talk about **neutral equilibrium**.

As an example let us consider the body shown in Figure 2.1. If this body floats freely at the surface we conclude from Eq. (2.17) that the total volume is larger than the weight divided by the specific gravity of the fluid. This body floats in stable equilibrium as to draught. To show this let us imagine that some force causes it to move downwards so that its draught increases by the quantity δT. Archimedes' principle tells us that a new force, $\gamma LB\delta T$, appears and that it is directed upwards. Suppose now that the cause that moved the body downwards decreases slowly. Then, the force $\gamma LB\delta T$ returns the body to its initial position. In fact, as the body moves (slowly) upwards, δT decreases until it becomes zero and then the motion ceases. If the force that drove the body downwards ceases abruptly, the body oscillates around its initial position and, if damping forces are active—they always exist in nature—the body will eventually come to rest in its initial position.

Next, we assume that some force moved the body upwards so that its draught decreases by δT. A force $-\gamma LB\,\delta T$ appears now and it is directed downwards. Therefore, if the body is released slowly it will descend until $\delta T = 0$. This completes the proof that the body floating freely **at the surface** is in stable equilibrium with regard to its draught. We mention "with regard to draught" because, as shown in the next section, the body may be unstable with regard to heel.

When a body floats freely, but is completely submerged, its weight equals exactly its volume multiplied by the specific gravity of the liquid. This body is in neutral equilibrium because it can float **at any** depth. Any small perturbation will move the body from a depth to another one.

If the weight of the body is larger than its total volume multiplied by the specific gravity of the liquid, then the body will sink.

Summing-up, we may distinguish three cases:

1. The total volume of a body is larger than its weight divided by the specific gravity of the water:

$$V_{total} > W/\gamma$$

   The body floats at the surface and we can control the draught by adding or reducing weights.

2. The weight of the body exactly equals the total volume multiplied by the specific gravity of the liquid:

$$V_{total} = W/\gamma$$

   The body can float at any depth and we cannot control the position by adding or reducing weights. Any additional weight would cause the body to sink bringing it into case 3. Reducing even slightly its weight will cause the body to come to the surface; its situation changes to case 1.

3. The weight of the body is larger than its volume multiplied by the specific gravity of the water:

$$V_{total} < W/\gamma$$

   The body will sink. To change its position we must either reduce the weight until we reach at least situation 2, or add buoyancy in some way.

In the above analysis we assumed that the specific gravity of the liquid, γ, is constant throughout the liquid volume. This assumption may not be correct if large variations in temperature or salinity are present, or if the liquid volume consists of layers of different liquids. Interesting situations can arise in such cases. Other situations can arise at depths at which the water density increases while the volume of the floating body shrinks because of the compressibility of its structure. These cases are beyond the scope of this book.

2.5 Initial Stability

Figure 2.9a is a transverse section through a ship in *upright condition*, that is unheeled. If this section passes through the centre of buoyancy, B, we know from Stevin's law that it contains the centre of gravity, G. The waterline is W_0L_0. The weight force, W, acts through the centre

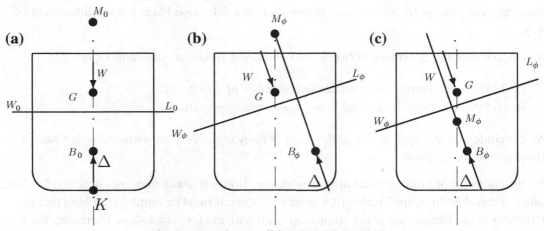

Figure 2.9 The condition of initial stability

of gravity, G; the buoyancy force, Δ, through the centre of buoyancy, B_0. The forces W and Δ are equal and collinear and the ship is in an equilibrium condition. Let the ship heel to the starboard with an angle ϕ. For reasons that will become clear in Section 2.8, we assume that the *heel angle is small*. As previously explained, we leave the ship as she is and draw the waterline as inclined to port, with the same angle ϕ. This is done in Figure 2.9b where the new waterline is $W_\phi L_\phi$. If the weights are fixed, as they should be, the centre of gravity remains in the same position, G. Because a volume submerges at starboard, and an equal volume emerges at port, the centre of buoyancy moves to starboard, to a new position, B_ϕ. Both forces W and Δ are vertical, that is perpendicular to the waterline $W_\phi L_\phi$. These two forces form a moment that tends to return the ship toward port, that is to her initial condition. We say that the ship is stable.

Figure 2.9c also shows the ship heeled toward starboard with an angle ϕ. In the situation shown in this figure the moment of the two forces W and Δ heels the ship further toward starboard. We say that the ship is unstable.

The difference between the situations in Figure 2.9b and c can be described elegantly by the concept of **metacentre**. This abstract notion was introduced by Pierre Bouguer (French, 1698–1758) in 1746, in his *Traité du Navire*. Around the same time Euler was also looking for a criterion of stability. He recognized Bouguer's priority. The fascinating story of the quest for a stability criterion can be read in Ferreiro (2010). Let us refer again to Figure 2.9b. For a ship, the dot-point line is the trace of the port-to-starboard symmetry plane, that is the *centreline*. More generally, for any floating body, the dot-point line is the *line of action of the buoyancy force before heeling*. The new line of action of the buoyancy force passes through the new centre of buoyancy and is perpendicular to $W_\phi L_\phi$. The two lines intersect in the point M_ϕ.

Bouguer called this point *metacentre*. Remember, this definition holds for small heel angles only.

We can see now the difference between the two heeled situations shown in Figure 2.9:

- in (b) the metacentre is situated *above the centre of gravity, G*;
- in (c) the metacentre is situated *below the centre of gravity, G*.

We conclude that *the equilibrium of the floating body is stable if the metacentre is situated above the centre of gravity.*

For his contributions of overwhelming importance, Bouguer was sometimes described as "the father of naval architecture" (quotation in Stoot, 1959). It must be emphasized here that the definition of the metacentre is not connected at all with the form of a ship. Therefore, the fact that in the above figures the metacentre is the intersection of the new line of action of the buoyancy force and the centreline is true only for symmetrical hulls heeled from the upright condition. For a general floating body we can reformulate the definition as follows:

> *Let us consider a floating body and its centre of buoyancy B_ϕ. Let the line of action of the buoyancy force be R. If the body changes its inclination by an angle $\delta\phi$, the centre of buoyancy changes its position to $B_{\phi+\delta\phi}$ and the new line of action of the buoyancy force will be, say, S. When $\delta\phi$ tends to zero, the intersection of the lines R and S tends to a point that we call metacentre.*

Readers familiar with elementary differential geometry will recognize that, defined as above, the metacentre is the **the centre of curvature of the curve of centres of buoyancy**. The notion of curvature is defined in Chapter 13.

2.6 Metacentric Height

In the preceding section we learnt that a surface ship is initially stable if its initial metacentre is above the centre of gravity. For actual calculations we must find a convenient mathematical formulation. We do this with the help of Figure 2.9a. We choose a reference point, K, at the intersection of the centreline and the baseline and we measure vertical coordinates from it, upwards. Thus defined, K is the origin of z-coordinates. A good recommendation is to choose K as the lowest point of the ship keel; then, there will be no negative z-coordinates. We remember easily the chosen notation because K is the initial letter of the word *keel*.

In the same figure M_0 is the **initial metacentre**, that is the metacentre corresponding to the upright condition. Dropping the subscripts 0 we can write

$$\overline{GM} = \overline{KB} + \overline{BM} - \overline{KG} \tag{2.20}$$

and the condition of initial stability is expressed as

$$\overline{GM} > 0 \qquad (2.21)$$

The vector $\overline{GM}$ is called **metacentric height**. The vector $\overline{KB}$ is the z-coordinate of the centre of buoyancy; it is calculated as the z-coordinate of the centroid of the submerged hull as one of the results of *hydrostatic calculations*. The vector $\overline{BM}$ is the **metacentric radius** whose calculation we are going to discuss in Section 2.8.2. The vector $\overline{KG}$ is the z-coordinate of the centre of gravity of the floating body; it results from **weight calculations**. The quantities $\overline{KB}$ and $\overline{BM}$ depend upon the ship geometry, the quantity $\overline{KG}$ depends upon the distribution of masses.

It is regrettable that certain authors of textbooks of fluid dynamics treat superficially the problem and state that for stability the centre of gravity, G, should be situated below the centre of buoyancy, B. The condition $\overline{KG} < \overline{KB}$ is certainly *sufficient*, but not *necessary*, while the condition $\overline{GM} > 0$ is both *necessary* and *sufficient*. Moreover, to lower the centre the gravity below the centre of buoyancy can require so much ballast that no cargo can be taken aboard. Practically, this is unacceptable. As we are going to learn in Chapter 6, a very low centre of gravity means a very high metacentric height and this may lead to unacceptably short roll periods.

2.7 A Lemma on Moving Volumes or Masses

Figure 2.10 shows a system of two masses, m_1 and m_2. Let the x-coordinate of the mass m_1 be x_1; that of the mass m_2, x_2. The centre of gravity of the system is G and its x-coordinate is given by

$$x_G = \frac{x_1 m_1 + x_2 m_2}{m_1 + m_2} \qquad (2.22)$$

Figure 2.10 Moving a mass in a system of masses

Let us move the mass m_2 a distance d in the x-direction. The new centre of gravity is G^* and its x-coordinate,

$$x_G^* = \frac{x_1 m_1 + (x_2 + d)m_2}{x_1 + m_1} = x_G + \frac{dm_2}{m_1 + m_2} \tag{2.23}$$

The product dm_2 is the change of moment caused by the translation of the mass m_2. The centre of gravity of the system moved a distance equal to the change of moment divided by the total mass of the system. A formal statement of this Lemma is:

> *Given a system of masses, if one of its components is moved in a certain direction, the centre of gravity of the system moves in the same direction, a distance equal to the change of moment divided by the total mass.*

A similar Lemma holds for a system of volumes in which one of them is moved to a new position. The reader is invited to solve Exercise 2.6 and prove the Lemma for a three-dimensional system of masses.

2.8 Small Angles of Inclination

In this section we prove two very important theorems for bodies that incline at **constant displacement**. This is usually the case of floating bodies that change their inclination without the addition or loss of weights. Constant displacement means **constant volume of displacement**. In Chapter 1 we mentioned that Romance languages use for the *submerged volume* terms derived from the Latin word *carina*, for instance *carène* in French, *carena* in Italian. Correspondingly, the theory of floating bodies inclined at constant volume of displacement is called *Théorie des isocarènes* in French, *Teoria delle isocarene* in Italian, and *Teoria de las isocarenas* in Spanish. The prefix "iso" comes from Greek and means "equal." Thus, Romance languages use one single term to mean "bodies inclining at constant volume of displacement."

A second assumption in this section is that the **angle of inclination is small**. The results developed under this assumption are valid for any floating body. The results are valid for any angle of inclination only for floating bodies belonging to a particular class of forms called **wall sided**, a concept explained in Section 6.12.

2.8.1 A Theorem on the Axis of Inclination

Let us assume that the initial waterplane of the body shown in Figure 2.11 is $W_0 L_0$. Next we consider the same body inclined by a *small* angle ϕ, such that the new waterplane is $W_\phi L_\phi$. The weight of the body does not change; therefore, also the submerged volume does not change. If so, the volume of the "wedge" that submerges at right, between the two planes

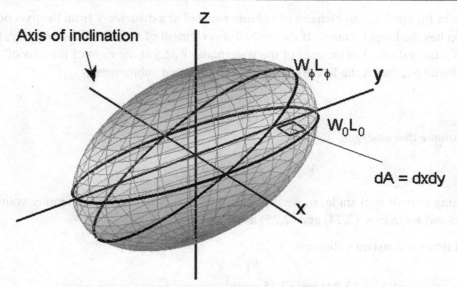

Figure 2.11 Euler's theorem on the axis of inclination, 1

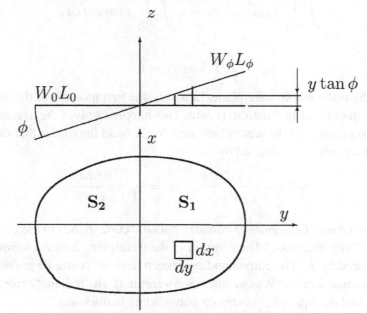

Figure 2.12 Euler's theorem on the axis of inclination, 2

W_0L_0 and $W_\phi L_\phi$, equals the volume of the wedge that emerges at left, between the same two planes. Let us express this mathematically. We take the intersection of the two planes as the x-axis. This is the *axis of inclination*.

As shown in Figure 2.12, an element of volume situated at a distance y from the axis of inclination has the height $y \tan \phi$. If the base of this element of volume is $dA = dxdy$, the volume is $y \tan \phi \, dxdy$. Let the area of the waterplane $W_0 L_0$ at the right of the axis of intersection be $\mathbf{S_1}$; that at the left, $\mathbf{S_2}$. Then, the volume that submerges is

$$V_1 = \int \int_{\mathbf{S_1}} y \tan \phi \, dxdy \qquad (2.24)$$

and the volume that emerges,

$$V_2 = - \int \int_{\mathbf{S_2}} y \tan \phi \, dxdy \qquad (2.25)$$

By assuming a small heel angle, ϕ, we can consider the submerging and emerging volumes as wall sided and write Eqs. (2.24) and (2.25) as we did.

The condition for constant volume is

$$V_1 = V_2$$

Combining this with Eqs. (2.24) and (2.25) yields

$$\int \int_{\mathbf{S_1}} y \tan \phi \, dxdy = - \int \int_{\mathbf{S_2}} y \tan \phi \, dxdy \qquad (2.26)$$

and, finally

$$\int \int_{\mathbf{S}} y \, dxdy = 0 \qquad (2.27)$$

where $\mathbf{S} = \mathbf{S_1} + \mathbf{S_2}$ is the whole waterplane. In words, the first moment of the waterplane area, with respect with the axis of inclination, is zero. This happens only if the axis of inclination passes through the *centroid* of the waterplane area. We remind the reader that the coordinates of the centroid of an area A are defined by

$$x_C = \frac{\int \int_A x \, dxdy}{\int \int_A dxdy}, \quad y_C = \frac{\int \int_A y \, dxdy}{\int \int_A dxdy}$$

Or, as the *Webster's Ninth Collegiate Dictionary* puts it, "corresponds to the centre of mass of a thin plate of uniform thickness." The centroid of the waterplane area is known as **centre of flotation** and is noted by F. The corresponding French term is "centre de gravité de la flottaison," the German term is "Wasserlinien-Schwerpunkt," the Italian, "centro del galleggiamento," and the Spanish, "centro de gravedad de la flotación."

A statement of this property is:

> Let the initial waterplane of a floating body be $W_0 L_0$. After an inclination, at constant volume of displacement, with an angle ϕ, the new waterplane is $W_\phi L_\phi$. The intersection of the two waterplanes is the axis of inclination. If the angle of inclination tends to zero, the axis of inclination tends to a straight line passing through the centroid of the waterplane area.

In practice, this property holds if the angle of inclination is sufficiently small. For heeling of a vessel, this can mean a few degrees, 5° for some forms, even 15° for others. If the inclination is the trimming of an intact vessel, the angles are usually small enough and this property always holds. The property also holds for larger heel angles if the floating body is **wall sided**. This is the name given to floating bodies whose surface includes a cylinder (in the broader geometrical sense), with generators perpendicular to the initial waterplane. An illustration of such a case is given in Example 2.5. In French and Italian, for example, the term used for wall-sided bodies is *cylindrical floating bodies*.

The term used in some languages, such as French, Italian, or Spanish, for an axis passing through the centroid of an area is **barycentric axis**. This term is economic and we shall use it whenever it will help us to express ideas more concisely.

2.8.2 *Metacentric Radius*

Let us refer again to Figure 2.9. As we shall see, the vector $\overline{B_\phi M_\phi}$ plays an important role in stability. Leaving the subscript ϕ, we generically call $\overline{BM}$ *metacentric radius*; in this section we calculate its magnitude. To do so we must find the displacement of the centre of buoyancy, B, for a small angle of inclination ϕ. Here we use the Lemma on moving volumes and we calculate

$$change\ of\ coordinate = \frac{change\ of\ moment\ of\ volume}{total\ volume}$$

As seen from Figures 2.11 and 2.12, the elemental change of volume is $y \tan \phi \, dx dy$. To find the changes of moment respective to the coordinate planes we must multiply the elemental volume by the coordinates of its centroid. To make things easier, we take the origin of coordinates in the initial centre of buoyancy, B_0, measure the x-coordinate parallel to the axis of inclination, positive forwards, the y-coordinate transversally, positive leftwards, and the z-coordinate vertically, positive upwards. The coordinates of the centre of buoyancy B_ϕ are obtained by integrating the changes of moment of the elemental volume, over the waterplane area **S**. The results are

$$x_B = \frac{\int \int_{\mathbf{S}} xy \tan \phi \, dx dy}{\nabla} = \frac{I_{xy}}{\nabla} \tan \phi \qquad (2.28)$$

$$y_B = \frac{\int \int_{\mathbf{S}} y^2 \tan \phi \, dx dy}{\nabla} = \frac{I}{\nabla} \tan \phi \qquad (2.29)$$

$$z_B = \frac{\int \int_{\mathbf{S}} \frac{1}{2} y^2 \tan^2 \phi \, dx dy}{\nabla} = \frac{1}{2} \frac{I}{\nabla} \tan^2 \phi \qquad (2.30)$$

Above, I is the moment of inertia of the waterplane area about the axis of inclination (remember, it is a barycentric axis), and I_{xy}, the **product of inertia** of the same area about the axes x and y. In German and some other languages I_{xy} is called *centrifugal moment of inertia*.

As we assumed that the angle ϕ is small, we can further write

$$x_B = \frac{I_{xy}}{\nabla}\phi$$

$$y_B = \frac{I}{\nabla}\phi$$

$$z_B = \frac{1}{2}\frac{I}{\nabla}\phi^2 \tag{2.31}$$

The coordinate z_B is of second order and we can neglect it if ϕ is small. As to the x-coordinate let us remember that conventional ships in upright condition enjoy a port-to-starboard symmetry. This means that for such ships, in upright condition, the product of inertia is zero so that x_B is zero too. Then $\overline{B_0 B_\phi}$ is essentially equal to y_B. For other floating bodies there is a three-dimensional theory that is beyond the scope of this book (see, for example, Appel, 1921; Hervieu, 1985). For our purposes it is sufficient to consider the projection of the curve of centres of buoyancy, B, on the plane that contains the initial centre of buoyancy, B_0, and is perpendicular to the axis of inclination. In this plane the length of the arc connecting B_0 to B_ϕ equals $\overline{BM}\phi$ (see Figure 2.13). We can write

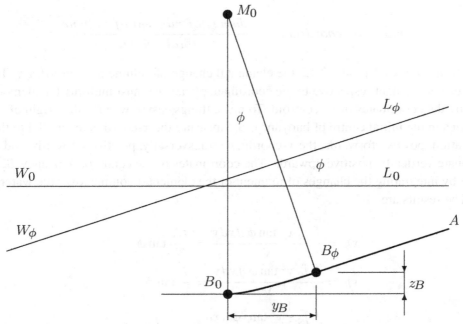

Figure 2.13 Calculation of metacentric radius

$$\frac{I}{\nabla}\phi = \overline{BM}\phi$$

and hence,

$$\overline{BM} = \frac{I}{\nabla} \tag{2.32}$$

A statement of this important theorem is:

> *The magnitude of the metacentric radius, $\overline{BM}$, is equal to the ratio of the waterplane moment of inertia, about the axis of inclination, to the volume of displacement.*

Returning to the third Eq. (2.31) we can see that z_B is always positive. This means that the curve of centres of buoyancy presents its concavity toward the waterline.

2.9 The Curve of Centres of Buoyancy

Figure 2.14 shows a floating body inclined by some angle; the corresponding waterline is W_1L_1 and the centre of buoyancy, B_1. Let us assume that the inclination increases by an additional, small angle, ϕ. Let the new waterline be W_2L_2 and the corresponding centre of

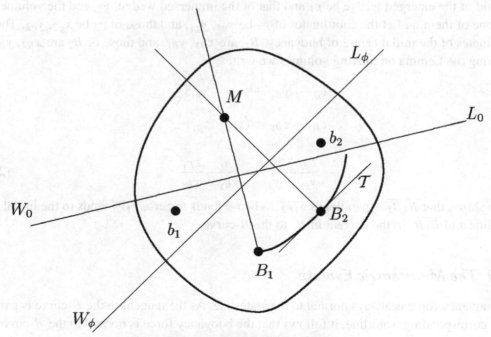

Figure 2.14 Properties of B and M curves

buoyancy, B_2. For a small angle ϕ we can write the coordinates of the new centre of buoyancy as

$$y_B = \frac{I}{\nabla}\phi$$

$$z_B = \frac{1}{2}\frac{I}{\nabla}\phi^2$$

Differentiation of these equations yields

$$dy_B = \frac{I}{\nabla}d\phi$$

$$dz_B = \frac{I}{\nabla}\phi d\phi$$

which shows that the slope of the tangent to the B-curve in B_2 is

$$\left.\frac{dz_B}{dy_B}\right|_{B_2} = \phi$$

This is the assumed angle of inclination. We reach the important conclusion that the tangent to the B-curve, in a point B_ϕ, is parallel to the waterline corresponding to the centre of buoyancy B_ϕ.

We could reach the same conclusion by the following reasoning. In Figure 2.14 let the centroid of the emerged wedge be g_1 and that of the immersed wedge, g_2, and the volume of each one of them, v. Let the coordinates of g_1 be x_{g1}, y_{g1}, and those of g_2 be x_{g2}, y_{g2}. The coordinates of the initial centre of buoyancy, B_1, are x_{B_1}, y_{B_1}, and those of B_2 are x_{B2}, y_{B_2}. Applying the Lemma on moving volumes we write

$$y_{B_2} - y_{B_1} = (y_{g2} - y_{g1})\frac{v}{\nabla}$$

$$z_{B_2} - z_{B_1} = (z_{g2} - z_{g1})\frac{v}{\nabla}$$

or

$$\frac{z_{g2} - z_{g1}}{y_{g2} - y_{g1}} = \frac{z_{B_2} - z_{B_1}}{y_{B_2} - y_{B_1}} \tag{2.33}$$

which shows that $\overline{B_1 B_2}$ is parallel to $\overline{g_1 g_2}$. When ϕ tends to zero, $\overline{g_1 g_2}$ tends to the initial waterline and $\overline{B_1 B_2}$ to the tangent in B_1 to the B-curve.

2.10 The Metacentric Evolute

The buoyancy force is always normal to the waterline. As the tangent to the B-curve is parallel to the corresponding waterline, it follows that the buoyancy force is normal to the B-curve. In

Figure 2.14 the normals to the B-curve in the points B_1 and B_2 intersect in a point M. In some languages this point is called *metacentric point*. When $B_1 \longrightarrow B_2$, the metacentric point tends to the metacentre.

Let the curve M be the locus of the metacentres corresponding to a given curve B. The curve M is the locus of centres of curvature of the curve B; it is also the **envelope** of the normals to the curve B. By definition, the curve M is the *evolute* of the curve B (see, for example Struik, 1961; Thuillier et al., 1991, pp. 223–224); it is called **metacentric evolute**. The term used in French is *développée métacentrique*, in German, *Metazentrische Evolute*, in Italian, *evoluta metacentrica*, and in Spanish, *evoluta metacéntrica*.

Conversely, the curve B intersects at right angles the tangents to the metacentric evolute. Then, by definition, the curve B is the *involute* of the curve M. The term used in French is *développante*; in German, *Evolvente*, in Italian, *evolventa*, and in Spanish, *evolvente*.

The concepts of B and M curve are illustrated in Examples 2.5 and 2.6. Some readers may be familiar with another example of a pair of curves that stay one to the other in the relationship *evolute-involute*. The shape of the tooth flanks used today in most gears is that of an involute of circle.

2.11 Metacentres for Various Axes of Inclination

In Eq. (2.32) the moment of inertia, I, is calculated about the axis of inclination. This axis passes through the centroid, F, of the waterplane and so does any other axis of inclination. It can be shown that there is a pair of orthogonal axes such that the moment of inertia about one of them is minimum and about the other maximum. Then, the metacentric radius corresponding to the former axis is minimum, and the moment about the latter axis is maximum (see, for example, Hahn, 1992; Meriam and Kraige, 2008; or Rumpl and Sondershausen, 1994). Correspondingly, one of the metacentric radii is minimum and the other maximum. In some European countries the smallest radius is denoted by r and is called *small metacentric radius*, while the largest radius is denoted by R and is called *large metacentric radius*.

The fact that the minimum and the maximum metacentric radii are measured in planes that are perpendicular one to the other is a consequence of an important theorem in the differential geometry of curves and surfaces. This theorem, due to Euler, is mentioned in some detail in Chapter 13. The theory of *principal curvatures* it treated, for example, in Struik, 1961, Sections 2–6; Davies and Samuels, 1996, Section 4.3; Banchoff and Lovett, 2010, Section 6.4; or Marsh, 2000, Theorem 10.10. In fact, the minimum and the maximum metacentric radii are the inverses of the principal curvatures, that is they are the *principal radii of curvature* of the surface of the centre of buoyancy.

In the theory of moments of inertia the two axes for which we obtain the extreme values of moments of inertia are called **principal axes** and the corresponding moments, **principal moments of inertia**. When the waterplane area has an axis of symmetry, this axis is one of the principal axes; the other one is perpendicular to the first. The waterplane area of ships in upright condition has an axis of symmetry: the intersection of the waterplane and the centreline plane. The moment of inertia about this axis is the smallest one; it is used to calculate the **transverse metacentric radius**. The moment of inertia about the axis perpendicular in F to the centreline is the largest; it enters in the calculation of the **longitudinal metacentric radius**.

To give an idea of the relative orders of magnitude of the transverse and longitudinal metacentric radii, let us consider a parallelepipedic barge whose length is L, breadth, B, and draught, T. The volume of displacement equals $\nabla = LBT$. The transverse metacentric radius results from

$$\overline{BM} = \frac{LB^3/12}{LBT} = \frac{B^2}{12T}$$

The longitudinal metacentric radius is given by

$$\overline{BM_L} = \frac{BL^3/12}{LBT} = \frac{L^2}{12T}$$

The ratio of the two metacentric radii is

$$\frac{\overline{BM_L}}{\overline{BM}} = \left(\frac{L}{B}\right)^2$$

The length-breadth ratio ranges from 3.1, for some motor boats, to 10.5, for fast cruisers. Correspondingly, the ratio of the longitudinal to the transverse metacentric radius varies roughly between 10 and 110. As a rule of thumb, the longitudinal metacentric radius is of the same order of magnitude as the ship length.

2.12 Summary

A body submersed in a fluid is subjected to an upwards force equal to the weight of the displaced fluid. This is *Archimedes' principle*. The hydrostatic force predicted by this principle passes through the centroid of the displaced fluid volume; we call that point *centre of buoyancy* and denote it by the letter B.

For a floating body the mass of the displaced fluid equals the mass of that body. The symbol for the immersed volume is ∇; that for the displaced mass, Δ. If the density of the fluid is ρ, we can write

$$\Delta = \rho\nabla$$

Values of the density of water in different navigation ways are given in the Appendix of this chapter.

If a floating body is inclined by a small angle, the new waterplane intersects the initial one along a line that passes through its centroid, that is, through the centre of flotation, F.

If the floating body is a ship, using the notations described in Chapter 1 we write

$$\rho C_B LBT = \sum_{i=1}^{n} m_i$$

where m_i is the mass of the ith item aboard and n, the total number of ship items. Δ is called *displacement mass* and ∇, *displacement volume*. The above equation expresses the condition of equilibrium of forces. The condition of equilibrium of moments requires that the centre of gravity, G, of the floating body and its centre of buoyancy, B, lie on the same vertical. This conditions is known as *Stevin's law*.

We say that a floating body is initially "stable" if after a small perturbation of its position of equilibrium, that body returns to its initial position when the perturbation disappears. To study initial stability, Bouguer introduced the notion of *metacentre*. Let the line of action of the buoyancy force in the initial position be R. If the floating body is inclined by a small angle, $\delta\phi$, the buoyancy force acts along a new line, say S. When $\delta\phi$ tends to zero, the intersection of the two lines, R, and S, tends to a point, M, called *metacentre*.

The equilibrium of a floating body is stable if its metacentre lies above its centre of gravity. The distance from the centre of gravity to the metacentre, $\overline{GM}$, is called *metacentric height* and is considered positive upwards. The condition of initial stability can be expressed as

$$\overline{GM} > 0$$

The distance from the centre of buoyancy to the metacentre, $\overline{BM}$, is called *metacentric radius*. Its value is given by

$$\overline{BM} = \frac{I}{\nabla}$$

where I is the moment of inertia of the waterplane about the axis of inclination, a line that passes through the centroid of the area, that is through the centre of flotation, F. Let K be the origin of vertical coordinates. We can write

$$\overline{GM} = \overline{KB} + \overline{BM} - \overline{KG}$$

By its definition, the metacentre is the centre of curvature of the curve described by B for different angles of inclination. The curve described by the metacentre is the evolute of the curve of centres of buoyancy. The normals to the curve B are tangents to the curve M.

2.13 Examples

Example 2.1 (Melting ice cube).   The following problem is sometimes presented as an intelligence quiz. We describe it here as a fine application of Archimedes' principle.

Let us suppose that somebody wants to cool a glass of water by putting in it a cube of ice made of the **same** water. Should he fear that when the cube melts the level of the water will rise?

We look for a property that does not change between the two states of the cube. This is the mass of the cube, M. Let the density of water be δ. The volume of water displaced by the cube equals M/δ. After meltdown the cube becomes a volume of water equal to M/δ. Conclusion: the water volume in the glass does not change and neither does the water level.

Example 2.2 (Designing a floating body).   To show how Archimedes' principle can be turned into a design equation let us assume that we are interested in a cylindrical float that should carry a load of mass M. To simplify the exercise let the float be made of homogeneous material, for example wood, of density ρ_{wood}. Let the density of the surrounding water be ρ_{water}. As usual in ship design, we must assume certain relationships between the main dimensions. Referring to Figure 2.15 our choice is

Cylinder diameter d
Float depth $D = d/2$
Float draught $T = D/2$
vcg of mass M 0.1 D above the float deck

Above *vcg* means *vertical centre of gravity*, that is the z-coordinate of the centre of gravity of the respective mass.

The design equation says that the displacement mass equals the float mass plus the given mass, M

$$\rho_{water}\frac{\pi d^2}{4}T = \rho_{wood}\frac{\pi d^2}{4}D + M$$

Figure 2.15   Designing a buoy

Substituting in this equation the relationships assumed above we obtain

$$d^3 = \frac{8M}{\pi(\rho_{water}/2 - \rho_{wood})}$$

This yields the condition

$$\frac{\rho_{water}}{2} - \rho_{wood} > 0$$

The downloadable file `FloatDesign.m` contains a MATLAB function that solves the problem and checks the stability of the float. For $\rho_{water} = 1.025 \text{ t m}^{-3}$ and $\rho_{wood} = 0.5 \text{ t m}^{-3}$ the results are

Float diameter, d	5.884 m
Displacement mass, Δ	41.000 t
Wood mass	40.000 t
Vertical centre of buoyancy, $\overline{KB}$	0.736 m
Metacentric radius, $\overline{BM}$	1.471 m
Vertical centre of gravity, $\overline{KG}$	1.514 m
Metacentric height, $\overline{GM}$	0.692 m

Other simple examples of the use of Archimedes' principle in writing a design equation can be found in Biran and Breiner (1995, pp. 289–290), and in Biran (2011, pp. 140–143).

Example 2.3 (Cone floating with vertex down). Figure 2.16a shows a cone floating top down in water. The diameter of the base is D; the height, H, and the diameter of the waterplane area, d. We assume that the cone is made of homogeneous material with specific gravity γ_c. Let the specific gravity of the water be γ_w. We want to find out under which conditions the cone can float as shown in the figure. As explained in this chapter, the conditions to be fulfilled are Archimedes' principle and $\overline{GM} > 0$.

We begin by finding the draught, T. Archimedes' principle allows us to write

$$\gamma_c \frac{\pi D^2}{3 \cdot 4} H = \gamma_w \frac{\pi d^2}{3 \cdot 4} T \tag{2.34}$$

Geometrical similarity between the submerged cone and the whole cone yields

$$\frac{d}{D} = \frac{T}{H} \tag{2.35}$$

Substituting d from Eq. (2.35) into Eq. (2.34), and noting

$$\alpha = \left(\frac{\gamma_c}{\gamma_w}\right)^{1/3}$$

Figure 2.16 A floating cone

we obtain

$$T = \alpha H \tag{2.36}$$

Other quantities necessary for checking the initial stability are derived by knowing that the centroid of a right, circular cone of height H is situated on its axis, at a distance $H/4$ from the base:

$$\overline{KG} = \frac{3}{4}H$$

$$\overline{KB} = \frac{3}{4}T = \frac{3}{4}\alpha H$$

$$\overline{BM} = \frac{I}{\nabla}$$

With

$$I = \frac{\pi d^4}{64}, \quad \nabla = \frac{\pi d^2}{3 \cdot 4}T$$

the metacentric radius is

$$\overline{BM} = \frac{3d^2}{16T} = \frac{3\alpha}{16}\left(\frac{D}{H}\right)^2 H$$

and the metacentric height

$$\overline{GM} = \overline{KB} + \overline{BM} - \overline{KG}$$

$$= \frac{3}{4}\alpha H + \frac{3\alpha}{16}\left(\frac{D}{H}\right)^2 H - \frac{3}{4}H$$

The cone is stable if

$$\frac{3}{4}\alpha + \frac{3\alpha}{16}\left(\frac{D}{H}\right)^2 - \frac{3}{4} > 0 \tag{2.37}$$

From Eq. (2.37) we can deduce a condition for the specific gravity of the cone material

$$\alpha > \frac{1}{1 + \frac{1}{4}\left(\frac{D}{H}\right)^2} \tag{2.38}$$

or, a condition for the D/H ratio:

$$\left(\frac{D}{H}\right)^2 > 4 \cdot \frac{1-\alpha}{\alpha} \tag{2.39}$$

Obviously, for the cone to float, the ratio α must be smaller than one. Thus, the complete condition for the cone material is

$$\frac{1}{1 + \frac{1}{4}\left(\frac{D}{H}\right)^2} < \alpha < 1 \tag{2.40}$$

Example 2.4 (Cone floating with vertex up). Figure 2.16b shows a cone floating top up. Noting by F_b the *freeboard*, that is the difference $H - T$, Archimedes' principle yields the equation

$$\gamma_c \frac{\pi D^2}{3 \cdot 4} H = \gamma_w \frac{\pi}{3 \cdot 4}(D^2 H - d^2 F_b) \tag{2.41}$$

We obtain

$$F_b = \frac{\gamma_w - \gamma_c}{\gamma_w} \cdot \frac{D^2 H}{d^2} \tag{2.42}$$

Geometrical similarity gives us

$$d = \frac{D}{H} F_b \tag{2.43}$$

Combining Eqs. (2.42) and (2.43) we obtain

$$F_b = \left(\frac{\gamma_w - \gamma_c}{\gamma_w}\right)^{1/3} H$$

With

$$\beta = \left(\frac{\gamma_w - \gamma_c}{\gamma_w}\right)^{1/3}$$

we write for the freeboard

$$F_b = \beta H \tag{2.44}$$

and for the draught,

$$T = H - F_b = (1 - \beta)H \tag{2.45}$$

The diameter of the waterplane section is given by

$$d = \frac{D}{H} F_b = \beta D \tag{2.46}$$

To find the vertical coordinate of the centre of buoyancy we calculate the volumes and moments about base of the whole cone and of the emerging cone. The differences of the respective values yield the values or the submerged, truncated cone. Thus, from Eq. (2.41) we obtain the submerged volume:

$$\nabla = \frac{\gamma_c}{\gamma_w} \cdot \frac{\pi D^2}{3 \cdot 4} H = (1 - \beta^3) \frac{\pi D^2}{3 \cdot 4} H \tag{2.47}$$

The emerging volume is

$$V_E = \beta^3 \frac{\pi D^2}{3 \cdot 4} H \tag{2.48}$$

The moment of the whole cone about its basis is

$$M_c = \frac{\pi D^2 H^2}{3 \cdot 4^2} \tag{2.49}$$

and the moment of the emerging cone about the same base is

$$M_E = \frac{\pi D^2 \beta^3 H^2}{3 \cdot 4} \left[(1 - \beta)H + \frac{\beta H}{4} \right] = \frac{\pi \beta^3 D^2 H^2}{3 \cdot 4^2} (4 - 3\beta) \tag{2.50}$$

The moment of the submerged volume is

$$M_\nabla = M_c - M_E = \frac{\pi D^2 H^2}{3 \cdot 4^2} (1 - 4\beta^3 + 3\beta^4) \tag{2.51}$$

The vertical centre of buoyancy is given by

$$\overline{KB} = \frac{M_\nabla}{\nabla} = \frac{1 - 4\beta^3 + 3\beta^4}{1 - \beta^3} \cdot \frac{H}{4} \tag{2.52}$$

We calculate the metacentric radius as

$$\overline{BM} = \frac{I}{\nabla} \tag{2.53}$$

With

$$I = \frac{\pi \beta^4 D^4}{64} \tag{2.54}$$

and Eq. (2.47) we obtain

$$\overline{BM} = \frac{3}{16} \cdot \frac{\beta^4}{1 - \beta^3} \cdot \frac{D^2}{H} \tag{2.55}$$

The height of the centre of gravity is

$$\overline{KG} = H/4 \qquad (2.56)$$

and the resulting metacentric height is

$$\overline{GM} = \overline{KB} + \overline{BM} - \overline{KG}$$
$$= \frac{1 - 4\beta + 3\beta^4}{1 - \beta^3} \cdot \frac{H}{4} + \frac{3}{16} \cdot \frac{\beta^4}{1 - \beta^3} \cdot \frac{D^2}{H} - \frac{H}{4} \qquad (2.57)$$

The cone is stable if

$$1 - 4\beta^3 + 3\beta^4 + \frac{3}{4}\beta^4 \left(\frac{D}{H}\right)^2 > 1 - \beta^3 \qquad (2.58)$$

We obtain a condition for the D-to-H ratio:

$$\left(\frac{D}{H}\right)^2 > 4\frac{1 - \beta}{\beta} \qquad (2.59)$$

The condition for the specific gravity of the cone material is

$$\beta > \frac{4}{(D/H)^2 + 4} \qquad (2.60)$$

In addition β must also fulfil the inequalities

$$0 < \beta < 1 \qquad (2.61)$$

Example 2.5 (A parallelepipedic barge). Let us consider a parallelepipedic barge; it has a constant, rectangular transverse section as shown in Figure 2.17. Let L be the length, B the breadth, H the depth, and T the draught. For this simple body form we can calculate analytically the positions of the centre of buoyancy and of the metacentre. We shall do this in two ways:

- Starting from known principles of mechanics and elementary results of differential geometry;
- Using the theorems developed in this chapter.

We begin this example by discussing the case in which the waterline reaches first the deck and later the bottom. Formally, this condition is expressed by

$$H - T < T \qquad (2.62)$$

that is $H < 2T$. In upright condition the centre of buoyancy, B_0, is situated in the centreline plane and its height above the bottom equals $T/2$. As shown in Figure 2.18, we use a coordinate system with the origin in B_0, and measure y horizontally, positive rightwards, and z vertically, positive upwards.

In Figure 2.17 we consider that the barge heels to starboard by an angle ϕ and the new waterline is $W_\phi L_\phi$. We distinguish several phases:

Figure 2.17   A barge with simple geometrical form

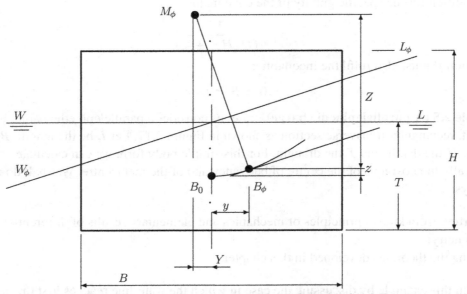

Figure 2.18   Centre of buoyancy and metacentre of simple barge

1. The new waterline is situated between the original waterline, $W_0 L_0$, and the waterline passing through the corner of the deck. Formally, this case is defined by

$$0 \geq \phi \leq \arctan \frac{H - T}{B/2} \tag{2.63}$$

2. The waterline is situated between the waterline that passes through the starboard deck corner and the waterline that passes through the port-side bottom corner. Formally, this

means

$$\arctan \frac{H-T}{B/2} < \phi \le \arctan \frac{2T}{B} \tag{2.64}$$

3. As the angle ϕ increases, two other phases can be distinguished. However, it is easier to consider those phases as being symmetric to the first two.

Phase 1

For the simple form considered in this example we can start from the principles of statics. We first observe that within the whole heel range defined by Eq. (2.63) the two waterlines W_0L_0 and $W_\phi L_\phi$ intersect in the centreline plane. Indeed, the submerging and the emerging wedges thus defined are equal, that is the volume of displacement is constant (*isocarène* heeling). In other words we are dealing with a wall-sided barge.

To calculate the change of moment we multiply the volume of each wedge by the coordinate of its centroid measured from a convenient coordinate plane. Then, the coordinates of the centre of buoyancy, B_ϕ, are obtained by means of the Lemma on moving volumes (Section 2.7). The calculations for the y-coordinate are shown in Table 2.2.

This is the place to stop for a short digression on this *tabular form of calculations*. Let us refer to Table 2.2. Column 2 contains the volumes of the initial hull, of the submerged wedge and of the emerged wedge. Column 3 contains the y-coordinates of the volumes entered in column 2. As said, these coordinates are measured from the centreline plane; we call them tcb, an acronym for *transverse centre of buoyancy*. We use lower-case letters and reserve the upper-case notation, TCB, for the y-coordinate of the whole body. Column 4 contains the moments of the initial body and of the wedges, about the centreline plane. These moments are calculated as products of the terms in column 2, by those in column 3. The procedure is described symbolically by the expression $4 = 2 \times 3$ written in the subheading of column 4.

The sum of the terms in column 2 equals the total volume of the heeled barge; it is written in the cell identified by the entries *Volume* and *Total*. Similarly, the sum of the partial moments in column 4 is the moment of the heeled barge about the centreline plane; it appears in the cell corresponding to the entries *Moment* and *Total*. Dividing the moment of the heeled barge by

Table 2.2 Calculating the transverse centre of buoyancy of the heeled barge

Solid	Volume	tcb	Moment
1	2	3	$4 = 2 \times 3$
Initial	LBT	0	0
Submerged wedge	$LB^2 \tan\phi/8$	$2B/(3 \cdot 2)$	$LB^3 \tan\phi/(3 \cdot 8)$
Emerged wedge	$-LB^2 \tan\phi/8$	$-2B/(3 \cdot 2)$	$LB^3 \tan\phi(3 \cdot 8)$
Total	LBT	$B^2 \tan\phi/(12T)$	$LB^3 \tan\phi/12$

its volume yields the y-coordinate of the heeled barge:

$$TCB = \frac{LB^3 \tan \phi / 12}{LBT} = B^2 \tan \phi / (12T)$$

This result is written in the cell identified by the entries *tcb* and *Total*.

A similar procedure is used to find the z-coordinate of the heeled barge; it is shown in Table 2.3. Calculations in tabular form are standard in Naval Architecture. More about them is written in Chapters 3 and 7 and we expect the reader to discover gradually the advantages of this way of solving problems. Obviously, Tables 2.2 and 2.3 can be consolidated. Then, the volumes are entered only once.

Tables 2.2 and 2.3 yield the **parametric equations** of the curve of centres of buoyancy:

$$y = \frac{1}{12} \cdot \frac{B^2}{T} \tan \phi$$

$$z = \frac{1}{24} \cdot \frac{B^2}{T} \tan^2 \phi \qquad (2.65)$$

We call the curve of centres of buoyancy **B curve**. From Eqs. (2.65) we can derive

$$z = \frac{6T}{B^2} x^2 \qquad (2.66)$$

This is the equation of a parabola.

The slope of the curves of centres of buoyancy is given by

$$\frac{dz}{dy} = \frac{dz/d\phi}{dy/d\phi} = \tan \phi \qquad (2.67)$$

where

$$\frac{dy}{d\phi} = \frac{B^2}{12T} \cdot \frac{1}{\cos^2 \phi} d\phi \qquad (2.68)$$

and

$$\frac{dz}{d\phi} = \frac{B^2}{12T} \cdot \frac{\tan \phi}{\cos^2 \phi} d\phi \qquad (2.69)$$

Table 2.3   Calculating the vertical centre of buoyancy of the heeled barge

Solid	Volume	vcb	Moment Change
1	2	3	4 = 2 × 3
Initial	LBT	0	0
Submerged wedge	$LB^2 \tan \phi / 8$	$B \tan \phi / (3 \cdot 2)$	$LB^3 \tan^2 \phi / (8 \cdot 3 \cdot 2)$
Emerged wedge	$-LB^2 \tan \phi / 8$	$-B \tan \phi / (3 \cdot 2)$	$LB^3 \tan^2 \phi / (8 \cdot 3 \cdot 2)$
Total	LBT	$B^2 \tan^2 \phi / (24T)$	$LB^3 \tan^2 \phi / (3 \cdot 8)$

Equation (2.67) shows that the tangent in B_ϕ has the slope ϕ, meaning that it is parallel to the corresponding waterline.

To find the radius of curvature of the B curve we calculate

$$\frac{d^2 z}{dy^2} = \frac{1}{\cos^2 \phi} \frac{d\phi}{dy} = \frac{12T}{B^2} \tag{2.70}$$

and use a formula that can be found in many books on calculus or classic differential geometry (see, for example, Stoker, 1969, p. 26; Taillé, 1975, p. 73; Gray, 1993, p. 11):

$$R = \frac{(1 + (dz/dy)^2)^{3/2}}{d^2 z/dy^2} = \frac{B^2}{12T} \cdot \frac{1}{\cos^3 \phi} \tag{2.71}$$

Now, let us use the theorems developed in this chapter. The volume of displacement of the barge is

$$\nabla = LBT$$

Equations (2.31) yield

$$x_b = \frac{I_{xy}}{\nabla} \tan \phi = \frac{0}{LBT} = 0$$

$$y_B = \frac{I}{\nabla} \tan \phi = \frac{1}{12} \cdot \frac{B^2}{T} \tan \phi$$

$$z_B = \frac{1}{2} \cdot \frac{I}{\nabla} \tan^2 \phi = \frac{1}{24} \cdot \frac{B^2}{T} \tan^2 \phi \tag{2.72}$$

These are exactly the results obtained in Tables 2.2 and 2.3. As to the metacentric radius, we calculate from Eq. (2.32)

$$\overline{BM_0} = \frac{I}{\nabla} = \frac{LB^3/12}{LBT} = \frac{1}{12} \cdot \frac{B^2}{T}$$

and, for any heel angle ϕ,

$$\overline{BM_\phi} = \frac{L(B/\cos\phi)^3/12}{LBT} = \frac{1}{12} \cdot \frac{B^2}{T} \cdot \frac{1}{\cos^3 \phi} \tag{2.73}$$

This is exactly the length of the radius of curvature obtained from Eq. (2.71).

Phase 2

In this phase the waterline passed the starboard deck corner and approaches the port-side bottom corner. If we consider the barge heeled by 90°, so that the starboard side becomes the new bottom, the barge is again a wall-sided floating body. This observation allows us to continue the calculations in the same manner as for Phase 1. However, they would be more complex so that algebraic technicalities could obscure insight. To avoid this, we make a simplifying assumption (Hervieu, 1985): $T = H/2$. Then, the angle defining the limit between Phase 1 and Phase 2 is given by

$$\tan \phi = \frac{H}{B}$$

Substituting this value into Eq. (2.72) we find that at this angle the coordinates of the centre of buoyancy are

$$y_B = \frac{1}{12} \cdot \frac{B^2}{T} \tan \phi = \frac{B}{6}$$

$$z_B = \frac{1}{24} \cdot \frac{B^2}{T} \tan^2 \phi = \frac{H}{12}$$

It is easy to see, in Figure 2.19 that these are the expected coordinates.

To continue the calculations in Phase 2, we use a new system of coordinates, η, ζ, with the origin in the centre of buoyancy, B_{90}, of the barge heeled by 90°. The relationships between the two systems of coordinates can be derived from Figure 2.20. We obtain thus

$$y_B = B/4 - \zeta_B$$
$$z_B = H/4 + \eta_B \tag{2.74}$$

Figure 2.19   Simple barge

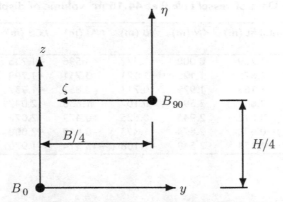

Figure 2.20 Coordinate systems for simple barge

B = 10, H = 6

Figure 2.21 B and M curves of simple barge

The equations shown above are implemented in a MATLAB function called BARGE1 that can be found on the website provided for this book. The results of running the function with $B = 10$ and $H = 6$ are shown in Figure 2.21.

The reader is invited to experiment with various values of B and H and see how they influence the shape of the B and M curves. A more general treatment of the same problem can be found in Krappinger (1960).

Example 2.6 (B and M curves of *Lido 9*). Table 2.4 contains hydrostatic data of the vessel *Lido 9* for a volume of displacement equal to 44.16 m³ and the heel angles $0°, 15°, 30°, \ldots, 90°$. As shown in Figure 2.22, all the data are measured in a system of coordinates ξ, η, ζ. In this example, the axes $K\eta$ and $K\zeta$ rotate with an angle ϕ with respect to the axes Ky, Kz in which the hull surface is defined. The angle ϕ is the *heel angle*. The draught, T, is measured perpendicularly to the waterline; in our figure it is $T = \overline{KQ}$. As we see, $\overline{KN}$ is parallel to the waterline. The centre of buoyancy corresponding to the heel angle ϕ is marked B_ϕ and the respective metacentre, M_ϕ. In the table we dropped the subscripts ϕ. The height of the centre of buoyancy, $\overline{NB_\phi}$, is measured perpendicularly to the waterline and so is the height of the metacentre, $\overline{NM_\phi}$.

Table 2.4   Data of vessel *Lido 9* at 44.16 m³ volume of displacement

Heel Angle (°)	Draught (m)	$\overline{KN}$ (m)	$\overline{NB}$ (m)	$\overline{NM}$ (m)	LCB (m)	$\overline{NM_L}$ (m)
0	1.729	0.000	1.272	4.596	−1.735	23.371
15	1.575	1.122	1.121	3.711	−1.799	23.730
30	1.163	1.979	0.711	2.857	−1.932	23.154
45	0.600	2.595	0.107	1.830	−2.047	23.133
60	−0.012	2.945	−0.625	0.479	−2.072	17.473
75	−0.693	2.874	−1.393	−0.869	−2.008	14.298
90	−1.354	2.539	−2.108	−13.314	−1.970	12.792

Figure 2.22   The coordinates of the points *B* and *M*

In this example we want to draw the curve of centres of buoyancy, B, and the metacentric evolute, M, at the given volume of displacement. With the data in Table 2.4 and the definitions shown in Figure 2.22 it is possible to draw manually these curves. Instead of this it is possible to use an M-file to draw the B and M curves for any ship we may want. The data is written in a convenient way, on an M-file named after the vessel we are studying. Thus, the contents of the file `lido9.m` can be found on the website of this book.

Next, we project all points we are interested in on a *transverse* plane, that is a plane for which the longitudinal coordinate, x, is constant. We do this as in Figure 2.22. Let our plane be the midship section. For *Lido 9* this section is described by the points $\mathbf{P_1}, \mathbf{P_2}, \ldots, \mathbf{P_{15}}$ whose coordinates are given in Table 2.5.

Table 2.5 Points defining the midship section of the ship *Lido 9*

Point	y	z	Point	y	z
P_1	0.000	0.50	P_9	3.176	2.250
P_2	0.240	0.50	P_{10}	3.200	2.500
P_3	0.240	0.58	P_{11}	3.218	2.750
P_4	1.100	1.00	P_{12}	3.230	3.000
P_5	1.787	1.25	P_{13}	3.230	3.360
P_6	2.460	1.50	P_{14}	2.099	3.425
P_7	2.902	1.75	P_{15}	0.000	3.489
P_8	3.100	2.00			

Let x_B, y_B be the coordinates of the centre of buoyancy, B, and x_M, y_M those of the metacentre, M. With the help of Figure 2.22 we can write

$$y_B = \overline{KN}\cos\phi - \overline{KB}\sin\phi$$
$$z_B = \overline{KN}\sin\phi + \overline{KB}\cos\phi \tag{2.75}$$

and

$$y_M = \overline{KN}\cos\phi - \overline{KM}\sin\phi$$
$$z_M = \overline{KN}\sin\phi + \overline{KM}\cos\phi \tag{2.76}$$

Equation (2.75) can be rewritten in matrix form as

$$\begin{bmatrix} x_B \\ y_B \end{bmatrix} = \begin{bmatrix} \cos\phi & -\sin\phi \\ \sin\phi & \cos\phi \end{bmatrix} \begin{bmatrix} \overline{KN} \\ \overline{KB} \end{bmatrix} \tag{2.77}$$

Similarly, Eq. (2.76) can be written in matrix form as

$$\begin{bmatrix} x_M \\ y_M \end{bmatrix} = \begin{bmatrix} \cos\phi & -\sin\phi \\ \sin\phi & \cos\phi \end{bmatrix} \begin{bmatrix} \overline{KN} \\ \overline{KM} \end{bmatrix} \tag{2.78}$$

The **transformation matrix**

$$\begin{bmatrix} \cos\phi & -\sin\phi \\ \sin\phi & \cos\phi \end{bmatrix} \tag{2.79}$$

performs counterclockwise rotation, around the origin, with the angle ϕ. In this example we need twice this rotation. As we may need it for more calculations in the future, it is worth programming a MATLAB function that evaluates the matrix and add this function to our toolbox. A possible listing of a file called `rotate.m` is given on the website of this book.

To draw the waterline we need a point on it. The easiest to calculate is the point Q shown in Figure 2.22. Here $\overline{KQ}$ corresponds to the draught T calculated by the programme ARCHIMEDES. The equation of the waterline passing through this point is

$$y - T\cos\phi = \tan\phi(x - T\sin\phi) \tag{2.80}$$

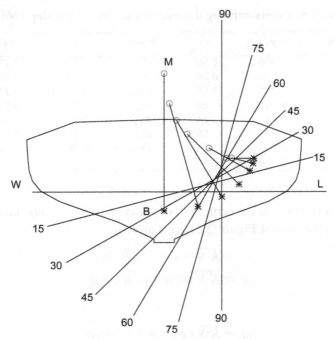

Figure 2.23   B and M curves of vessel *Lido 9*

The M-file, called b_curve, provided on the website of the book, performs all the calculations. The resulting plot is shown in Figure 2.23.

Table 2.4 contains a column that we did not use until now: the LCB values. We included these data to show that at finite angles of heel the centre of buoyancy can leave its initial transverse plane and move along the ship. This is the case of ships that do not have a fore-to-aft symmetry. Then, when the heel changes, the trim also changes until centre of gravity and centre of buoyancy lie again on the same vertical (Stevin's law).

Example 2.7 (Catamaran stability).   Up to this point we have considered floating bodies whose buoyancy is provided by one submerged volume. If the floating body is a ship, we say that she is a **monohull ship**. In the example that follows we are going to show that stability can be greatly improved by distributing the buoyancy in two hulls. Then we talk about a *twin-hull ship*, but more often we use the term **catamaran**, a word derived from the Tamil "kattumarum" composed of two words meaning "to tie" and "tree." As the etymology indicates, catamarans have been in use for centuries in the Indian and Pacific Oceans. Today, many competition sailing boats and fast ferries are of the catamaran type.

Let us consider in Figure 2.24 a barge of breadth B and length L. Assuming the draught T, the displacement volume is

Figure 2.24 Monohull versus catamaran

$$\nabla = LBT$$

and the metacentric radius,

$$\overline{BM} = \frac{B^3L/12}{LBT} = \frac{1}{12} \cdot \frac{B^2}{T}$$

We can obtain the same displacement volume with two hulls of breadth $B/2$, the same length, L, and the same draught, T. Assuming that the distance between the centrelines of the two hulls is $3B/2$, the resulting metacentric radius is

$$\overline{BM} = \frac{2}{LBT}\left[L \cdot \frac{(B/2)^3}{12} + \frac{B}{2} \cdot L \cdot \left(\frac{3B}{2 \cdot 2}\right)^2 \right] = \frac{7B^2}{12T}$$

The first term between parentheses represents the sum of the moments of inertia of the waterlines about their own centrelines. The second term accounts for the parallel translation of the hulls from the plane of symmetry of the catamaran. The second term is visibly the greater. The ratio of the catamaran $\overline{BM}$ to that of the monohull is 7. The improvement in stability is remarkable.

Catamarans offer also the advantage of larger deck areas and, under certain conditions, improved hydrodynamic performances. On the other hand, the weight of structures increases and the overall performance in waves must be carefully checked. It may be worth mentioning that also many vessels with three hulls, that is *trimarans*, have been built. Moreover, *Nigel Gee and Associates* of Southampton developed a remarkable concept of a large ship with a main, slender hull, and four side hulls; that is a *pentamaran*.

Example 2.8 (Submerged bodies).   Completely submerged bodies have no waterplane; the surface of the centre of buoyancy, B, is reduced to one point. Therefore, the metacentric radii are equal to zero. Then the condition of initial stability is reduced to

$$\overline{GM} = \overline{KB} - \overline{KG} > 0$$

In simple words, the centre of gravity, G, must be situated under the centre of buoyancy, B. We invite the reader to draw a sketch showing the two mentioned points and derive the condition of stability by simple mechanical considerations. Submerged bodies do not develop hydrostatic moments that oppose inclinations, as they do not develop hydrostatic forces that oppose changes of depth.

Example 2.9 (An offshore platform).   Figure 2.25 is a sketch of an *offshore platform* of the *semi-submersible* type. The buoyancy is provided by four pontoons, each of diameter b and length ℓ. The platform deck is supported by four columns. The depth of the platform is H and the draught, T.

Our problem is to find a condition for the height of the centre of gravity, $\overline{KG}$, for given platform dimensions. To do this, we calculate the limit value of $\overline{KG}$ for which the metacentric height, $\overline{GM}$, is zero. The metacentric radius is given by

$$\overline{BM} = \frac{I}{\nabla}$$

Figure 2.25   A semi-submersible platform

Table 2.6 Calculation of $\overline{BM}$

	Volume	Vertical Arm	Moment
Pontoons	$\pi b^2 \ell$	$b/2$	$\pi b^3 \ell$
Columns	$\pi b^2 (T - b/2)$	$(T+b)/2$	$\frac{\pi b^2 (2T^2 + bT - b^2)}{4}$
Total	$\pi b^2 (\ell + T - b/2)$	$\frac{2T^2 + bT + 4b\ell}{4(\ell + T)}$	$\frac{\pi b^2 (4b\ell + 2T^2 + bT - b^2)}{4}$

where the moment of inertia of the waterplane, I, and the volume of displacement, ∇, are

$$I = 4 \left[\frac{\pi b^4}{64} + \left(\frac{\ell - b}{2} \right)^2 \frac{\pi b^2}{4} \right] = \frac{\pi b^2}{4} \left[5 \left(\frac{b}{2} \right)^2 - 2\ell b + \ell^2 \right]$$

$$\nabla \simeq 4 \frac{\pi b^2}{4} \ell + 4 \frac{\pi b^2}{4} T = \pi b^2 (\ell + T)$$

In calculating the volume of displacement, ∇, we did not take into account the overlapping between column and pontoon ends. In conclusion

$$\overline{BM} = \frac{5 (b/2)^2 - 2\ell b + \ell^2}{4(\ell + T)} \simeq \frac{\ell(\ell - 2b)}{4(\ell + T)} \tag{2.81}$$

where we neglected the term in b^2, usually small in comparison with other terms.

The height of the centre of buoyancy above the baseline is calculated in Table 2.6. Neglecting the terms in $-b^2$ and $b/2$ we obtain

$$\overline{KB} = \frac{2T^2 + bT + 4b\ell}{4(\ell + T)} \tag{2.82}$$

The height of the metacentre above the baseline is given by

$$\overline{KM} = \overline{KB} + \overline{BM} \approx \frac{2T^2 + bT + \ell^2}{4(l + T)} \tag{2.83}$$

The condition for initial stability is

$$\overline{KG} < \frac{2T^2 + bT + \ell^2}{4(\ell + T)} \tag{2.84}$$

To rewrite Eq. (2.84) in non-dimensional form we define

$$\alpha = b/\ell, \quad \beta = T/\ell$$

and obtain

$$\frac{\overline{KG}}{\ell} = \frac{2\alpha + 2\beta^2 + \alpha\beta + 1}{4(1 + \beta)} \tag{2.85}$$

2.14 Exercises

Exercise 2.1.   (Melting icebergs) In Example 2.1 we learnt that if an ice cube melts in a glass of water, the level of water does not change. Then, why do people fear that the meltdown of all icebergs would cause a water-level rise and therefore the flooding of lower coasts? Show that they are right.

Hint: Icebergs are formed from compressed snow; their average density is 0.89 t m^{-3}. The density of ocean water can be assumed equal to 1.025 t m^{-3}.

Exercise 2.2.   (The tip of the iceberg) Calculate what part of an iceberg's volume can be seen above the water and explain the meaning of the expression *"The tip of the iceberg."*

Hint: See Exercise 2.1.

Exercise 2.3.   (Whisky on the rocks) Instead of considering a cube of ice floating in a glass of water, as in Example 2.1, let us think of a cube of ice floating in a glass of whisky. What happens when the cube melts?

Exercise 2.4.   (Perpetual motion?) This exercise is proposed by Salin and Martin (1997). In Figure 2.26 we consider a cylinder mounted in the wall of a tank filled with water. The axis of the cylinder lies in the wall plane. The cylinder can rotate freely and the watertightness between it and the tank is ensured by some means. As we learnt, the static pressure of the surrounding water produces a force exercised on the cylinder. At first glance this force could

Figure 2.26   A solution for perpetual motion

cause the rotation of the cylinder. Would this be a case of *perpetual motion* (*Perpetuum mobile*)? Explain why it isn't and why Archimedes' principle does not hold here.

Exercise 2.5.   (Draughts of a parallelepipedic barge) Consider a parallelepipedic (or, with another term, a *box-shaped*) barge characterized by the following data:

Length, L 10 m
Breadth, B 3 m
Displacement mass, Δ 30 t

Find the draught, T_1, in fresh water, and the draught, T_2, in ocean water. See the Appendix of this chapter for various water densities.

Exercise 2.6.   (A Lemma about moving masses in 3D) Prove the Lemma in Section 2.7 for a three-dimensional system of masses and a three-dimensional displacement of one of the masses.

Exercise 2.7.   (Area properties) Figure 2.27 shows schematically a ship waterline. Your data are:

L 120 m
B 20 m
e 30 m

Your tasks are:

1. to calculate the waterplane area, A_{wl};
2. to find the coordinates, x_F, y_F, of the centre of flotation (centroid of the waterplane area);
3. to calculate the moments of inertia (second moments) of the waterplane area about the given coordinate axes;
4. to calculate the centroidal moments of inertia (moments about the centroidal axes);
5. to draw the waterline at a standard scale and show on it the centre of flotation and the centroidal axes.

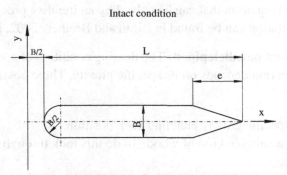

Figure 2.27   A schematic waterline

Figure 2.28   A cylindrical float

It is recommended to carry on the calculations in an Excel spreadsheet.

Exercise 2.8.   (A cylindrical float) Figure 2.28 shows a cylindrical body made of homogeneous wood of density 0.67 t m^{-3} and floating in sea water of density 1.025 t m^{-3}. Your data are

L   1.70 m
B   0.55 m

Your tasks are:

1. to calculate the draught, T;
2. to calculate the metacentric height, $\overline{GM}$;
3. to draw your conclusion. Can you give a general proof that the stability of any cylinder floating as in this exercise is neutral?
4. to calculate the longitudinal metacentric height, $\overline{GM}_L$;
5. to calculate the block coefficient, C_B, the midship coefficient, C_M, the waterline coefficient, C_{WL}, the prismatic coefficient, C_P, and the vertical prismatic coefficient, C_{VP}.

Hint: To calculate the draught you need the angle subtended by the waterline breadth. This leads to a transcendental equation that can be solved by an iterative procedure. A solution and its MATLAB implementation can be found in Biran and Breiner (2002, pp. 537–540).

Exercise 2.9.   (A wooden parallelepiped) The floating condition of a wooden, homogeneous block of square cross-section depends on its specific gravity. Three possible positions are shown in Figure 2.29:

1. Find the ranges of specific gravity enabling each position.
2. For each range find a suitable kind of wood. To do this look through tables of wood properties.
3. Can you imagine other floating positions? In the affirmative, calculate the corresponding wood-density range.

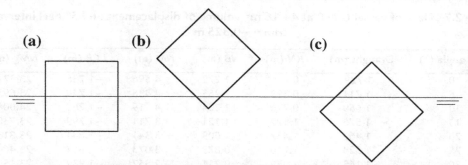

Figure 2.29 Different floating conditions of a wooden, parallelepipedic block

Hint: A floating position is possible if the corresponding metacentric height is positive.

Exercise 2.10. (Cone floating vertex down) Consider a cone floating with vertex down as in the left-hand side of Figure 2.16. Let us assume that the cone is made of red cedar of density 0.35 t m^{-3}, has the height $H = 1$ m, the base diameter $D = 2$ m, and floats in sweet water. Check if the data fulfil inequality (2.39). Calculate directly the metacentric height, $\overline{GM}$, and verify that the result corresponds to that predicted by the inequality.

Exercise 2.11. (Cone floating with vertex up) Consider a cone floating with its vertex up, as in the right-hand part of Figure 2.16. Let us assume that the cone floats in sea water of density 1.025 t m^{-3}, is made of red cedar of density 0.38 t m^{-3}, and has the height $H = 1$ m.

- a. Assuming that $D/H = 0.8$. Is the cone stable according to Eq. (2.60)? Verify your answer by direct calculation of $\overline{GM}$.
- b. Assuming that $D/H = 0.9$. Is the cone stable according to Eq. (2.60)? Verify your answer by direct calculation of $\overline{GM}$.

Exercise 2.12. (B and M curves—variable heel) Table 2.7 contains the same data items as Table 2.4, but calculated at 5-degree intervals. With this "resolution" it is possible to plot smooth B and M curves. First, write the data on a file lido9a similar to file lido9. Next, modify the programme shown in Example 2.6 to plot only the B and M curves of the vessel whose data are called from the keyboard. Run the programme with the data given at 5-degree intervals and print a hardcopy of the resulting plot.

Exercise 2.13. (B and M curves—variable trim) Table 2.8 contains data of the vessel *Lido 9* for constant volume of displacement equal to 44.16 m^3, upright condition, and trim varying between −1.000 and 1.1 m. The LCB values in column 5 are equivalent to the $\overline{KN}$ values in Example 2.6, Figure 2.22. The $\overline{NM}_L$ values refer to the longitudinal metacentre.

Write the data on an M-file, lido9b.m, and use the programme b_curve to plot the B- and M-curves in the longitudinal plane (xz-plane). Here the M-curve is the locus of the longitudinal metacentre.

Table 2.7   Data of vessel *Lido 9* at 44.16 m³ volume of displacement and 5° heel intervals, trim = −0.325 m

Heel angle (°)	Draught (m)	$\overline{KN}$ (m)	$\overline{NB}$ (m)	$\overline{NM}$ (m)	LCB (m)	$\overline{NM}_L$ (m)
0	1.729	0	1.272	4.596	−1.735	23.371
5	1.711	0.399	1.255	4.438	−1.740	23.693
10	1.659	0.776	1.204	4.119	−1.761	24.008
15	1.575	1.122	1.121	3.711	−1.799	23.730
20	1.462	1.432	1.009	3.341	−1.841	23.813
25	1.324	1.716	0.872	3.073	−1.887	23.464
30	1.163	1.979	0.711	2.857	−1.932	23.154
35	0.985	2.215	0.528	2.464	−1.971	22.822
40	0.796	2.419	0.326	2.105	−2.002	22.810
45	0.600	2.595	0.107	1.830	−2.047	23.133
50	0.402	2.749	−0.126	1.537	−2.106	21.837
55	0.198	2.870	−0.372	1.082	−2.113	19.757
60	−0.012	2.945	−0.625	0.479	−2.072	17.473
65	−0.235	2.960	−0.883	−0.185	−2.041	16.162
70	−0.464	2.931	−1.140	−0.543	−2.025	15.117
75	−0.693	2.874	−1.393	−0.869	−2.008	14.298
80	−0.919	2.788	−1.640	−1.171	−1.994	13.633
85	−1.140	2.678	−1.878	−1.446	−1.981	13.121
90	−1.354	2.539	−2.108	−13.314	−1.970	12.792

Table 2.8   Data of vessel *Lido 9* at 44.16 m³ volume of displacement, 0.1 m trim intervals, upright condition

Trim (m)	Draught (m)	$\overline{NB}$ (m)	$\overline{NM}$ (m)	LCB (m)	$\overline{NM}_L$
−1.000	1.673	1.174	4.536	−2.777	23.681
−0.900	1.653	1.192	4.550	−2.623	23.904
−0.800	1.668	1.208	4.564	−2.468	24.069
−0.700	1.683	1.224	4.577	−2.313	24.163
−0.600	1.697	1.238	4.585	−2.157	24.145
−0.500	1.709	1.251	4.589	−2.001	23.954
−0.400	1.721	1.24	4.592	−1.848	23.584
−0.300	1.732	1.276	4.598	−1.697	23.293
−0.200	1.742	1.286	4.604	−1.548	22.951
−0.100	1.750	1.295	4.610	−1.401	22.556
0.000	1.758	1.304	4.614	−1.257	22.108
0.100	1.765	1.311	4.615	−1.114	22.137
0.200	1.772	1.319	4.614	−0.971	22.115
0.300	1.777	1.324	4.612	−0.829	22.046
0.400	1.782	1.329	4.610	−0.690	21.910
0.500	1.786	1.333	4.606	−0.546	21.707
0.600	1.789	1.336	4.603	−0.407	21.431
0.700	1.792	1.338	4.599	−0.270	21.116
0.800	1.793	1.339	4.599	−0.135	20.895
0.900	1.794	1.340	4.597	0.000	20.871
1.000	1.795	1.340	4.594	0.135	20.834
1.100	1.795	1.338	4.590	0.269	20.829

Appendix—Water Densities

	Density ($t\,m^{-3}$)
Fresh water	1.000
Eastern Baltic Sea	1.003
Western Baltic Sea	1.015
Black Sea	1.018
Oceans	1.025
Red Sea	1.044
Caspian Sea	1.060
Dead Sea	1.278

CHAPTER 3

Numerical Integration in Naval Architecture

Chapter Outline

3.1 Introduction

In Chapter 2, we have learnt that the evaluation of ship properties, such as displacement and stability, requires the calculation of areas, centroids, and moments of plane figures, and of volumes and centres of volumes. Such properties are calculated by integration. In the absence of an explicit definition of the hull surface, in terms of calculable mathematical functions, the integrations cannot be carried out by analytic methods. The established practice has been to describe the hull surface by tabulated data, as shown in Chapter 1, and to use these data in numerical calculations.

Two methods for numerical integration are described in this chapter: the **trapezoidal** and **Simpson's rules**. The treatment is based on Biran and Breiner (2002). The rules are exemplified on integrands defined by explicit mathematical expressions; this is done to convince the reader that the two methods of numerical integration are efficient, and to allow an evaluation of errors. The first examples are followed in Chapter 4 by Naval-Architectural applications to real ship data presented in tabular form.

Many Naval-Architectural problems require the calculation of the definite integral

$$\int_a^b f(x)dx,$$

of a function bounded in the finite interval $[a, b]$. We approximate the definite integral by the weighted sum of a set of function values, $f(x_1 1)$, $f(x_2)$, $\ldots$, $f(x_n)$, evaluated, or measured, at n points $x_i \in [a, b]$, $i = 1, 2, \ldots, n$, i.e.

$$\int_a^b f(x)dx \approx \sum_{i=1}^n a_i f(x_i) \qquad (3.1)$$

In Naval Architecture the coefficients a_i are called **multipliers**, in some books on Numerical Methods they are called **weights**.

There are several ways of deriving formulae for numerical integration—also called **quadrature formulae**—of the form shown in Eq. (3.1); three of them are mentioned below.

1. By geometrical reasoning, considering $\int_a^b f(x)dx$ as the area under the curve $f(x)$, between $x = a$ and $x = b$.
2. By approximating the function $f(x)$ by an interpolating polynomial, $P(x)$, and integrating the latter instead of the given function, so that

$$\int_a^b f(x)dx \approx \int_a^b P(x)dx$$

3. By developing the given function into a Taylor or MacLaurin series and integrating the first terms of the series.

The first approach yields a simple intuitive interpretation of the rules for numerical quadrature and of the errors involved. This interpretation enables the user to derive the rules whenever required, and to adapt them to particular situations, for instance, when changing the subintervals of integration. On the other hand, each rule must be derived separately. The advantages of the other approaches are:

- The derivation is common to a group of rules which thus appear as particular cases of a more general method.
- The derivation yields an expression of the error involved.

In the next two sections we shall use the geometrical approach to derive the two most popular rules, namely the **trapezoidal** and **Simpson's rules**. These two methods are sufficient for solving most problems encountered in Naval Architecture. The error terms will be given without derivation; however, interpretations of the error expressions will follow their presentation.

3.2 The Trapezoidal Rule

Let us consider the function $f(x)$ represented in Figure 3.1. We assume that we know the values $f(x_1)$, $f(x_2)$, ..., $f(x_5)$ and we want to calculate the definite integral

$$I = \int_{x_1}^{x_5} f(x)dx \tag{3.2}$$

The integral in Eq. (3.2) represents the area under the curve $f(x)$. Let us connect the points $f(x_1)$, $f(x_2)$, ..., $f(x_5)$ by straight-line segments (the dot-dash lines in the figure). We approximate the area under the curve by the sum of the areas of four trapezoids, that is, the area of the trapezoid with the base $\overline{x_1x_2}$ and the heights $f(x_1)$, $f(x_2)$, plus the area of the trapezoid with the base $\overline{x_2x_3}$ and heights $f(x_2)$, $f(x_3)$, and so on. We obtain

$$I \approx (x_2 - x_1)\frac{f(x_1) + f(x_2)}{2} + (x_3 - x_2)\frac{f(x_2) + f(x_3)}{2} + \cdots \tag{3.3}$$

For constant x-spacing, $x_2 - x_1 = x_3 - x_2 = \cdots = h$, Eq. (3.3) can be reduced to a simpler form:

$$I \approx h\left[\frac{1}{2}f(x_1) + f(x_2) + f(x_3) + \cdots + f(x_{n-1}) + \frac{1}{2}f(x_n)\right] \tag{3.4}$$

We call the intervals $[x_1, x_2]$, $[x_2, x_3]$, and so on, *subintervals*.

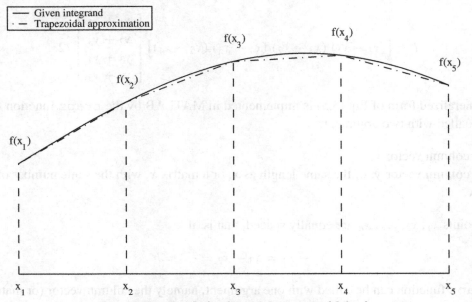

Figure 3.1 The derivation of the trapezoidal rule

As an example let us calculate

$$\int_{0^0}^{90^0} \sin x \, dx$$

The calculation presented in tabular form is as follows:

Angle Degrees	$\sin x$	Multiplier	Product
0	0.0000	1/2	0.0000
15	0.2588	1	0.2588
30	0.5000	1	0.5000
45	0.7071	1	0.7071
60	0.8660	1	0.8660
75	0.9659	1	0.9659
90	1.0000	1/2	0.5000
Sum	–	–	3.7979

The calculations were performed with MATLAB and the precision of the display in the *short format*, that is four decimal digits, was retained. To obtain the approximation of the integral we multiply the sum in column 4 by the constant subinterval, h:

$$(\pi * 15/180) * 3.7979 = 0.9943$$

Above we measured the interval in radians, as we should do in such calculations.

Equation (3.3) in matrix form yields

$$I \approx \left[(x_2 - x_1) \ (x_3 - x_2) \ (x_4 - x_3) \ (x_5 - x_4) \right] \begin{bmatrix} y_1 + y_2 \\ y_2 + y_3 \\ y_3 + y_4 \\ y_4 + y_5 \end{bmatrix} /2 \tag{3.5}$$

The generalized form of Eq. (3.5) is implemented in MATLAB by the `trapz` function that can be called with two arguments:

1. the column vector x,
2. the column vector y, of the same length as x, or a matrix y, with the same number of rows as x.

If the points $x_1, x_2, \ldots, x_n$ are equally spaced, that is, if

$$x_2 - x_1 = x_3 - x_2 = \cdots = h$$

the `trapz` function can be called with one argument, namely the column vector (or matrix) y. In this case, the result must be multiplied by the common x-interval, h.

3.2.1 Error of Integration by the Trapezoidal Rule

In any subinterval $[x_i, x_{i+1}]$, the error of the approximation, I_i, obtained by the trapezoidal rule equals

$$I_i - \int_{x_i}^{x_{i+1}} f(x)dx = h^3 \frac{d^2 f(\xi_i)}{dx^2}/12, \tag{3.6}$$

where ξ_i is a point in the subinterval (x_i, x_{i+1}) and $h = x_{i+1} - x_i$. Usually, the interval of integration, $[x_1, x_m]$, is divided into several subintervals; if we assume that they are equal, and note by I the trapezoidal approximation over the whole interval, we can write

$$\left| I - \int_{x_1}^{x_m} f(x)dx \right| = \left| \sum_{i=1}^{m-1} \frac{h^3}{12} \frac{d^2 f(\xi)}{dx^2} \right|$$

$$\leq \frac{x_m - x_1}{12} h^2 \max_{\xi \in [x_1, x_m]} \left| \frac{d^2 f(\xi)}{dx^2} \right| \tag{3.7}$$

We do not know the maximum value of the derivative in Eq. (3.7); otherwise we would have been able to calculate the exact value of the integral. We can, however, say the following:

- By substituting in Eq. (3.7) the maximum value of $df^2(x)/dx^2$ in the interval $[x_1, x_m]$ we can calculate an upper boundary of the error.
- The error is proportional to the square of h: if we halve the subinterval, the error is reduced approximately in the ratio $1/4$.
- The method is exact if $df^2(x)/dx^2 = 0$. This is the case for linear functions. As a matter of fact, the derivation of the trapezoidal rule was based on a linear approximation of $f(x)$.

Example 3.1.   In this example we consider the integral

$$\int_0^{\pi/2} \{1 + \sin(x)\}dx = [x - \cos(x)]_0^{\pi/2}$$

$$= \pi/2 + 1 = 2.570\,796\,326\,794\,90$$

To calculate the same integral numerically, by means of the trapezoidal rule, we begin by dividing the interval $[0, \pi/2]$ into two subintervals and obtain the value $2.518\,855\,775\,763\,42$. The error equals -2.02% of the correct value. We can reduce the error by halving the subinterval h. Experimenting with subintervals equal to $\pi/8, \pi/16, \ldots, \pi/128$, we obtain the results shown in Table 3.1 where they are compared with the results yielded by Simpson's rule (see the following sections). For $h = \pi/8$, Figure 3.1 shows the error as the sum of the small areas contained between the dash-dot line (the trapezoids) and the solid line (the given curve). This area looks really small. The errors in percent of the true values are shown in Table 3.2. As predicted by Eq. (3.7), each time we divide the subinterval h by 2, the error is divided by approximately 4. It is easy to see that as $h \to 0$, the trapezoidal approximation of the integral tends to the true value.

Table 3.1 Results by trapezoidal and by Simpson's rule

Subinterval	Integral	
	Trapezoidal Rule	*Simpson's Rule*
$\pi/4$	2.51885577576342	2.57307620428711
$\pi/8$	2.55791212776767	2.57093091176909
$\pi/16$	2.56758149868107	2.57080462231886
$\pi/32$	2.56999300727997	2.57079684347960
$\pi/64$	2.57059552111492	2.57079635905990
$\pi/128$	2.57074612688700	2.57079632881103

Table 3.2 Percent error by trapezoidal and by Simpson's rule

Subinterval	Percent Error	
	Trapezoidal Rule	*Simpson's Rule*
$\pi/4$	−2.0204	0.08868371
$\pi/8$	−0.5012	0.00523515
$\pi/16$	−0.1251	0.00032268
$\pi/32$	−0.0312	0.00002010
$\pi/64$	−0.0078	0.00000126
$\pi/128$	−0.0195	0.00000008

In this example, by reducing the size of the subinterval h we could make the error negligible. This was easy because we had an explicit expression for $f(x)$, and we could evaluate as many values of $f(x)$ as we wanted. When there is no explicit mathematical definition, as it happens when the ship lines are defined only by drawings or tables of offsets, the number of function values that can be measured, or evaluated, is restricted by practical limitations. In such cases, we must be satisfied if the precision of the integration is consistent with the precision of the measurements, or of calculations involving the same constants and variables. To understand this point better, let us suppose that we want to calculate the ship displacement mass as $\Delta = \rho \nabla$, where ρ is the density of the surrounding water. It makes no sense to be very precise in the calculation of the displacement volume, ∇, if we multiply it afterward by a conventional value of the density, ρ. The density varies from sea to sea (see table in Appendix of Chapter 2), and in the same sea it varies with temperature. In most calculations it would be impossible to take into account these variations, and the Naval Architect or the ship Master has to use the value prescribed by the regulations relevant to the ship under consideration. For example, for oceans and the Mediterranean sea, various regulations specify the value $1.025 \, \text{t m}^{-3}$. An exception is the *inclining experiment*, a case in which the actual density must be measured. But, even in that case the precision of the measurement is limited and not better than that of the ∇-value calculated with the rules described in this chapter.

3.3 Simpson's Rule

In Figure 3.2 the solid line passing through the points B, C, and D represents the integrand $f(x)$. We want to calculate the integral of $f(x)$ between $x = A$ and $x = E$, that is, the area $ABCDEFA$. This time we shall approximate $f(x)$ by a parabola whose equation has the form

$$f(x) = a_0 + a_1 x + a_2 x^2 \tag{3.8}$$

The parabola is represented by a dashed-dotted line in Figure 3.2. We need three points to define this curve; therefore, in addition to the values of $f(x)$ calculated at the two extremities, i.e., at the points B and D, we shall also evaluate $f(x)$ at the half-interval, obtaining the point C. Let

$$\overline{AB} = f(x_1), \quad \overline{FC} = f(x_2), \quad \overline{ED} = f(x_3)$$
$$h = \overline{AE}/2 = (x_3 - x_1)/2$$

We divide the total area under $f(x)$ into two partial areas:

1. the trapezoid $ABDEA$;
2. the parabolic segment $BCDGB$.

The first area equals

$$\overline{AE} \cdot \frac{\overline{AB} + \overline{ED}}{2} = 2h \cdot \frac{f(x_1) + f(x_3)}{2}$$

Figure 3.2 The derivation of Simpson's rule

For the second area we use a result from geometry that says that the area of a parabolic segment equals two-thirds of the area of the circumscribed parallelogram. Correspondingly, we calculate the second area as 2/3 of the circumscribed parallelogram $BHID$, i.e.

$$\frac{2}{3} \cdot \overline{AE} \cdot \overline{CG} = \frac{2}{3} \cdot 2h \left(f(x_2) - \frac{f(x_1) + f(x_3)}{2} \right)$$

Adding the two partial sums yields

$$\int_{x_1}^{x_3} f(x)dx \approx \frac{h}{3} \left[f(x_1) + 4f(x_2) + f(x_3) \right] \tag{3.9}$$

which is the elementary form of **Simpson's rule**.

Usually we have to integrate the function $f(x)$ over a larger interval $[a, b]$. Then, we achieve a better approximation by dividing the given interval into more subintervals. From the way we derived Eq. (3.9) we see that the number of subintervals must be even, say $n = 2k$, where k is a natural number. Let

$$h = \frac{a - b}{n} = x_2 - x_1 = x_3 - x_2 = \cdots = x_{n+1} - x_n$$

Applying Eq. (3.9) for each pair of subintervals, and adding all partial sums, we get

$$\int_{x_1}^{x_{n+1}} f(x)dx = \frac{h}{3}[f(x_1) + 4f(x_2) + 2f(x_3) + 4f(x_4) + \cdots + 4f(x_n) + f(x_{n+1})] \tag{3.10}$$

which is the extended form of Simpson's rule, for equal subintervals. This form is very helpful when calculations are carried on manually. As an example, let us calculate

$$\int_{0°}^{90°} \sin x \, dx$$

In tabular form the calculation is

Angle degrees	$\sin x$	Multiplier	Product
0	0.0000	1	0.0000
15	0.2588	4	1.0353
30	0.5000	2	1.0000
45	0.7071	4	2.8284
60	0.8660	2	1.7321
75	0.9659	4	3.8637
90	1.0000	1	1.0000
Sum	–	–	11.4595

To obtain the approximation of the integral, we multiply the sum in column 4 by the constant subinterval:

$$(\pi * 15/180) * 11.4595/3 = 1.0000$$

When a computer is used, there is no need to have all subintervals equal and it is sufficient to have **pairs of equal intervals**. A MATLAB function called `simp` that implements Eq. (3.9) is described in Biran and Breiner (2002), Chapter 10.

As an example, let us calculate by Simpson's rule the same integral that we exemplified in Section 3.2. As shown in Tables 3.1 and 3.2, the results are much better than those obtained with the trapezoidal rule.

3.3.1 Error of Integration by Simpson's Rule

Denoting by I_i the approximation obtained by Simpson's rule in the subinterval $[x_1, x_3]$, the error equals

$$I_i - \int_{x_1}^{x_3} f(x)dx = h^5 \frac{1}{90} \frac{d^4 f(\xi)}{dx^4}, \tag{3.11}$$

where $x_1 \leq \xi \leq x_3$. Summing up the errors in all pairs of subintervals, and denoting by I the approximation calculated with Simpson's rule, we obtain

$$\left| I - \int_{x_1}^{x_n} f(x)dx \right| = \left| \sum_{1}^{n/2} \frac{d^4 f(\xi)}{dx^4} \cdot \frac{h^5}{90} \right| \tag{3.12}$$

$$\leq \frac{x_n - x_1}{180} h^4 \max_{\xi \in [x_1, x_n]} \left| \frac{df^4(\xi_m)}{dx^4} \right| \tag{3.13}$$

At this point we can say the following about Simpson's rule:

1. If we divide h by 2, the error decreases approximately in the ratio $1/16$.
2. Simpson's rule yields the exact result if $d^4 f/dx^4 = 0$. This is certainly true for second-degree parabolas, which is not surprising because we assumed such a curve when we developed the rule. It is interesting that the method is also exact for cubics (third-degree curves).
3. For an equal number of subintervals, Simpson's rule yields better results than the trapezoidal rule. On the other hand, Simpson's rule imposes a serious constraint: the number of subintervals must be even, or, equivalently, $f(x)$ must be evaluated at an uneven

number of equally spaced points, or, in other words, an uneven number of ordinates. If, for example, we calculate the area of a waterline, we need an uneven number of equally spaced stations.

Example 3.2.   We refer again to the Example 3.1 using this time Simpson's rule. We can experiment with decreasing subintervals and obtain the results shown in Table 3.1, where they are compared with the results yielded by the trapezoidal rule. The convergence is considerably faster than that obtained in the case of the trapezoidal rule. The percent errors are shown in Table 3.2. As predicted by Eq. (3.13), each time we divide the subinterval h by 2, the error decreases approximately in the ratio 1/16. Note also that only two subintervals yield better results with Simpson's rule than eight with the trapezoidal rule.

3.4 Calculating Points on the Integral Curve

The trapezoidal and Simpson's rule produce one number for an interval of ordinates, that is

$$I(a, b) = \int_a^b f(x)dx$$

Sometimes we are interested not in one number only, but in a sequence of numbers that describe the integral as a function within the given interval

$$I(x) = \int_a^x f(x)dx, \quad a \le x \le b \tag{3.14}$$

Thus, in certain hydrostatic calculations we may need to know the areas of transverse sections (stations) as functions of draught (see Chapter 4). Another example is that of calculations of dynamic stability which require the knowledge of the area under the curve of the righting arm as function of the heel angle. The latter subject is discussed in Chapters 6 and 8. An appropriate name for a procedure that yields such an integral is **integral with variable upper limit**.

Let us consider a sequence of points, $x_1, x_2, \ldots, x_n$, and a sequence of values $f(x_1), f(x_2), \ldots, f(x_n)$. In the first example above, the values of the independent variable, x_i, represent draught, the functions $f(x_i)$, half-breadth, and the integral, the area of the station up to that draught.

In the second example, x_i is a heel angle, $f(x_i)$, the righting arm, $\overline{GZ}$, and the integral, the area under the righting-arm curve up to the respective angle. We could calculate the integral in Eq. (3.14) by applying one of the integration rules over the interval $[x_1, x_2]$, then over $[x_1, x_3]$, and so on. This procedure would be awkward. Table 3.3 illustrates an algorithm that yields the integral with variable upper limit in a "continuous" calculation. Let us detail the algorithm.

In column 1, we write the current number of the points at which we know the values of the function to be integrated. In column 2, we write the x_i-values; i.e., the draughts in the first

Table 3.3   The algorithm for integration with variable upper limit

No.	Position	Function	Sums	Integrals
1	*2*	*3*	*4*	*5*
1	x_1	$f(x_1)$	0	0
		$\downarrow$	$\hookleftarrow$	
2	x_2	$f(x_2)$		
		$\hookrightarrow$	$f(x_1) + f(x_2)$	$h\left(\frac{f(x_1)}{2} + \frac{f(x_2)}{2}\right)$
		$\downarrow$	$\hookleftarrow$	
3	x_3	$f(x_3)$		
		$\hookrightarrow$	$f(x_1) + 2f(x_2) + f(x_3)$	$h\left(\frac{f(x_1)}{2} + f(x_2) + \frac{f(x_3)}{2}\right)$
		$\downarrow$	$\hookleftarrow$	
...	...	...	...	...
n	x_n	$f(x_n)$	$f(x_1) + 2f(x_2) + \cdots + 2f(x_{n-1}) + f(x_n)$	$h\left(\frac{f(x_1)}{2} + f(x_2) + \cdots + f(x_{n-1}) + \frac{f(x_n)}{2}\right)$

example given above, or the heel angles in the second example. In column 3, we write the values of the functions $f(x_i)$ at the points x_i shown in column 2. For columns 3 and 4 the algorithm is

> Write 0 in column 4, line 1
> For $i = 1: (n-1)$
> - Pick up the value in column 4, line i.
> - Go left and add the value in column 3, line i.
> - Go down and add the value in column 3, line $i + 1$.
> - Write the result in column 4, line $i + 1$.
>
> End

In column 5, line i, we write the result of the product of the content of column 4, line i, by the half of the subinterval of integration. Visual inspection of column 5 shows that the expressions appearing there are exactly those yielded by the trapezoidal rule over the intervals $[x_1, x_2], [x_1, x_2], \ldots, [x_1, x_n]$.

Let us illustrate the above procedure with the example of $\int_0^x \sin x\, dx$ in the interval $[0, 2\pi]$. As known

$$\int_0^x \sin \xi\, d\xi = -\cos \xi \,|_0^x = 1 - \cos x \qquad (3.15)$$

Figure 3.3 shows the implementation of the algorithm in Microsoft's Excel. A MATLAB function based on the same algorithm, called `intvar`, can be found on the website of this book. The algorithm is used in the following MATLAB lines that solve the particular problem exemplified above:

Figure 3.3 An excel spreadsheet for $\int_0^x \sin x\, dx$ in the interval $[0, 2\pi]$

```
x = 0: 10: 180;        % angles in degrees
phi = pi * x/180;      % angles in radians
y = sin(phi);          % function to be integrated
l = length(y);
y1 = y(1:(l-1));
s1 = [0 cumsum(y1)];
y2 = y(2:l);
s2 = [0 cumsum(y2)];
% now multiply by half subinterval
S = (pi * 10/(2 * 180)) * (s1 + s2);
```

In our experience, the MATLAB procedure is slightly more exact than the Excel spreadsheet. Table 3.4 compares the result yielded by Eq. (3.15) with those obtained with the MATLAB function. The agreement between the results obtained analytically, in Excel, and in MATLAB is remarkable.

Table 3.4 Integral with variable upper limit—comparing the analytic result with that obtained in MATLAB

Angle Degrees	Analytic Result	Numerical Result	Angle Degrees	Analytic Result	Numerical Result
0	0.0000	0.0000	100	1.1736	1.1707
10	0.0152	0.0152	110	1.3420	1.3386
20	0.0603	0.0602	120	1.5000	1.4962
30	0.1340	0.1336	130	1.6428	1.6386
40	0.2340	0.2334	140	1.7660	1.7616
50	0.3572	0.3563	150	1.8660	1.8613
60	0.5000	0.4987	160	1.9397	1.9348
70	0.6580	0.6563	170	1.9848	1.9798
80	0.8264	0.8243	180	2.0000	1.9949
90	1.0000	0.9975			

3.5 Intermediate Ordinates

The integration rules developed in Sections 3.2 and 3.3 were based on a subdivision into equal subintervals. This procedure is not always the best one. Let us consider, for example, the waterline shown in Figure 3.4. We may appreciate that the shape of the curve between Stations 0 and 1 suits neither the trapezoidal nor Simpson's rule; applying either of them would yield large errors. We learnt that reducing the intervals would also reduce the errors. Therefore, let us introduce an **intermediate** station between Stations 0 and 1 and appropriately call it Station $\frac{1}{2}$. We introduce another intermediate station between Stations 9 and 10 and call it $9\frac{1}{2}$. We invite the reader to check that the corresponding sequence of trapezoidal multipliers is now

$$1/4, 2/4, 3/4, 4/4, 1, \ldots 1, 4/4, 3/4, 2/4, 1/4 =$$
$$1/4, 1/2, 3/4, 1, \ldots, 1, 3/4, 1/2, 1/4$$

The subdivision illustrated in Figure 3.4 suits Simpson's rule too, because we have a pair of equal subintervals $\delta L/2$, four pairs of equal subintervals δL, and a pair of equal subintervals $\delta L/2$.

3.6 Reduced Ordinates

We present in this section another way of overcoming the problem described in the preceding section. In continuation, we show how the same method can be adapted for a more difficult case.

Figure 3.4 Intermediate ordinates at station $\frac{1}{2}$ and station $9\frac{1}{2}$

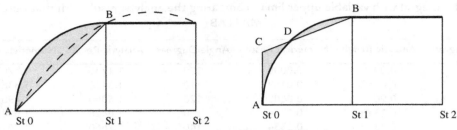

Figure 3.5 Reduced ordinates—a simple case

Let us consider the thick, solid-line curve shown in the left-hand side of Figure 3.5; it may be, for example, the after part of a waterline. If we calculate the area under the curve by the trapezoidal rule, and enter 0 for the half-breadth at Station 0, and the actual half-breadth at Station 1, we miss the whole shaded area. If we use Simpson's rule with the same values, plus the actual half-breadth at Station 2 (remember, for Simpson's rule we must take two equal subintervals), we obtain, in fact, the area under the dashed line, and this can be again less than the actual area.

The right-hand part of Figure 3.5 shows a simple way of improving the result. Let us draw the line BC so that the two shaded areas look equal. Our intention is to rely upon visual appreciation because we are looking for a quick procedure. Then, we take the length of the segment $\overline{AC}$ as the **reduced ordinate** at Station 0.

Above, the curve we are interested in begins exactly at one station. Frequently it happens that the curve begins or ends between stations. Such a case is illustrated in Figure 3.6, which may represent the forward part of a waterline.

To obtain a **reduced ordinate** we begin by applying the procedure described above, and substitute the given curve arc by the straight-line segment $\overline{AB}$. Next, we connect the point A

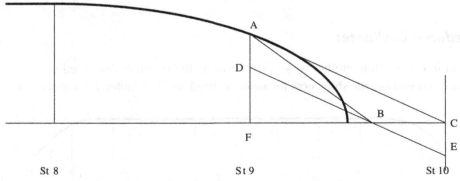

Figure 3.6 Reduced ordinates—a more complex case

to the point C and draw $\overline{BE}$ parallel to $\overline{AC}$. The reduced ordinate is $\overline{CE}$ and we use it with a minus sign. To prove that the proposed procedure yields the correct result, we extend the segment $\overline{BE}$ until it intercepts the Station 9 at point D. We are looking for the area of the triangle ABF, but this area equals the area of the triangle ACF minus that of the triangle ABC. Now, the area of the triangle ABC is half the area of the parallelogram ACED. Noting $\overline{AF} = y_9$, the half-breadth at Station 9, and $FC = \delta L$, we can write

$$Area = \frac{y_9 \cdot \delta L}{2} - \frac{\overline{CE} \cdot \delta L}{2} \qquad (3.16)$$

This is exactly the result we would obtain by applying the trapezoidal rule with the value y_9 for Station 9 and the length of the segment $\overline{CE}$ taken with the minus sign.

3.7 Other Procedures of Numerical Integration

We described in this section two rules for numerical integration: the trapezoidal and Simpson's rule. Additional methods of integration have been developed and employed. For example, a third rule popular in English-language literature is *Simpson's second rule* in which the given integrand is approximated by a third-degree parabola. This rule is applied on sets of three equal subintervals, or, in other words, sets of four equally spaced ordinates. This is a very serious constraint.

As shown in Chapter 13, CAD programmes used today in Naval Architecture describe the hull surface by *piecewise polynomials*, that is they fit polynomials and combination of polynomials to curve segments and surface *patches*. Then, it is possible to use the polynomial coefficients to obtain the integrals by simple algebraic formulae. For example, if a segment of a waterline is described by the equation

$$y = c_1 x^2 + c_2 x + c_3 \qquad (3.17)$$

then the area enclosed between the curve segment, the centreline, and the stations $x = a, x = b$ is

$$\int_a^b y dx = \frac{c_1}{3}x^3 + \frac{c_2}{2}x^2 + c_3 x \Big|_a^b = \frac{c_1}{3}(b^3 - a^3) + \frac{c_2}{2}(b^2 - a^2) + c_3(b - a) \qquad (3.18)$$

An example of fitting a MATLAB spline to points along a ship station and integrating the area under the curve is described in Biran (2011), Section 4.5. Similar equations can be derived for other properties, namely moments and moments of inertia. A convenient way of doing this is by means of Green's theorem, as shown in a programme included in the companion software of this book.

3.8 Summary

Naval Architecture requires the calculation of areas, moments of areas, moments of inertia of areas, volumes, and moments of volumes. Such calculations involve definite integrals. Usually, the hull surface is defined by lines drawings or tables of offsets, and not by explicit mathematical expressions. Then, the integrals can be obtained only by numerical methods. In a numerical method we approximate the integral by a weighted sum of a finite set of function values, that is

$$\int_a^b f(x)dx \approx \sum_{i=1}^n a_i f(x_i) \tag{3.19}$$

Two methods that implement such approximations are introduced in this chapter: the *trapezoidal rule* and *Simpson's rule*. The trapezoidal rule approximates the given curve by straight-line segments, while Simpson's rule approximates it by a parabola. The rules are exemplified on integrands for which we know the exact solutions. Thus, it is possible to show convincingly that the approximations yield satisfactory results. Also, it is possible to see that, as the number of ordinates—i.e., the number of points at which the integral is evaluated—increases, the error decreases. The number of ordinates must be limited for practical reasons. This is possible because it is sufficient to maintain a precision consistent with measurements or other calculations. Simpson's rule yields, on one hand, results closer to the exact value. On the other hand, it imposes a serious constraint: the number of subintervals must be even.

By applying a rule of integration over one interval we obtain one number. In Naval Architecture, it is sometimes necessary to have a set of numbers that describe the integral curve as a function of the independent variable, i.e.

$$I(x) = \int_a^x f(x)dx, \quad a \leq x \leq b$$

This *integral with variable upper limit* can be obtained with the aid of an elegant algorithm described in this chapter.

The shape of curves encountered in Naval Architecture can be such that over certain intervals, generally toward their ends, it may be necessary to use smaller subintervals of integration. We then use *intermediate ordinates*. In the case of a waterline, these ordinates are *intermediate stations*.

In the lines plan, some lines can terminate within a subinterval, and not at the end of the subinterval. For example, by construction the design waterline usually begins at the aft perpendicular AP, and ends at the forward perpendicular FP. Most other waterlines can begin and end between stations. For good approximations of the areas under such curves, while using the initially given subdivision into subintervals, the lines must be corrected yielding *reduced ordinates* that will be used in the integration.

3.9 Examples

EXAMPLE 3.1. Calculate the integral

$$\int_0^{45} x^3 dx$$

by the following methods:

(a) analytic, **(b)** trapezoidal rule, five ordinates, **(c)** trapezoidal rule, nine ordinates, **(d)** Simpson's rule, five ordinates, **(e)** Simpson's rule, nine ordinates.

Solution.

(a)

$$\int_0^{45} x^3 dx = \left.\frac{x^4}{4}\right|_0^{45} = 1025156.25$$

(b)   The following values were calculated in Microsoft's Excel:

No. of Ordinate	Trapezoidal Multiplier	x	f(x)	Products
1	*2*	*3*	*4*	*5 = 2 × 4*
1	1/2	0.00	0.00	0.00
2	1	11.25	1423.83	1423.83
3	1	22.50	11390.63	11390.63
4	1	33.75	38443.36	38443.36
5	1/2	45.00	91125.00	45562.50
Sum				96820.31
Integral		(45/4) Sum =		1089228.52

The error is

$$E = 1025156.25 - 1089228.52 = -64072.27$$

and the relative error is

$$E_r = 100 \times \frac{E}{1025156.25} = -6.25\%$$

(c)   The following values were calculated in Microsoft's Excel:

No. of Ordinate	Trapezoidal Multiplier	x	f(x)	Products
1	2	3	4	5 = 2 × 4
1	1/2	0.00	0.00	0.00
2	1	5.63	177.98	177.98
3	1	11.25	1423.83	1423.83
4	1	16.88	4805.42	4805.42
5	1	22.50	11390.63	11390.63
6	1	28.13	22247.31	22247.31
7	1	33.75	38443.36	38443.36
8	1	39.38	61046.63	61046.63
9	1/2	45.00	91125.00	45562.50
Sum	–	–	–	185097.66
Integral		(45/8)Sum/3 =		1041174.32

The error is

$$E = 1025156.25 - 1041174.32 = -16018.07$$

and the relative error is

$$E_r = 100 \times \frac{E}{1025156.25} = -1.56\%$$

(d) The following values were calculated in Microsoft's Excel:

No. of Ordinate	Simpson's Multiplier	x	f(x)	Products
1	2	3	4	5 = 2 × 4
1	1	0.00	0.00	0.00
2	4	11.25	1423.83	5695.31
3	2	22.50	11390.63	22781.25
4	4	33.75	38443.36	153773.44
5	1	45.00	91125.00	91125.00
Sum				273375.00
Integral		(45/4)Sum/3		1025156.25

The error is

$$E = 1025156.25 - 1025156.25 = 0$$

and the relative error is

$$E_r = 100 \times \frac{E}{1025156.25} = 0\%$$

(e) The following values were calculated in Microsoft's Excel:

No. of Ordinate	Simpson's Multiplier	x	f(x)	Products
1	*2*	*3*	*4*	*5 = 2 × 4*
1	1	0.00	0.00	0.00
2	4	5.63	177.98	711.91
3	2	11.25	1423.83	2847.66
4	4	16.88	4805.42	19221.68
5	2	22.50	11390.63	22781.25
6	4	28.13	22247.31	88989.26
7	2	33.75	38443.36	76886.72
8	4	39.38	61046.63	244186.52
9	1	45.00	91125.00	91125.00
Sum	–	–	–	546750.00
Integral		(45/8)Sum/3 =		1025156.25

The error is

$$E = 1025156.25 - 1025156.25 = 0$$

and the relative error is

$$E_r = 100 \times \frac{E}{1025156.25} = 0\%$$

MATLAB solution

(a) Analytic.

```
format long
a = 45^4/4 = 1.025156250000000e + 006
```

(b) Trapezoidal rule, five ordinates.

```
x = 0: 45/4: 45;
y = x. ^3;
b = trapz(x, y) = 1.089228515625000e + 006
error = a −b = −6.4072e + 004
percent_error = 100 * (a − b)/a = −6.2500%
```

(c) Trapezoidal rule, nine ordinates.

```
x = 0: 45/8: 45;
y = x.^3;
c = trapz(x, y) = 1.041174316406250e + 006
```

```
error = a - c = -1.6018e + 004
percent_error = 100 * (a - c)/a = -1.5625%
```

(d) Simpson's rule, five ordinates.

```
x = 0: 45/4: 45;
y = x.^3;
d = simp(x', y') = 1.025156250000000e + 006
error = a - d = 0
percent_error = 100 * (a - d)/a = 0%
```

(e) Simpson's rule, nine ordinates.

```
x = 0: 45/8: 45;
y = x.^3;
e = simp(x', y') = 1.025156250000000e + 006
error = a - e = 0
percent_error = 100 * (a - e)/a = 0%
```

3.10 Exercises

Exercise 3.1. Calculate the integral

$$\int_{-\pi/2}^{\pi/2} \sin x \, dx$$

by the following methods: (a) analytic, (b) trapezoidal rule, five ordinates, (c) trapezoidal rule, nine ordinates, (d) Simpson's rule, five ordinates, (e) Simpson's rule, nine ordinates. Analyze the errors and explain your results.

Exercise 3.2. Find the trapezoidal multipliers corresponding to integration over the set of stations

$$0, \frac{1}{2}, 1, 1\frac{1}{2}, 2, 3, \ldots 8, 8\frac{1}{2}, 9, 9\frac{1}{2}, 10$$

Exercise 3.3. Find the Simpson's multipliers corresponding to integration over the set of stations

$$0, \frac{1}{2}, 1, 2, 3, \ldots 8, 9, 9\frac{1}{2}, 10$$

CHAPTER 4

Hydrostatic Curves

Chapter Outline

4.1 Introduction

In the preceding chapter we learnt several methods of numerical integration used in Naval Architecture. In this chapter, we are going to apply them to the calculation of areas, centroids, moments of inertia of areas, volumes, and centres of volume. We call these properties **hydrostatic data** and show how to plot them, as functions of draught, in curves that allow further calculations.

Another set of plots consists of **Bonjean curves**; they enable the user to calculate the displacement and the centres of buoyancy for a given waterline, in an upright condition. The waterline can be not only a straight line, as is the case in still water, but also a curve. The latter case can arise when the hull is deflected because of a longitudinal bending moment or thermal expansion, or when the vessel floats in waves. The vessel is said to be in a **hogging condition** if the keel is concave downwards, and in a **sagging condition** if the keel is concave upwards.

All the properties mentioned above are represented as functions of draught. Certain functional relationships exist between some of those curves. Three such properties are described in this chapter.

Another subject dealt with in this chapter is that of **affine hulls**, that is hulls obtained from given ship lines by multiplying by the same **scale factor** all dimensions parallel to an axis of coordinates. The properties of an affine hull can be derived by simple formulae from the properties of the parent hull.

Within this chapter we use the following notations:

i	station number, as in the lines drawing;
j	station number defined such that the distance from the origin of x-coordinates is $j\delta L$;
x_i	x-coordinate of station i;
y_i	half-breadth of station i on a given waterline;
α_i	integration multiplier for station i. For Simpson's rule we assume that the common factor $1/3$ is included in α_i;
δL	subinterval of integration along the x-axis;
δT	subinterval of integration along the z-axis.

For the above definitions we have, obviously, $j = 0$ in the origin of coordinates.

4.2 The Calculation of Hydrostatic Data

4.2.1 Waterline Properties

In this section, we refer to Figure 4.1 and assume that all waterlines are symmetric about the centreline. This assumption is true for almost all ships in upright condition.

We calculate the **waterplane area,** of a given waterline, as

$$A_W = 2 \int_a^b y\,dx \approx 2 \left(\sum_{i=n_1}^{n_n} \alpha_i y_i \right) \delta L \tag{4.1}$$

where the waterline begins at station n_1, with $x = a$, and ends at station n_n, with $x = b$.

The **moment of the waterplane area about a transverse axis** passing through the origin of coordinates is

$$M_x = 2 \int_a^b xy\,dx \approx 2 \left(\sum_{i=n_1}^{n_n} \alpha_i x_i y_i \right) \delta L = 2 \left(\sum_{i=n_1}^{n_n} \alpha_i j_i y_i \right) \delta L^2 \tag{4.2}$$

Leaving the indexes n_1, n_n we write the x-**coordinate of the centre of flotation** of the given line as

$$x_F = \frac{M_x}{A_W} = \frac{2 \left(\sum \alpha_i j_i y_i \right) \delta L^2}{2 \left(\sum \alpha_i y_i \right) \delta L} = \frac{\left(\sum \alpha_i j_i y_i \right)}{\left(\sum \alpha_i y_i \right)} \delta L \tag{4.3}$$

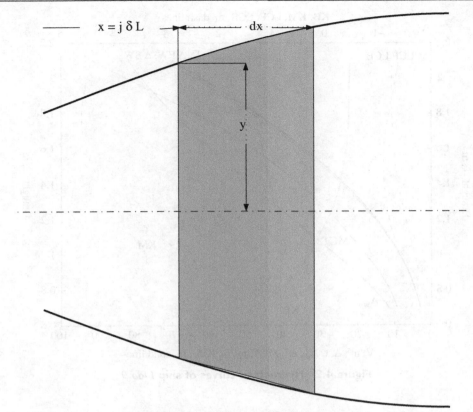

Figure 4.1 An element of waterline area

The notation x_F corresponds to the DIN 81209 standard. The notation used in English-language texts is **LCF**, an acronym for **longitudinal centre of flotation**. The corresponding curve is shown in Figure 4.2. To calculate the transverse **moment of inertia of the waterplane area**, i.e., the moment of inertia about the centreline, we first write the moment of inertia of the elemental area shown in grey in Figure 4.1:

$$dI_T = \frac{(2y)^3 dx}{12} = \frac{2}{3} y^3 dx \tag{4.4}$$

Then, the moment of inertia of the whole waterplane equals

$$I_T = \int_a^b \frac{2}{3} y^3 dx \approx \frac{2}{3} \left(\sum_{i=n_1}^{n_n} \alpha_i y_i^3 \right) \delta L \tag{4.5}$$

Figure 4.2 Hydrostatic curves of ship *Lido 9*

The **moment of inertia of the waterplane area about a transverse axis** passing through the origin of coordinates is calculated as

$$I_y = 2 \int_a^b x^2 y \, dx \approx 2 \left(\sum_{i=n_1}^{n_n} \alpha_i x_i^2 y_i \right) \delta L = 2 \left(\sum_{i=n_1}^{n_n} \alpha_i j_i^2 y_i \right) \delta L^3 \tag{4.6}$$

In Section 2.8.1, we learnt that, for small angles of inclination, the initial and the inclined waterline intersect themselves along a line passing through the centre of flotation (barycentric axis). For longitudinal inclinations, that is trim, in intact condition, this is almost always true. Therefore, we are interested in finding the moment of inertia of the waterplane area about the transverse barycentric axis. We find this moment, called **longitudinal moment of inertia,** by using a theorem on the parallel translation of the axes of coordinates

$$I_L = I_y - x_f^2 A_W \tag{4.7}$$

The geometrical properties of the waterplane area can be conveniently calculated in a spreadsheet, such as that shown in Table 4.1. The table contains the data of the lowest waterline in Figure 1.11 and it was calculated in Microsoft's Excel.

Table 4.1 A waterline sheet

Station No.	Trapezoidal Multiplier	Half-Breadth m	Levers	Functions of Area	Functions of Moments	Functions of I_x	Cubes of Half-Breadth	Functions of I_T
	α_i	y_i	j_i	$\alpha_i y_i$	$\alpha_i j_i y_i$	$\alpha_i j_i^2 y_i$	y_i^3	$\alpha_i y_i^3$
1	*2*	*3*	*4*	*5=2×3*	*6=5×4*	*7=6×4*	*8=3³*	*9=2×8*
0	1/2	0.000	−5	0.000	0.000	0.000	0.000	0.000
1	1	0.900	−4	0.900	−3.600	14.400	0.729	0.729
2	1	1.189	−3	1.189	−3.567	10.701	1.681	1.681
3	1	1.325	−2	1.325	−2.650	5.300	2.326	2.326
4	1	1.377	−1	1.377	−1.377	1.377	2.611	2.611
5	1	1.335	0	1.335	0.000	0.000	2.379	2.379
6	1	1.219	1	1.219	1.219	1.219	1.811	1.811
7	1	1.024	2	1.024	2.048	4.096	1.074	1.074
8	1	0.749	3	0.749	2.247	6.741	0.420	0.420
9	1	0.389	4	0.389	1.556	6.224	0.059	0.059
10	1/2	0.000	5	0.000	0.000	0.000	0.000	0.000
Sums	–	–	–	9.507	−4.124	50.058	–	13.091

The final results are obtained by using the sums in Table 4.1 as follows:

$$\delta L = 0.893\text{ m}$$
$$A_W = 2 \times 0.893 \times 9.507 = 16.98\text{ m}^2$$
$$LCF = \frac{-4.124}{9.507} \times 0.893 = -0.387\text{ m}$$
$$I_T = \frac{2}{3} \times 13.091 \times 0.893 = 7.79\text{ m}^4$$
$$I_y = 2 \times 50.058 \times 0.893^3 = 71.29\text{ m}^4$$
$$I_L = 71.29 - (-0.387)^2 \times 16.98 = 68.75\text{ m}^4$$

We recommend the reader to check the plausibility of the results by comparing them with the data of the circumscribed rectangle. For example, the area of this rectangle is

$$2 \times 1.377 \times 8.928 = 24.588\text{ m}^2$$

that is greater than the waterplane area, and so it should be.

Table 4.1 requires a few explanations. The fourth line contains column numbers. An expression like $5 = 2 \times 3$ means that the numbers in column 5 are the products of numbers in column 2 by numbers in column 3. Similarly, the numbers in column 7 are the products of numbers in column 6 by numbers in column 4. This means that the number 14.400, for example, is obtained by one multiplication, namely $-3.600 \times (-4)$, and not by two multiplications and a squaring operation, $1 \times (-4)^2 \times (0.900)$. Proceeding in this way we spare computer resources and reduce the possibilities of errors. Another simplification results from the use of the factors j, most of them integers. Multiplication by integers is easier when carried on manually, and it is not affected by numerical errors. Thus, instead of multiplying the products in column 5 by x-distances that are "real" numbers (fractional values) and introduce numerical errors at each station, we multiply by integers. Then, the sums of the products in columns 6 and 7 are multiplied only once by the length, and the square of the length of the subinterval of integration, δL, which can be a real number.

Let us make a final comment on the use of electronic spreadsheets for calculations such as those in Table 4.1. The values of half-breadths, y_i, are entered only once, in column 3, although they are repeatedly used in all calculations. In this way we reduce the possibilities of errors that can occur when entering a number. Moreover, if we must change the value of a half-breadth, we do it in one place only, and the change spreads automatically over the whole table.

Instead of using an electronic spreadsheet, such as Microsoft's Excel, one can write a programme in a suitable language, for example MATLAB. Such a programme can be useful if the calculations are chained with other computer operations. For the reasons explained above we recommend to write the programme following the principles used in the waterline sheet shown in Table 4.1.

4.2.2 Volume Properties

We can obtain the displacement volume corresponding to a given draught, T_0, by integrating "vertically" the waterplane areas from the lowest hull point to the given draught:

$$\nabla = \int_0^{T_0} A_W dz \approx \left(\sum_{i=1}^{i_T} \alpha_i A_{Wi} \right) \delta T \tag{4.8}$$

The moment of the displacement volume above the baseline can be also obtained by "vertical" integration:

$$M_B = \int_0^{T_0} T A_W dz \approx \left(\sum_{i=1}^{i_T} \alpha_i z_i A_{Wi} \right) \delta T = \left(\sum_{i=1}^{i_T} \alpha_i j_i A_{Wi} \right) \delta^2 T \tag{4.9}$$

where z_i is the z-coordinate of the ith waterline, and j_i the number of the waterline counted from the baseline.

From Eqs. (4.8) and (4.9) we calculate the vertical coordinate of the centre of buoyancy, z_B, as

$$z_B = \frac{M_B}{\nabla} \approx \frac{\left(\sum_{i=1}^{i_{T0}} \alpha_i j_i A_{Wi} \right) \delta^2 T}{\left(\sum_{i=1}^{i_{T0}} \alpha_i A_{Wi} \right) \delta T} = \frac{\left(\sum_{i=1}^{i_{T0}} \alpha_i j_i A_{Wi} \right)}{\left(\sum_{i=1}^{i_{T0}} \alpha_i A_{Wi} \right)} \delta T \tag{4.10}$$

The notation z_B is that prescribed in the DIN 81209 standard. The notations common in English-language books are $\overline{KB}$, or VCB, the latter being the acronym of **vertical centre of buoyancy**. The procedure used with Eq. (4.10) yields bad approximations for the lowest waterlines. Therefore, we recommend to neglect the results for the first waterlines. As shown in Section 4.4, we can also calculate the displacement and the vertical centre of buoyancy by "longitudinal" integration of values read in Bonjean curves.

4.2.3 Derived Data

Let us suppose that we know the displacement, Δ_0, corresponding to a given draught, T_0, and we want to find by how many tonnes that displacement will change if the draught changes by δT, centimetres. Let the waterplane area be A_W m^2, and the water density, ρ_W t m^{-3}. For a small draught change we may neglect the slope of the shell (in other words we assume a wall-sided hull) and we write

$$\delta\Delta = \rho_W A_W \delta T$$

If we measure Δ in tonnes, and δT in centimetres, we obtain

$$\delta\Delta = \rho_W \frac{A_W}{\delta T} \tag{4.11}$$

We call the quantity $\rho_W \frac{A_W}{100}$ **tonnes per centimetre immersion**, where, as explained previously, the *tonne* is a unit of mass, and use for it the notation **TPC**. In older, English-language books we find the notation TPI as an acronym for **tonnes per inch** where the *ton* is a unit of weight. This quantity is calculated from an expression similar to Eq. (4.11), but adapted for English and American units. For SI units

$$TPC = \frac{A_W}{100} \times \rho_W \tag{4.12}$$

where ρ_W should be taken from the Appendix of Chapter 2. The problem posed above can be inverted: find the change in draught, δT, corresponding to a change of displacement, $\delta\Delta$. The obvious answer is

$$\delta T = \frac{\delta\Delta}{TPC}$$

The above calculations yield good approximations as long as the changes $\delta\Delta$, δT are small. In fact, Eq. (4.11) is a linearization of the relationship between displacement volume and waterplane area.

Trim calculations will be discussed in more detail in Chapter 7. However, as one quantity required for those calculations is derived from hydrostatic data and is usually presented with the latter, we introduce this quantity here. Let us calculate the moment necessary to change the trim by 1 m. If the length between perpendiculars is L_{pp} and is measured in m, the corresponding angle of trim is defined by

$$\tan\theta = \frac{1}{L_{pp}} \tag{4.13}$$

The notation θ for the angle of trim corresponds to the standards ISO 7463 and DIN 81209-1. At the angle of trim given by Eq. (4.13), the displacement and buoyancy forces are separated by a distance $\overline{GM_L} \sin \theta$, where $\overline{GM_L}$ is the **longitudinal metacentric height** calculated as

$$\overline{GM_L} = \overline{KB} + \overline{BM_L} - \overline{KG}$$

The couple formed by the displacement and buoyancy forces is

$$\Delta \overline{GM_L} \sin \theta$$

For small angles of trim we assume $\tan \theta \approx \sin \theta$ and then the **moment to change trim by 1 m** is equal to

$$M_{CT} = \frac{\Delta \overline{GM_L}}{L_{pp}} \tag{4.14}$$

where M_{CT} is measured in t m/m, Δ in t, and $\overline{GM_L}$ and L_{pp}, in m. Although the SI unit is the metre, some design offices use the "moment to change trim by 1 cm." Then, the value of M_{CT} given by Eq. (4.14) should be divided by 100.

In the first design stages $\overline{KG}$ is not known. As $\overline{BM_L} \gg \overline{KB} - \overline{KG}$, we can assume the approximation $\overline{GM_L} \approx \overline{BM_L}$.

In Table 4.2, calculated with the ARCHIMEDES programme, the moment to change trim is based on the displacement volume, ∇, and is measured in m^4/m. Let us check, for example, the value corresponding to the draught 1.9 m. We rewrite Eq. (4.14) as

$$M_{CT} = \frac{\nabla \overline{BM_L}}{L_{pp}} \tag{4.15}$$

and calculate

$$M_{CT} = \frac{54.197 \times (22.242 - 1.401)}{15.5} = 72.872 \mathrm{m}^4/\mathrm{m}$$

This is exactly the value appearing in Table 4.2.

4.2.4 Wetted Surface Area

We call **wetted surface** area the hull area in contact with the surrounding water. When we speak about a certain value of the wetted surface area we mean the value corresponding to a given draught. We need this quantity when we calculate the **ship resistance**, that is the force by which the water opposes the forward motion of the ship. Besides this, the protection against corrosion, be it active or passive, depends on the value of the wetted surface area. The methods used to calculate the wetted surface area can be extended to the evaluation of the shell area up to any given height. The total shell area is needed for a preliminary estimation of the weight of shell plates and the weight of paint.

Table 4.2 Hydrostatic data of ship *Lido 9*

Data	Units		Draught								
		m	0.700	0.900	1.100	1.300	1.500	1.700	1.900	2.100
Trim difference (by head > 0)	m	0.000	0.000	0.000	0.000	0.000	0.000	0.000	0.000	0.000
Volume of displacement	m³	2.998	6.090	11.212	18.669	28.379	40.314	54.197	69.825
LCB fwd of midship	m	−1.599	−1.747	−1.600	−1.446	−1.329	−1.268	−1.246	−1.266
KB	m	0.506	0.660	0.819	0.973	1.120	1.263	1.401	1.536
Waterline area	m²	11.529	20.221	31.449	42.998	54.183	64.708	74.088	81.810
LCF	m	−1.973	−1.648	−1.298	−1.150	−1.092	−1.137	−1.259	−1.388
Long mom of inertia	m⁴	144.830	218.207	334.093	469.420	642.827	857.657	1129.524	1416.003
Moment to change trim	m⁴/m	9.344	14.078	21.554	30.285	41.473	55.333	72.872	91.355
Transverse mom of inertia	m⁴	2.950	9.364	25.814	55.665	93.061	134.428	171.925	201.990
Longitudinal, KM	m	48.813	36.491	30.615	26.117	23.772	22.538	22.242	21.815
Transverse, KM	m	1.490	2.198	3.121	3.955	4.400	4.598	4.574	4.429
Block coefficient, CB	–	0.110	0.126	0.149	0.177	0.216	0.261	0.301	0.342
Waterline coefficient, CW	–	0.296	0.377	0.461	0.531	0.620	0.712	0.783	0.841
Midship coefficient, CM	–	0.069	0.124	0.172	0.220	0.280	0.344	0.398	0.444
Prismatic coefficient, CP	–	–	–	0.870	0.807	0.773	0.758	0.770

In the past the wetted surface area was calculated as the area of the hull **expansion**. In simple terms, to do this one has to "open" the hull surface and lay it flat on a plane. This operation can be done exactly for certain surfaces called **developable** (see Chapter 13), such as the surfaces of cubes, cylinders, or cones. Many hull surfaces are not developable, for some only the **middlebody** is developable. Then, the Naval Architect must be satisfied with an approximation, such as described in Comstock (1967, pp. 39–41). Recent computer programmes for Naval Architecture calculate the wetted surface area by methods of differential geometry. Approximate formulae for calculating the wetted surface area of many ship types can be found in the literature of speciality. If the chosen hull belongs to a series of models tested in a towing tank, the wetted surface area is usually included in the data supplied by the experimenting institution.

4.3 Hydrostatic Curves

Table 4.2 shows the hydrostatic data of the ship *Lido 9*, for draughts between 0.7 and 2.1 m, as calculated by the ARCHIMEDES programme. The data appear at discrete draught intervals. It is usual to represent those data also as **hydrostatic curves** that allow interpolation at any required draught. Such curves are part of the documentation that must be on board, for use by deck officers in calculations required for the operation of the vessel. Many ships are provided today with board computers that store the input data of the vessel and enable the officers to calculate immediately any data they need. Even in those cases the hydrostatic curves and the knowledge to use them should be present for emergency cases in which the computer fails.

There are no universally accepted standards for plotting hydrostatic data and we can find a wide variety of "styles." For our purposes we choose a simple model that can be accommodated in the space of a textbook page, but still shows the major features common to all representations. The curves are plots of functions of the draught, T, at constant trim and heel. In general, the trim equals zero (ship on even keel), but it is possible to plot hydrostatic curves for any given, non-zero trim. The heel is almost always zero. The hydrostatic curves represent data calculated for parallel waterplanes. Romance languages use a short, elegant term for this situation. For instance, in French one talks about "carènes isoclines," while Italian uses the term "carene isocline" and Spanish "carenas isoclinas."

Let us refer to Figure 4.2. The draught axis is vertical, positive upwards. The various properties are measured horizontally, each at its own scale, so that all curves can be contained in the same paper format. In our example, the curves of volume of displacement, ∇, displacement in fresh water, ΔFW, displacement in salt water, ΔSW, waterplane area, A_W, moment to change trim by 1 m, M_{CT}, and longitudinal metacentre above keel, KM_L, are measured along the lower scale that is to be read as 0–100 m^3, 0–100 t, 0–100 m^2, 0–100 m^4/m, or 0–100 m, respectively. The vertical centre of buoyancy, $\overline{KB}$, the transverse metacentre above keel, $\overline{KM}$,

Figure 4.3 Coefficients of form of ship *Lido 9*

the longitudinal centre of flotation, *LCF*, and the longitudinal centre of buoyancy, *LCB*, are measured along the upper scale graduated from −2 to 5 m. To simplify things, we plot the coefficients of form, C_B, C_M, C_P, and C_{WL} in another graph shown in Figure 4.3.

Let us return to the volume and displacement values represented in hydrostatic curves. The displacement volume, ∇, is usually the volume of the moulded hull. The displacements in fresh and in salt water should be **total displacements** that include the displacements of shell plates and appendages. Appendages found in all kinds of ships include rudders, propellers, propeller shafts and struts, bilge keels, and roll fins. The sonar domes of warships are also appendages if they do not appear in the lines drawing and are not directly taken into account in hydrostatic calculations. The volumes of tunnels that accommodate bow thrusters should be subtracted from the volume of the moulded, submerged hull when calculating total displacements.

American literature recommends to calculate separately the volumes and moments of shell plates and appendages, and to add them to those of the moulded hull. This procedure requires detailed knowledge of all appendages and shell plates, an information not available in early design stages. An approximate, simple method consists in adding a certain percentage to the moulded displacement volume. This amounts to multiplying the moulded volume by a **displacement factor** that is the sum of surrounding-water density and the relative part of appendages and shell plates. Examples of values found in European projects are

$$\Delta_{FW} = (1.000 + 0.008)\nabla = 1.008\nabla$$

for a vessel displacing a few hundred tonnes, and

$$\Delta_{FW} = (1.000 + 0.005)\nabla = 1.005\nabla$$

for larger vessels. The corresponding displacements in salt water of density 1.025 t m^{-3} are

$$\Delta_{SW} = (1.025 + 0.008)\nabla = 1.033\nabla$$

$$\Delta_{SW} = (1.025 + 0.005)\nabla = 1.030\nabla$$

To understand why the additional percentage decreases with increasing volume let us remember that volumes increase like the cubes of dimensions, while surfaces, such as those of plates and rudders, increase like the square of dimensions.

4.4 Bonjean Curves and their Use

Figure 4.4 shows the midship section of the ship *Lido 9* in solid, thick line. Its equation is of the form

$$z = f(y)$$

The **Bonjean curves** are defined by the equations

$$A = \int_{keel}^{T} y\,dz \qquad (4.16)$$

$$M = \int_{keel}^{T} zy\,dz \qquad (4.17)$$

The first integral yields the **sectional area** as function of draught, while the second integral is the moment of the sectional area about the baseline, also as function of draught.

Figure 4.5 shows the Bonjean curves of the ship *Lido 9*. The ship outline appears in solid line. The scales along the x-axis and the T-axis are different, otherwise the drawing format would be too long. The waterline appearing in the figure corresponds to the mean draught 2 m and the trim 0.5 m. The data corresponding to this line are written in Table 4.3; they are read along horizontal lines starting from the intersection of the waterline with the corresponding station. For example, the midship station is intersected by the waterline a small distance below 2 m. On the horizontal corresponding to that draught we read the sectional area

$$A = 2 \times 1.34 = 2.68 \text{ m}^2$$

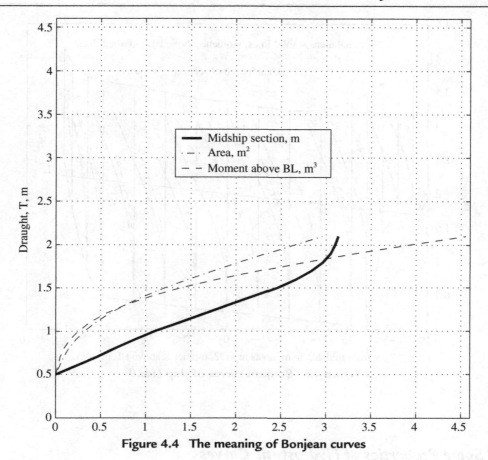

Figure 4.4 The meaning of Bonjean curves

and the moment about BL

$$M = 5 \times 0.79 = 3.95 \text{ m}^3$$

To simplify the example we neglect the data corresponding to the ship volumes aft of Station 0 and forward of Station 10. The respective values are indeed very small and by not including them we can integrate by either trapezoidal or Simpson's rule without having to correct multipliers.

The final results are calculated as follows:

$$\delta L = 1.55 \text{ m}$$
$$\nabla = 2 \times 1.55 \times 18.57 = 57.57 \text{ m}^3$$
$$LCB = \frac{-6.21}{18.57} \times 1.55 = -0.518 \text{ m}$$
$$\overline{KB} = \frac{27.20}{18.57} = 1.465 \text{ m}$$

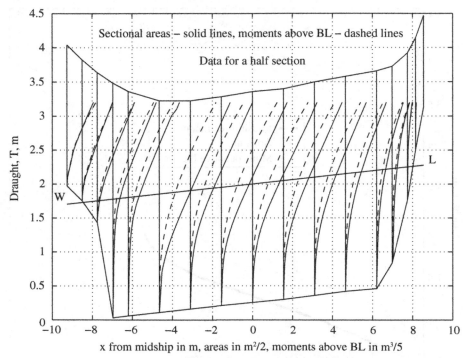

Figure 4.5 Bonjean curves of ship *Lido 9*

4.5 Some Properties of Hydrostatic Curves

In Section 4.2 we have learnt how to calculate hydrostatic data and represent them as functions of draught, for constant trim and heel. In addition to the functional dependence of each variable on draught, certain relationships between the various curves hold true. In this section, we are going to show three of them. Relationships between the various hydrostatic curves have been used to check visually the correctness of hydrostatic calculations. Such checks were obviously very useful when calculations were carried on by tedious manual procedures, even if with the help of mechanical integrating devices. Today we rely on the correctness and accuracy of computer programmes, but errors can still occur when plotting the output of the programmes by means of procedures that are not part of the hydrostatic programme. Besides this, reading this section is a good exercise in understanding the meaning of hydrostatic data.

In Figure 4.6 we consider a floating body with the waterline WL. The centre of buoyancy is B, the displacement volume is ∇, and the waterplane area, A_W. The moment of the submerged volume about the plane zOy is $x_B\nabla$, the moment of the submerged volume about the plane xOz equals $y_B\nabla$, and the moment of the submerged volume about the plane yOx is $z_B\nabla$.

Table 4.3 A Bonjean sheet

Station No.	Trapezoidal Multiplier	Lever Arm	Sectional Area	Functions of Area	Moment from MS	Moment above BL	Functions of Moment
	α_i	j_i	A_i	$\alpha_i A_i$	$\alpha_i j_i A_i$	M_i	$\alpha_i M_i$
1	2	3	4	$5 = 2 \times 4$	$6 = 3 \times 5$	7	$8 = 2^7$
0	1/4	−5	0.23	0.06	−0.29	0.37	0.09
$\frac{1}{2}$	1/2	−4.5	0.68	0.34	−1.53	0.93	0.47
1	3/4	−4	1.04	0.78	−3.12	1.45	1.09
2	1	−3	2.99	2.99	−8.98	3.83	3.83
3	1	−2	2.21	2.21	−4.41	3.11	3.11
4	1	−1	2.62	2.62	−2.62	3.76	3.76
5	1	0	2.68	2.68	0.00	3.93	3.93
6	1	1	2.42	2.42	2.42	3.68	3.68
7	1	2	2.09	2.09	4.17	3.29	3.29
8	1	3	1.51	1.51	4.54	2.47	2.47
9	3/4	4	0.87	0.65	2.60	1.45	1.09
$9\frac{1}{2}$	1/2	4.5	0.43	0.21	0.97	0.77	0.38
10	1/4	5	0.03	0.01	0.04	0.06	0.01
Sum				18.57	−6.21		27.20

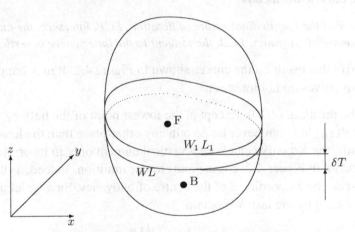

Figure 4.6 Properties of "isocline" floating bodies

Let us assume that the waterline rises by a draught change equal to δT. Then, the submerged volume increases by $\delta V = A_W \delta T$. Let the centre of the additional volume be F. When δT tends to zero, F tends to the centroid of the waterline, that is to the *centre of flotation*. The moments of the submerged volume change by

$$\delta(x_b \nabla) = x_F A_W \delta T$$
$$\delta(y_b \nabla) = y_F A_W \delta T$$
$$\delta(z_b \nabla) = z_F A_W \delta T \qquad (4.18)$$

Expanding the left-hand side of Eqs. (4.18) we obtain

$$\nabla \delta x_B + x_B \delta \nabla = x_F A_W \delta T$$
$$\nabla \delta y_B + y_B \delta \nabla = y_F A_W \delta T$$
$$\nabla \delta z_B + z_B \delta \nabla = z_F A_W \delta T \tag{4.19}$$

Dividing by $\delta \nabla = A_W \delta T$, rearranging terms and passing to infinitesimal quantities we rewrite Eqs. (4.19) as

$$x_F - x_B = \frac{d(x_B)}{dT} \cdot \frac{\nabla}{A_W}$$
$$y_F - y_B = \frac{d(y_B)}{dT} \cdot \frac{\nabla}{A_W}$$
$$z_F - z_B = \frac{d(z_B)}{dT} \cdot \frac{\nabla}{A_W} \tag{4.20}$$

Let us consider the first of Eqs. (4.20) and assume $x_F = x_B$. The left-hand side becomes zero and so must be the right-hand side. The displacement volume, ∇, can equal zero only at the lowest point of the hull. For any other point for which $x_f = x_B$ we must have $d(x_B)/dT = 0$. In the hydrostatic curves this means

> *Where the curve of the longitudinal centre of flotation, LCF, intersects the curve of the longitudinal centre of buoyancy, LCB, the tangent to the latter curve is vertical.*

We can easily verify this result on the curves shown in Figure 4.2. It may happen that for some ship forms the two curves do not intersect.

We turn now to the third Eq. (4.20). Except at the lowest point of the hull, z_F can never equal z_B. It results that $d(z_B)/dT$ can never be zero in any other place than the lowest point of the hull. In other words, the $\overline{KB}$ curve can have a vertical tangent only in its origin. This result, which can be checked in Figure 4.2, corresponds to our intuition. Indeed, as the draught increases, so must do the z-coordinate of the centre of buoyancy. Finally, let us divide, side by side, the first Eq. (4.20) by the last. We obtain

$$\frac{x_F - x_B}{z_F - z_B} = \frac{dx_B}{dz_B} \tag{4.21}$$

and remark that $z_F = T$. To discover the geometric significance of Eq. (4.21) let us examine Figure 4.7 built with data of the ship *Lido 9*; it contains a plot of z_B as function of x_B or, with alternative notations, $\overline{KB}$ values as function of *LCB*. The point F is the centre of flotation corresponding to a draught of 0.9 m, and the point B, the centre of buoyancy for the same draught. We can write

$$\tan(\angle OBF) = \frac{dx_B}{dz_B} = \frac{\overline{OF}}{\overline{BO}} = \frac{x_F - x_B}{z_F - z_B} \tag{4.22}$$

as predicted by Eq. (4.21).

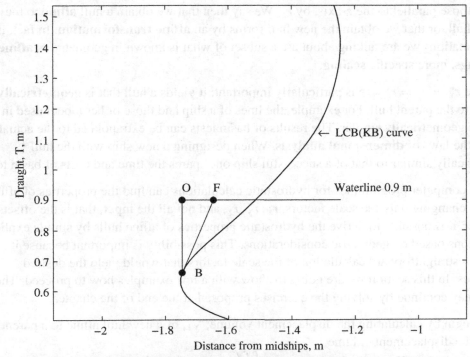

Figure 4.7   Relationship between centre of flotation and centre of buoyancy

Conventional ships are symmetric about their centrelines. Then, $y_f = y_B = 0$ and so is $d(y_B)/dT$. For floating bodies that have no port-to-starboard symmetry, it makes sense to divide the second of Eqs. (4.20) by the third and obtain

$$\frac{y_F - y_B}{z_F - z_B} = \frac{dy_B}{dz_B} \tag{4.23}$$

Then, a property similar to that derived for the $z_B(x_B)$-curve can be found for the $z_B(y_B)$-curve. Examples of floating bodies that have no port-to-starboard symmetry are ships with permanent list caused by unsymmetrical loading, by negative metacentric height or by flooding.

4.6 Hydrostatic Properties of Affine Hulls

One way of obtaining new ship lines is to derive them by a transformation, or mapping, of some suitable, given lines. The simplest transformation is that in which all dimensions parallel to one of the coordinate axes are multiplied by the same **scale factor**. Thus, let all dimensions parallel to the x-axis be multiplied by r_x, all dimensions parallel to the y-axis be multiplied by

r_y, and those parallel to the z-axis, by r_z. We say then that we obtain a hull **affine** to the **parent** hull, or that we obtain the new hull forms by an **affine transformation**. In fact, the transformations we are talking about are a subset of what is known in geometry as **affine mappings,** more specific **scaling**.

The case $r_x = r_y = r_z = r$ is particularly important; it yields a hull that is **geometrically similar** to the parent hull. For example, the lines of a ship and those of her model used in basin tests are geometrically similar. The results of basin tests can be extrapolated to the actual ship size by the laws of **dimensional analysis**. When designing a new ship with the hull geometrically similar to that of a successful ship one spares the time and costs of basin tests.

Modern computer programmes for hydrostatic calculations can find the properties of affine hulls by changing only the scale factors, r_x, r_y, r_z, and not all the input, that is the offsets. However, it is possible to derive the hydrostatic properties of affine hulls by simple explicit expressions based on geometric considerations. This possibility is important because it permits a straightforward calculation of the scale factors that would yield the desired properties. In this section we are going to show with a few examples how to proceed. The reader may continue by solving the exercises proposed at the end of the chapter.

Let us begin by calculating the displacement volume, ∇_1, of a new hull affine to a parent hull having the displacement volume

$$\nabla_0 = \iiint dx dy dz \tag{4.24}$$

The dimensions of the new hull change as $x_1 = r_x x$, $y_1 = r_y y$, $z_1 = r_z z$, so that the new displacement volume is

$$\nabla_1 = \iiint dx_1 dy_1 dz_1 = \iiint r_x dx \cdot r_y dy \cdot r_z dz = r_x r_y r_z \nabla_0 \tag{4.25}$$

For geometrically-similar hulls we obtain $\nabla_1 = r^3 \nabla_0$.

With a similar reasoning we can find that for scale factors r_x, r_y, r_z the new longitudinal centre of buoyancy is $LCB_1 = r_x LCB_0$, the new longitudinal centre of flotation is $LCF_1 = r_x LCF_0$, and the new vertical centre of buoyancy, $\overline{KB_1} = r_z \overline{KB_0}$.

4.7 Summary

The methods of numerical integration learnt in Chapter 3 can be applied to the calculation of hydrostatic data. The properties of waterplanes are the area, A_W, the longitudinal coordinate of the centre of flotation, LCF, the transverse moment of inertia, I_T, and the longitudinal moment of inertia, I_L. These properties can be conveniently calculated in an electronic spreadsheet. The input data, that is the half-breadths, are entered only once, but are used repeatedly in all calculations. The various quantities are calculated each in a separate column.

Table 4.4 A summary of hydrostatic calculations

Quantity	Notation	How to Calculate it
Waterplane area	A_W	Eq. (4.1)
Moment of waterplane area about a transverse axis	M_x	Eq. (4.2)
Longitudinal centre of flotation	x_f, LCF	Eq. (4.3)
Transverse moment of inertia of waterplane area	I_T	Eq. (4.5)
Moment of inertia of waterplane about a transverse axis	I_y	Eq. (4.6)
Longitudinal moment of inertia of waterplane area	I_L	Eq. (4.7)
Displacement volume	∇	Eq. (4.8) and Table 4.3
Moment of displacement volume above baseline	M_B	Eq. (4.9)
Vertical centre of buoyancy	z_B, $\overline{KB}$, VCB	Eq. (4.10) and Table 4.3
Longitudinal centre of buoyancy	x_B, LCB	Table 4.3
Tonnes per centimetre immersion	TCP	Eq. (4.12)
Moment to change trim by 1 m	M_{CT}	Eqs. (4.14) and (4.15)

In the same line, corresponding to one station, the calculations are chained in a way that reduces the number of required arithmetic operations.

The hydrostatic data are calculated at discrete intervals, as functions of draught, for constant trim and heel. These data are plotted in hydrostatic curves that allow interpolation. These curves are part of the documentation that must be present aboard the ship and are used in calculations related to the operation of the vessel. A summary of the data yielded by hydrostatic calculations is given in Table 4.4.

The Bonjean curves represent the areas of transverse sections, and the moments of these areas above the baseline, as functions of draught. Bonjean curves are used in the processing of the results of inclining experiment (see Chapter 7).

Certain relationships exist between some hydrostatic curves. They can be used for visual checks of the hydrostatic curves.

One method of deriving new ship lines consists in multiplying by the same scale factor all dimensions parallel to an axis of coordinates. Such transformations are called affine transformations. The properties of a new hull, affine to a parent hull, can be derived from the properties of the parent hull by simple algebraic expressions. An important case of affine transformation is that in which the three scale factors are equal. Two hulls related in this way are geometrically similar. Affine transformations do not change the coefficients of form.

4.8 Examples

Example—The Displacement of Geometrically Similar Hulls

Let us assume, for example, that we derive a geometrically similar hull by increasing the linear dimensions with the scale factor 10%. The displacement volume increases by the factor

$1.1^3 = 1.331$. For a quick estimate let us write

$$\nabla_1 = r^3 \nabla_0 \tag{4.26}$$

Taking natural logarithms of both sides yields

$$\ln \nabla_1 = 3 \ln r + \ln \nabla_0 \tag{4.27}$$

We differentiate both sides considering ∇_0 constant and obtain

$$\frac{d\nabla_1}{\nabla_1} = 3\frac{dr}{r} \tag{4.28}$$

We have now a rule for simple and quick approximation: the percent change of the displacement volume equals three times the percent ratio change.

4.9 Exercises

Exercise 4.1.   Modify Table 4.1 for a coordinate origin in AP and repeat the calculation. Check the results with those shown in the original table.

Exercise 4.2.   Modify Table 4.1 for use with Simpson's rule and repeat the calculations.

Exercise 4.3.   Verify the values of M_{CT} in Table 4.2, for the draughts 1.8 and 2.1 m, using the displacement volume, $\overline{KB}$ and $\overline{KM_L}$ values shown there.

Exercise 4.4.   Modify Table 4.3 for a coordinate origin in AP and repeat the calculation. Check the results with those shown in the original table.

Exercise 4.5.   Modify Table 4.3 for use with Simpson's rule and repeat the calculations.

Exercise 4.6.   Using the data of ship *Lido 9* plot a figure in which you can verify the property described by Eq. (4.21) for the draught values 1.7, 1.9, and 2.1 m.

Exercise 4.7.   Show that affine transformations leave the coefficients of form unchanged. In mathematical terminology, **the coefficients of form are invariants of affine transformations**.

Exercise 4.8.   Show that for affine hulls the metacentric radius, $\overline{BM}$, behaves like B^2/T.

Statical Stability at Large Angles of Heel

5.1 Introduction

Chapter 4 dealt with hull properties calculated as functions of draught, at constant trim and heel. We reminded then that the maritime terminologies of Romance languages have a concise term for the set of submerged hulls characterized as above. Thus, for example, the term in French is **carènes isoclines**. The first part of the term, "iso" derives from the Greek "isos" and means "equal." The meaning of the term "isocline" is "equal inclination" (see Figure 4.6 in previous chapter). In this chapter we are going to discuss the properties of submerged hulls as functions of heel, at constant displacement volume. Again, Romance languages have a concise term for the set of submerged hulls of a given vessel, having the same displacement volume. For example, the French term is **isocarènes**, while the Italian term is **isocarene**. The assumption of constant displacement volume recognizes the fact that while a ship heels and rolls her weight remains constant. By virtue of Archimedes' principle, constant weight implies constant displacement volume.

The central notion in this chapter is the **righting arm**. We shall show how to calculate and represent the righting arm in a set of curves known as **cross-curves of stability**. Another topic is the plot of the righting arm as function of the heel angle, for a given displacement volume and a given height of the centre of gravity. This plot is called **curve of statical stability** and it is used to assess the ship stability.

117

Figure 5.1 Definition of righting arm

5.2 The Righting Arm

In Figure 5.1 we consider a ship whose waterline in upright condition is $W_0 L_0$. The corresponding centre of buoyancy is B_0 and the centre of gravity, G. Let us assume that the ship heels to starboard by an angle ϕ. The new waterline is $W_\phi L_\phi$ and the centre of buoyancy moves toward the submerged side, to the new position B_ϕ. The weight force, equal to Δ, passes through G and is vertical, that is perpendicular to $W_\phi L_\phi$. The buoyancy force, also equal to Δ, passes through B_ϕ and is also perpendicular to $W_\phi L_\phi$. The perpendicular from G to the line of action of the buoyancy force intersects the latter line in Z. The forces of weight and buoyancy produce a **righting moment** whose value is

$$M_R = \Delta \overline{GZ} \tag{5.1}$$

As Δ is a constant for all angles of heel, we can say that the righting moment is characterized by the **righting arm**, $\overline{GZ}$. From Figure 5.1 we write

$$\overline{GZ} = \ell_k - \overline{KG} \sin \phi \tag{5.2}$$

For reasons to be explained several lines below, the distance ℓ_k is called **value of stability cross-curves**. This quantity results from hydrostatic calculations based on the ship lines. Such calculations are left today to the computer. The term $\overline{KG} \sin \phi$ depends on $\overline{KG}$, a quantity obtained from **weight calculations** as explained in Chapter 7. In European literature the term ℓ_k is often described as "lever arm of stability of form," while the term $\overline{KG} \sin \phi$ is called "lever arm of stability of weight."

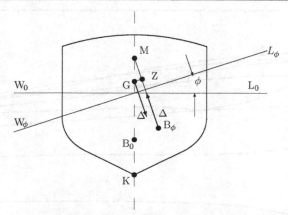

Figure 5.2 Righting arm, $\overline{GZ}$, at small angles of heel

It is important to note that ℓ_k is measured here from K, a point preferably chosen as the lowest keel point, or the projection of the lowest keel point on the midship section. The resulting ℓ_k value is thus always positive. This convention is practically standard in some European countries and, for its advantages, we follow it throughout this book. In American projects and computer programmes ℓ_k is often measured from one of the positions of the centre of gravity, G. For example, the reference point can be the centre of gravity for the full-load condition (see, for example, Lewis, 1988, pp. 78–9). When proceeding so, the designer must define in the clearest way the position of the reference point.

The relationship between the value of the stability cross-curves, ℓ_k, and the angle of heel, ϕ, is not linear and, in general, cannot be defined explicitly. For small angles of heel a linear expression for the righting arm, $\overline{GZ}$, can be derived from Figure 5.2:

$$\overline{GZ} = \overline{GM} \sin\phi \tag{5.3}$$

But, what do we mean by "small angle?" The answer is given by the same Figure 5.2. Equation (5.3) holds true as long as the metacentre, M, does not move visibly from its initial position. Thus, for many ships an angle equal to $5°$ is small, while for a few others even $15°$ may be a small angle. The value depends on both ship forms and loading condition. More insight on this point can be gained by looking at the metacentric evolutes shown in Chapter 2. A further criterion for the "smallness" of the heel angle will be given in the next section.

A useful way of plotting the ℓ_k values is shown in Figure 5.3. There, the ℓ_k curves are plotted as functions of the displacement volume, ∇, for a set of constant heel-angle values. Thus, we have a curve for $\phi = 10°$, one for $\phi = 20°$, and so on. To use Eq. (5.2) for a given displacement volume, say ∇_0, it is necessary to draw the vertical line $\nabla = \nabla_0$ and read the values where this line "crosses" the curves. Hence the term **cross-curves of stability**.

Figure 5.3   Cross-curves of stability of ship *Lido 9*

5.3 The Curve of Statical Stability

The plot of the righting arm, $\overline{GZ}$, calculated from Eq. (5.2), as function of the heel angle, ϕ, at constant ∇ and $\overline{KG}$ values is called **curve of statical stability**. Such diagrams are used to evaluate the stability of the ship in a given loading condition. For a full appreciation it is necessary to compare the righting arm with the various **heeling arms** that can endanger stability. We discuss several models of heeling arms in Chapter 6. An example of statical-stability curve is shown in Figure 5.4; it is based on Table 5.1. The table can be calculated in an electronic spreadsheet, or in MATLAB as shown in Biran and Breiner (2002), Example 2.9.

Let us identify some properties of the righting-arm curves. One important value is the maximum $\overline{GZ}$ value and the heel angle where this value occurs. For example, in Figure 5.4 the maximum righting-arm value is 1.009 m and the corresponding heel angle is 50°. Another important point is that in which the $\overline{GZ}$ curve crosses zero. The corresponding ϕ value is called **angle of vanishing stability**. In our example the righting-arm curve crosses zero at an angle greater than 90°, in a region outside the plot frame. The angle of vanishing stability can often occur at less than 90°, as shown, for example, in Figure 6.23.

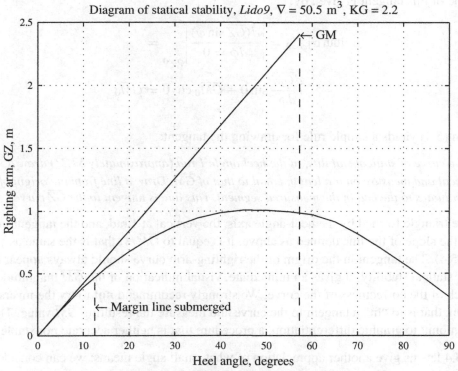

Figure 5.4 Statical-stability curve

Table 5.1 Ship *Lido 9*—righting arm, $\overline{GZ}$, for $\nabla = 50.5 \text{ m}^3$, $\overline{KG} = 2 \text{ m}$

Heel Angle (°)	ℓ_p, m (m)	$\overline{KG} \sin\phi$ (m)	$\overline{GZ}$ (m)	Heel angle (°)	ℓ_p, m (m)	$\overline{KG} \sin\phi$ (m)	$\overline{GZ}$ (m)
0	0.000	0.000	0.000	50	2.694	1.685	1.009
5	0.396	0.192	0.204	55	2.799	1.802	0.997
10	0.770	0.382	0.388	60	2.879	1.905	0.974
15	1.115	0.569	0.546	65	2.908	1.994	0.914
20	1.427	0.752	0.675	70	2.883	2.067	0.816
25	1.713	0.930	0.783	75	2.828	2.125	0.703
30	1.977	1.100	0.877	80	2.747	2.167	0.580
35	2.208	1.262	0.946	85	2.641	2.192	0.449
40	2.402	1.414	0.988	90	2.513	2.200	0.313
45	2.564	1.556	1.008				

$\overline{KM} = 4.608 \text{ m}$; $\overline{KG} = 2.200 \text{ m}$; $\overline{GM} = 2.408 \text{ m}$.

A very useful property refers to the tangent in the origin of the righting-arm curve. The slope of this tangent is given by:

$$|\tan \alpha|_{\phi=0} = \left| \frac{d(\overline{GZ}\sin\phi)}{d\phi = 0} \right|_{\phi=0} =$$

$$\frac{d\overline{GM}}{d\phi}\sin 0 + \overline{GM}_0 \cos 0 = \overline{GM}_0 \qquad (5.4)$$

Equation (5.4) yields a simple rule for drawing the tangent:

> *In the curve of statical stability, at the heel angle 1 rad (approximately 57.3°) draw a vertical and measure on it a length equal to that of $\overline{GM}$. Draw a line from the origin of coordinates to the end of the measured segment. This line is tangent to the $\overline{GZ}$ curve.*

From the triangle formed by the heel-angle axis, the vertical at 1 rad, and the tangent in origin, we find the slope of the line defined as above; it is equal to $\overline{GM}/1$, that is the same as yielded by Eq. (5.4). The tangent in the origin of the righting-arm curve should always appear in the curve of statical stability; it gives an immediate, visual indication of the $\overline{GM}$ magnitude, and it is a check of the correctness of the curve. We strongly recommend **not** to try the inverse operation, that is to "fit" a tangent to the curve and measure the resulting $\overline{GM}$ value. This would amount to graphic differentiation, a procedure that is neither accurate nor stable.

Figure 5.4 lets us give another appreciation of what small angle means: we can consider as small those heel angles for which the curve of the righting arm can be confounded with the tangent in its origin. In our example this holds true for angles up to 7–8°.

For any angle of heel, ϕ, we can rewrite Eq. (5.4) as

$$\frac{d\overline{GZ}}{d\phi} = \overline{ZM}_\phi \qquad (5.5)$$

where Z is as previously the foot of the perpendicular from G to the line of action of the buoyancy force, and M_ϕ is the metacentre corresponding to the heel angle ϕ. The geometric construction of this tangent is similar to that of the tangent in origin. For a proof of this result see, for example, Birbănescu-Biran (1979).

5.4 The Influence of Trim and Waves

Once it was usual to calculate the cross-curves of stability at constant trim, i.e. for the ship on even keel. This approach was justified before the appearance of computers and Naval-Architectural software. However, Eq. (2.28), developed in Chapter 2, shows that the longitudinal position of the centre of buoyancy changes if the heel angle is large. It happens so because at large heel angles the waterplane ceases to be symmetric about the centreline. If the centre of buoyancy moves along the ship, while the position of the centre of gravity is constant, the trim changes too. Therefore, cross-curves calculated at constant trim may not represent actual stability conditions. Jakić (1980) has shown that trim can greatly influence the values of cross-curves and, therefore, that influence should be taken into account. The old stability regulations, *BV 1033*, of the German Navy required, indeed, the calculation of the cross-curves at the trim induced by heel. More recently, the 2008 IMO code for intact stability recommends to calculate the cross-curves with the "free trim" option.

Modern computer programmes for Naval Architecture include this option. As we shall show in Chapter 9, waves perpendicular or oblique to the ship velocity influence the values of cross-curves and can cause a very dangerous effect called parametric resonance. This effect too must be taken into account and modern computer programmes can calculate cross-curves on waves. The stability regulations of the German Navy take into account the variation of the righting arm in waves (see Arndt, 1965, and Arndt et al., 1982).

5.5 Summary

In this chapter we dealt with the righting moment at large angles of heel, $M_R = \Delta \overline{GZ}$. The quantity $\overline{GZ}$, called righting arm, is the length of the perpendicular drawn from the centre of gravity, G, to the line of action of the buoyancy force. We assume that the ship heels at constant displacement. This is the desired situation in which the ship neither loses loads, nor takes water aboard. Then, the factor Δ is constant and the variation of the righting moment with heel is described by the variation of the righting arm $\overline{GZ}$. The value of the righting arm is calculated from

$$\overline{GZ} = \ell_k - \overline{KG} \sin \phi$$

where ℓ_k, called value of stability cross-curve, is the distance from the reference point K to the line of action of the buoyancy force, $\overline{KG}$, the distance of the centre of gravity from the same point K, and ϕ, the heel angle. It is recommended to take the point K as the lowest hull point.

The values of the stability cross-curves, ℓ_k, are usually represented as functions of the displacement volume, with the heel angle as parameter. One can read in this plot the values corresponding to a given displacement volume, calculate with them the righting arm and plot its values against the heel angle. This plot is called curve of statical stability and it is used to appreciate the stability of the ship, at a given displacement and height of the centre of gravity. To check the correctness of the righting-arm curve, it is recommended to draw the tangent in the origin. To do this, one should draw a vertical line at the angle of 1 rad and measure on the vertical a length equal to the metacentric height, $\overline{GM}$. The tangent is the line that connects the origin of coordinates to the point found as in the previous sentence.

Table 5.2   Terms related to stability at large angles of heel

English Term	Symbol	Computer Notation	Translations (Old European symbol)
Centre of buoyancy	B		Fr centre de carène (C) G Verdrängungsschwepunkt (F), I centro di carena, S centro de empuje, de carena
Centre of gravity	G		Fr centre de gravité, G Massenschwerpunkt, I centro di gravità, S centro de gravedad (del buque)
Curve of statical stability			Fr courbe de stabilité, G Stabilitätskurve, I curva di stabilitá, S curva de estabilidad
Heel angle (positive starboard down)	ϕ	HELANG	Fr angle de bande, angle de gîte G Krängungswinkel, I angolo di inclinazione trasversale, sbandamento, S ángulo de escora, ángulo de balance
Keel point—Reference point on BL	K		F point le plus bas de la carène, G Kielpunkt I intersezione della linea base con la sezione maestra S intersección de la línea base con la cuaderna maestra
Projected centre of gravity	Z		G projizierter Massenschwerpunkt
Righting lever	$\overline{GZ}$	GZ	F bras de levier (GK), G Aufrichtenden Hebelarm, I braccio radrizzante, S brazo adrizante
Value of stability cross-curve	ℓ_k	LK	Fr pantocarénes, bras de levier du couple de redressement, G Pantocarenenwert bezogen auf K S pantocarenas isoclinas
z-Coordinate of centre of gravity	$\overline{KG}$	ZKG	Fr distance du centre de gravité à la ligne d'eau zéro G z-Koordinate des Massenschwerpunktes I distanza verticale del centro di gravità, S altura del centro de gravedad sobre la quilla

The trim changes as the ship heels. That effect should be taken into account when calculating cross-curves of stability. Another influence to be taken into account is that of waves.

Table 5.2 summarizes the main terms related to stability at large angles of heel. As in Chapter 1, we note by "Fr" the French term, by "G" the German term, by "I" the Italian term,

Ship *Lido 9* – Cross–curves of stability

Figure 5.5 3D Cross-curves of stability of ship *Lido 9*

and by "S" the Spanish term. Old symbols used once in those languages are given between parentheses.

5.6 Example

Figure 5.5 is a 3D representation of the cross-curves of the ship *Lido 9*.

5.7 Exercises

Exercise 5.1. Plot in one figure the righting-arm curves and the tangents in origin of the ship *Lido 9*, for $\nabla = 50.5$ m^3 and $\overline{KG}$—values 1.8, 2.0, 2.4, and 2.6 m. Comment the influence of the centre-of-gravity height.

Exercise 5.2. Draw the curve of statical stability of the ship *Lido 9* for a displacement in sea water $\Delta = 35.3\ t$ and a height of the centre of gravity $\overline{KG} = 2.1$ m. Use data in Tables 4.2 and 5.1.

CHAPTER 6

Simple Models of Stability

Chapter Outline

6.1 Introduction

In Chapter 5 we learnt how to calculate and how to plot the righting arm in the curve of statical stability. It may be surprising that for a very long period the metacentric height and the curve of righting arms were considered sufficient for appreciating the ship stability. We do not proceed so in other engineering fields. As pointed out by Wendel (1965), one first finds out the resistance to ship advance and only afterwards dimensions of the engine. Also, we first calculate the load on a beam and only afterwards we dimension it. Similarly, we should determine the **heeling moments** and then compare them with the righting moment. It was only at the beginning of the 20th century that Middendorf proposed such a procedure for large sailing ships. His book, *Bemastung und Takelung der Schiffe* was first published in Berlin, in 1903, and it contained the first proposal for a ship-stability criterion. In 1933, Pierrottet (see Pierrottet, 1935) wrote in a publication of the test basin in Rome that the stability of a ship must be assessed by comparing the heeling moments with the righting moment. He detailed his proposal in 1935, in a meeting of INA, but had no immediate followers. In 1939 Rahola (see Rahola, 1939) published in Helsinki his doctoral thesis; it was based on extensive statistics and a very profound analysis of the qualities of stable and unstable vessels. Rahola proposed then a stability criterion that considered only the metacentric height and the curve of the righting arm. The Naval-Architectural community appreciated Rahola's work and his proposal was used, indeed, as a stability standard and stood at the basis of stability regulations issued later by national and international authorities.

It was only after the Second World War that the issue of comparing heeling and righting arms was brought up again. German researchers used then a very appropriate term: **Lever arm balance** (Hebelarm Bilanz). Eventually, newer stability regulations made compulsory the comparison of lever arms and we show in this chapter how to do it.

Heeling moments can be caused by wind, by the centrifugal force developed in turning, by transverse displacements of masses, by towing, or by the lateral pull developed in cables that connect two vessels during the transfer of loads at sea. In Chapter 5 we have shown that, when the ship heels at constant displacement, it is sufficient to consider the righting arm as an indicator of stability. Then, to assess the ship stability it is necessary to compare the righting arm with a **heeling arm**. According to the DIN-ISO standard, we note the heeling arm by the letter ℓ and indicate the nature of the righting arm by a subscript. To obtain a generic heeling arm, ℓ_g, corresponding to a generic heeling moment, M_g, we divide that moment by the ship weight

$$\ell_g = \frac{M_g}{g\Delta} \tag{6.1}$$

where Δ is the displacement mass, and g, the acceleration due to gravity. In older practice it has been usual to measure the displacement in units of force. Then, instead of Eq. (6.1) one had to use

$$\ell_g = \frac{M_g}{\Delta}$$

Much attention should be paid
to the system of units used in calculation. From now on we constantly use the displacement mass in calculations. At this point it may seem that we defined the heeling arm as above just to be able to compare the righting arm with a quantity having the same physical dimensions (and units!). In Section 6.7 we prove that this definition is mathematically justified.

In Figure 6.1 we superimposed the curve of a generic heeling arm, ℓ_g, over the curve of the righting arm, $\overline{GZ}$. For almost all positive heeling angles shown in the plot the righting arm is

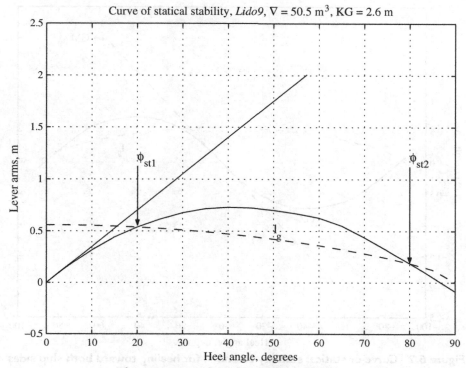

Figure 6.1 Angles of statical equilibrium

positive. We define the righting arm as positive if when the ship is heeled to starboard, the righting moment tends to return it toward port. In the same figure the heeling arm is also positive, meaning that the corresponding heeling moment tends to incline the ship toward starboard. What happens if the ship heels in the other direction, that is with the port side down? Let us extend the curve of statical stability by including negative heel angles, as in Figure 6.2. The righting arms corresponding to negative heel angles are negative. For a ship heeled toward port, the righting moment tends, indeed, to return the vessel toward starboard, therefore it has another sign than in the region of positive heel angles. The heeling moment, however, tends in general to heel the ship in the same direction as when the starboard is down and, therefore, it is positive. Summarizing, the righting-arm curve is symmetric about the origin, while the heeling-arm curves are symmetrical about the lever-arm axis.

In this chapter we present simplified models of the various heeling arms, models that allow reasonably fast calculations. Approximate as they may be, those models stand at the basis of regulations that specify the stability requirements for various categories of ships. In most cases, practice has shown that ships complying with the regulations were safe. The requirements themselves are explained in Chapters 8 and 10. By the end of this chapter we briefly explain why the simplifying assumptions are necessary in Naval-Architectural practice.

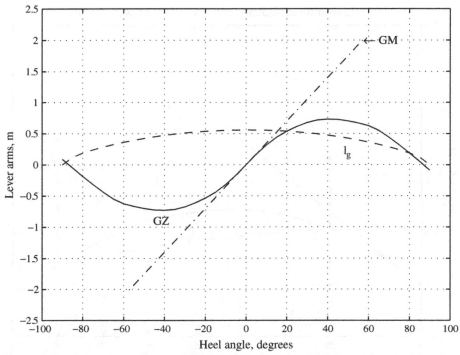

Figure 6.2 Curve of statical stability extended for heeling toward both ship sides

We can appreciate the stability of a vessel by comparing the righting arm with the heeling arm as long as the heeling moment is applied gradually and inertia forces and moments can be neglected. When the heeling moment appears suddenly, as caused, by example, by a gust of wind, one has to compare the heeling energy with the work done by the righting moment. Such situations are discussed in the section on dynamical stability. In continuation we show how moving loads, solid, or liquid, endanger the ship stability, and we develop formulae for calculating the corresponding reduction of stability. Other situations in which the stability is endangered are those of grounding or positioning in dock. We show how to predict the moment in which those situations may become critical. This chapter also discusses the situations in which a ship sails with a negative metacentric height.

6.2 Angles of Statical Equilibrium

Figure 6.1 shows the curve of a heeling arm, ℓ_g, superimposed on the curve of the righting arm, $\overline{GZ}$. In general, those curves intersect in two points; they are noted here as ϕ_{st1} and ϕ_{st2}. Both points correspond to positions of statical equilibrium because at both points the righting arm and the heeling arm are equal, and, therefore, the righting moment and the heeling moment are also equal. Only the first point corresponds to a position of **stable** equilibrium, while the second point corresponds to a situation of **unstable** equilibrium. In this section we give an intuitive proof of this statement; for a rigorous proof see Section 6.7.

Let us first consider the equilibrium at the first static angle, ϕ_{st1}, and assume that some perturbation causes the ship to heel further to starboard by a small angle, $\delta\phi$. When the perturbation ceases, at the angle $\phi_{st1} + \delta\phi$ the righting arm is larger than the heeling arm, returning thus the ship toward its initial position, at the angle ϕ_{st1}. Conversely, if the perturbation causes the ship to heel toward port, to an angle $\phi_{st1} - \delta\phi$, when the perturbation ceases the righting arm is smaller than the heeling arm, so that the ship returns toward the initial position, ϕ_{st1}. This situation corresponds to the definition of stable equilibrium given in Section 2.4.

Let us see now what happens at the second angle of equilibrium, ϕ_{st2}. If some perturbation causes the ship to incline further to starboard, the heeling arm will be larger than the righting arm and the ship will capsize. If the perturbation inclines the ship toward port, after its disappearance the righting arm will be larger than the heeling arm and the ship will incline toward port regaining equilibrium at the first static angle, ϕ_{st1}. We conclude that the second static angle, ϕ_{st2}, corresponds to a position of unstable equilibrium.

6.3 The Wind Heeling Arm

We use Figure 6.3 to develop a simple model of the heeling moment caused by a beam wind, that is a wind perpendicular to the centreline plane. In this situation the wind heeling arm is

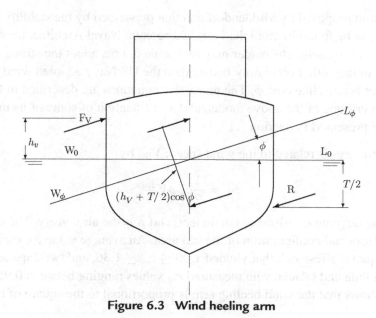

Figure 6.3 Wind heeling arm

maximal. In the simplest possible assumption the wind generates a force, F_V, that acts in the centroid of the lateral projection of the above-water ship surface, and has a magnitude equal to

$$F_V = p_V A_V$$

where p_V is the wind pressure, and A_V, the area of the above-mentioned projection of the ship surface. Let us call A_V **sail area**.

Under the influence of the force F_V the ship tends to drift, a motion opposed by the water with a force, R, equal in magnitude to F_V. To simplify calculations we assume that R acts at half-draught, $T/2$. The two forces, F_V and R, form a torque that inclines the ship until the heeling moment equals the righting moment. The value of the heeling moment in the upright condition is $p_V A_V (h_V + T/2)$, where h_V is the height of the sail-area centroid above $W_0 L_0$. The heeling arm in upright condition is

$$\ell_V(0) = \frac{p_V A_V (h_V + T/2)}{g\Delta}$$

How does the heeling arm change with the heeling angle? In the case of a "flat" ship, that is for $B = 0$, the area exposed to the wind varies proportionally to $\cos\phi$. In Figure 6.3 we show that for a flat ship the forces F_V and R would act in the centreline plane, both horizontally, that is parallel to the inclined waterline $W_\phi L_\phi$. Then, the lever arm of the torque would be proportional to $\cos\phi$. Summing up, the wind heeling arm equals

$$\ell_V(\phi) = \frac{p_V A_V \cos\phi}{g\Delta}(h_V + T/2)\cos\phi = \frac{p_V A_V (h_V + T/2)}{g\Delta}\cos^2\phi \qquad (6.2)$$

This is the equation proposed by Middendorf and that prescribed by the stability regulations of the US Navy; it can be found in more than one textbook on Naval Architecture where it is recommended for all vessels. The reader may feel some doubts about the strong assumptions accepted above. In fact, other regulatory bodies than the US Navy adopted wind heeling arm curves that do not behave like $\cos^2\phi$. The respective equations are described in Chapters 8 and 10. Our own critique of the above model, and a justification of some of its underlying assumptions, are presented in Section 6.13.

The wind pressure, p_V, is related to the wind speed, V_W, by

$$p_V = \frac{1}{2}c_w \rho V_W^2 \qquad (6.3)$$

where c_w is an aerodynamic resistance coefficient, and ρ is the air density. The coefficient c_w depends on the form and configuration of the sail area. An average value for c_w is 1.2. Wegner (1965) quotes a research that yielded $1.00 \leq c_w \leq 1.36$, and two Japanese researchers, Kinohita and Okada, who measured c_w values ranging between 0.95 and 1.24. Equation (6.3) shows that the wind heeling arm is proportional to the square of the wind speed. In this section we considered the wind speed as constant over all the sail area. This assumption is acceptable for a fast estimation of the wind heeling arm. However, we may know from our own experience that wind speed increases with height above the water surface. Some stability regulations recognize this phenomenon and we show in Chapters 8 and 10 how to take it into account. Calculations with variable wind speed, that is considering the **wind gradient**, yield lower, more realistic heeling arms for small vessels whose sail area lies mainly in the low wind-speed region. It may be worth mentioning that engineers take into account the wind gradient in the design of tall buildings and tall cranes.

6.4 Heeling Arm in Turning

When a ship turns with a linear speed V, in a circle of radius R_{TC}, a centrifugal force, F_{TC}, develops; it acts in the centre of gravity, G, at a height $\overline{KG}$ above the baseline. From mechanics we know that

$$F_{TC} = \Delta \frac{V^2}{R_{TC}}$$

Under the influence of the force F_{TC} the ship tends to drift, a motion opposed by the water with a reaction R. To simplify calculations, we assume again that the water reaction acts at half-draught, that is at a height $T/2$ above the baseline. The two forces, F_{TC} and R, form a torque whose lever arm in upright condition is $(KG - T/2)$. For a heeling, flat ship this lever arm is proportional to $\cos\phi$. Dividing by the displacement force, we obtain the **heeling lever of the centrifugal force in turning circle**:

$$\ell_{TC} = \frac{1}{g} \frac{V^2}{R_{TC}} (KG - T/2) \cos\phi \tag{6.4}$$

The speed V to be used in Eq. (6.4) is the speed in turning, smaller than the speed achieved when sailing on a straight-line path. The turning radius, R_{TC}, and the speed in turning, V, are not known in the first stages of ship design. If results of basin tests on a ship model, or of sea trials of the ship, or of a sister ship, are available, they should be substituted in Eq. (6.4). The stability regulations of the German Navy, BV 1030-1, provide a coefficient for the approximation of the turning radius to be used in the early design stages of naval ships (see Chapter 10). A discussion of this subject can be found in Wegner (1965). This author uses a non-dimensional factor

$$C_D = \left(\frac{V_D}{V_0}\right)^2 \cdot \frac{L_{pp}}{R_{TC}} \tag{6.5}$$

where V_D is the ship speed in turning, and V_0, the speed on a straight-line path. Substituting into Eq. (6.4) yields

$$\ell_{TC} = C_D \frac{V_0^2}{g L_{pp}} (\overline{KG} - T/2) \cos\phi \tag{6.6}$$

Quoting *Handbuch der Werften*, Vol. VII, Wegner shows that for 95% of 80 cargo ships the values of C_D ranged between 0.19 and 0.25. For a few trawlers the values ranged between 0.30 and 0.35.

6.5 Other Heeling Arms

A dangerous situation can arise if many passengers crowd on one side of the ship. There are two cases when passengers can do this: when attracted by a beautiful seascape, or when scared by some dangerous event. In the latter case passengers can also be tempted to go to upper decks. The resulting heeling arm can be calculated from

$$\ell_P = \frac{np}{\Delta}(y\cos\phi + z\sin\phi) \tag{6.7}$$

where n is the number of passengers, p, the average person mass, y, the transversal coordinate of the centre of gravity of the crowd, and z, the vertical translation of said centre. The second term between parentheses accounts for the virtual metacentric-height reduction. Wegner (1965) recommends to assume that up to seven passengers can crowd on a square metre, that the average mass of a passenger plus some luggage is 80 kg, and that the height of a passenger's centre of gravity above deck is 1.1 m. Similar values are prescribed by the regulations described in Chapters 8 and 10. The subject is again under discussion because the increasing prevalence of obesity in contemporary society. Wegner recommends to include in the deck area all areas that can be occupied by panicking people, e.g., tables, benches, and skylights. Other heeling moments can occur when a tug tows a barge. The barge can drift and then the tension in the towing cable can be decomposed into two components, one parallel to the tug centreline, the other perpendicular to the first. The latter component can cause capsizing of the tug. The process is very fast and there may be no survivors. To avoid this situation tugs must be provided with quick-release mechanisms that free instantly the towing cable. Lateral forces also appear when fishing vessels tow nets or when two vessels are connected by cables during replenishment-at-sea operations. Special provisions are made in stability regulations for the situations mentioned above. **Icing** is a phenomenon known to ship crews sailing in very cold zones. The accumulation of ice has a double destabilizing effect: it raises the centre of gravity and it increases the sail area. The importance of ice formation should not be underestimated. For example, Arndt (1960a) cites cases in which blocks of ice 1 m thick developed on a poop deck, or walls of 60 cm of ice formed on the front surface of a bridge. Therefore, stability regulations take into account the effect of ice.

6.6 Dynamical Stability

Until now we assumed that the heeling moments are applied gradually and that inertial moments can be neglected. Shortly, we studied **statical stability**. Heeling moments, however, can appear, or increase suddenly. For example, wind speed is usually not constant, but fluctuates. Occasionally, sudden bursts of high intensity can occur; they are called **gusts**. As another example, loosing a weight on one side of a ship can cause a sudden heeling moment that sends down the other side. In the latter cases we are interested in **dynamical stability**. It is no more sufficient to compare righting with heeling arms; we must compare the energy of the heeling moment with the work done by the opposing righting moment. It can be easily shown that the energy of the heeling moment is proportional to the area under the heeling-arm curve, and the work done by the righting moment is proportional to the area under the righting-arm curve. To prove this, let us remember that the work done by a force, F, which produces a motion from x_1 to x_2 is equal to

$$W = \int_{x_1}^{x_2} F dx \tag{6.8}$$

If the path of the force F is an arc of circle of radius r, the length of the arc that subtends an angle $d\phi$ is $dx = rd\phi$. Substituting into Eq. (6.8) yields

$$W = \int_{\phi_1}^{\phi_2} Frd\phi = \int_{\phi_1}^{\phi_2} Md\phi \tag{6.9}$$

where M is a moment.

A ship subjected to a sudden heeling moment M_h, applied when the roll angle is ϕ_1, will reach for an instant an angle ϕ_2 up to which the energy of the heeling moment equals the work done by the righting moment, so that

$$W = \int_{\phi_1}^{\phi_2} \frac{M_h}{g} d\phi = \int_{\phi_1}^{\phi_2} \overline{\Delta GZ} d\phi \tag{6.10}$$

or

$$\int_{\phi_1}^{\phi_2} \frac{M_h}{g\Delta} d\phi = \int_{\phi_1}^{\phi_2} \overline{GZ} d\phi \tag{6.11}$$

This condition is fulfilled in Figure 6.4 where the area under the heeling-arm curve is $A_2 + A_3$, and the area under the righting-arm curve is $A_1 + A_3$. As A_3 is common to both areas, the condition is reduced to $A_1 = A_2$. Moseley is quoted for having proposed the calculation of dynamical stability as early as 1850. It took several marine disasters and many years until the idea was accepted by the Naval-Architectural community.

In Figure 6.4 we marked with ϕ_{dyn} the maximum angle reached by the ship after being subjected to a gust of wind. An elegant way to find this angle is to calculate the areas under the curves as functions of the heel angle, ϕ, plot the resulting curves and find their points of intersection. The algorithm for calculating the integrals with variable upper limit is described in Section 3.4.

Figure 6.4 Dynamical stability

In Figure 6.4 we assumed that the gust of wind appeared when the ship was in an upright condition, that is $\phi_1 = 0$. As shown in Figure 6.5, the situation is less severe if $\phi_1 > 0$, and more dangerous if $\phi_1 < 0$. In both graphs the maximum dynamical angle is found by plotting the curve

$$\int_{\phi_1}^{\phi} \overline{GZ}d\phi - \int_{\phi_1}^{\phi} l_v d\phi$$

and looking for the point where it crosses zero. An analogy with a swing (or a pendulum) is illustrated in Figure 6.6. Many readers may have tried to accelerate a swing by pushing it periodically. Thus, they may know that a push given in position (a) sends the swing to an angle that is much larger than the angle achieved by pushing at position (b). Moreover, pushing the swing while it is in position (c) proves very difficult. The physical explanation is simple. In position (a) the energy transferred from the push is added to the potential energy accumulated by the swing, the latter energy acting to return the swing rightwards. In position (c) the potential energy accumulated by the swing tends to return it to position (b), opposing thus the energy impacted by the push. The influence of the roll angle on dynamical stability is taken into consideration by some stability regulations (see Chapter 8).

Figure 6.5 The influence of the roll angle on dynamical stability

Figure 6.6 Swing analogy

6.7 Stability Conditions—A More Rigorous Derivation

We describe the dynamics of heeling by Newton's equation for rotational motion

$$J\frac{d^2\phi}{dt^2} + g\Delta\overline{GZ} = M_H \tag{6.12}$$

where J is the **mass moment of inertia** of the ship, Δ, the mass displacement, and M_H, a heeling moment. The mass moment of inertia is calculated as the sum of the products of masses by the square of their distance from the axis of roll

$$J = \sum_{i=1}^{n} \left(y_i^2 + z_i^2\right) m_i \tag{6.13}$$

where y_i is the transverse and z_i is the height coordinate of the mass i. In the SI system we measure J in m^2 t. In Eq. (6.12) we neglected damping and added mass, terms briefly introduced in Section 6.13 and used in Chapter 12. We also neglect the **coupling** of heeling with other ship motions.

Let us multiply by $d\phi$ all terms of Eq. (6.12)

$$J\frac{d^2\phi}{dt^2}d\phi + g\Delta\overline{GZ}d\phi = M_H d\phi \tag{6.14}$$

We transform the factor that multiplies J as follows

$$\frac{d^2\phi}{dt^2}d\phi = \frac{d\dot\phi}{dt}d\phi = \frac{d\dot\phi}{d\phi}\cdot\frac{d\phi}{dt}\cdot d\phi = \dot\phi d\dot\phi \tag{6.15}$$

and integrate between an initial angle, ϕ_0, and a current angle, ϕ,

$$J\int_{\phi_0}^{\phi}\dot\phi d\dot\phi + g\Delta\int_{\phi_0}^{\phi}\overline{GZ}d\phi = \int_{\phi_0}^{\phi}M_H d\phi \tag{6.16}$$

The result is

$$\frac{1}{2}J\left(\dot\phi^2(\phi) - \dot\phi^2(\phi_0)\right) = \int_{\phi_0}^{\phi}M_H d\phi - g\Delta\int_{\phi_0}^{\phi}\overline{GZ}d\phi \tag{6.17}$$

The left-hand side of the above equation represents kinetic energy, K. In the position of stable equilibrium the potential energy has a minimum. As the sum of potential and kinetic energies is constant in a system such as that under consideration (it is a **conservative** system), the kinetic energy has a maximum in the position of statical equilibrium. The conditions for maximum are

$$\frac{dK}{d\phi} = 0$$
$$\frac{d^2K}{d\phi^2} < 0 \tag{6.18}$$

Substituting K by the right-hand side of Eq. (6.17) and differentiating we obtain

$$\frac{M_H}{g\Delta} = \overline{GZ}$$
$$\frac{d(M_H/(g\Delta))}{d\phi} < \frac{d\overline{GZ}}{d\phi} \tag{6.19}$$

The first Eq. (6.19) shows that at the point of statical equilibrium the righting arm equals the heeling arm. The second equation shows that at the point of stable statical equilibrium the slope of the righting arm must be greater than the slope of the heeling arm. This is a rigorous proof that the first static angle corresponds to a position of stable equilibrium, while the second static angle does not.

Until now we looked for the angles of statical equilibrium. Let us examine the dynamical phenomenon, that is the behaviour of the heeling angle, ϕ, as function of time. The conditions for maximum dynamic angle are

$$\dot{\phi} = 0, \quad \ddot{\phi} < 0 \tag{6.20}$$

Substituting the first part of Eq. (6.20) in Eq. (6.16) we obtain

$$\int_{\phi_0}^{\phi} \overline{GZ} d\phi = \int_{\phi_0}^{\phi} \frac{M_H}{g\Delta} d\phi \tag{6.21}$$

Equation (6.21) represents the condition of equality of the areas under the righting and the heeling arms. The second part of Eq. (6.20) when applied to Eq. (6.12) yields the condition

$$\overline{GZ} > \frac{M_H}{g\Delta} \tag{6.22}$$

Figure 6.7 shows two limiting cases. In the upper plot the first condition (6.19) is fulfilled, while the second is not. Therefore, in this case there is no angle of stable statical equilibrium and the ship is lost. In the lower Figure 6.7 the areas under the righting-arm and the heeling-arm curves are equal, but condition (6.22) is not fulfilled. Therefore, under the shown gust of wind the ship will capsize.

6.8 Roll Period

For small angles of heel, and assuming $M_H = 0$, we rewrite Eq. (6.12) as

$$J\frac{d^2\phi}{dt^2} + g\Delta\overline{GM}\phi = 0 \tag{6.23}$$

We say that this equation describes **unresisted roll**. We define the **mass radius of gyration**, i_m, by

$$J = i_m^2 \Delta \tag{6.24}$$

Substituting the above expression into Eq. (6.23) and rearranging yields

$$\frac{d^2\phi}{dt^2} + \frac{g\overline{GM}}{i_m^2}\phi = 0 \tag{6.25}$$

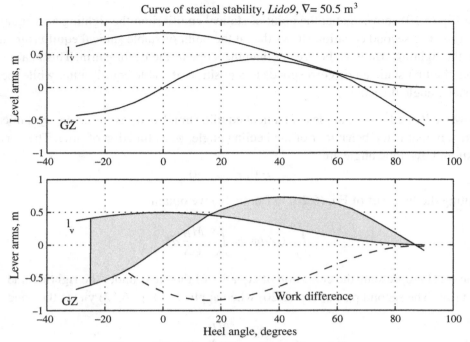

Figure 6.7   Two limiting cases of instability

With the notation

$$\omega_0 = \sqrt{\frac{g\overline{GM}}{i_m^2}} \qquad (6.26)$$

the steady-state solution of this equation is of the form $\phi = \Phi \sin(\omega_0 t + \epsilon)$, where ω_0 is the **natural angular frequency** of roll, and ϵ, the **phase**. The **natural period of roll** is the inverse of the roll frequency, f_0, defined by

$$\omega_0 = 2\pi f_0$$

Using algebra, we obtain

$$T_0 = 2\pi \frac{i_m}{\sqrt{g\overline{GM}}}, \qquad (6.27)$$

where the result is in seconds.

We conclude that the larger the metacentric height, $\overline{GM}$, the shorter the roll period, T_0. If the roll period is too short, the oscillations may become unpleasant for crew and passengers and can induce large forces in the transported cargo. Tangential forces developed in rolling are proportional to the angular acceleration, that is to

$$\frac{d^2\phi}{dt^2} = -\Phi\omega_0^2 \sin(\omega_0 t + \epsilon)$$

a quantity directly proportional to $\overline{GM}$.

Thus, while a large metacentric height is good for stability, it may be necessary to impose certain limits on it. IMO (2009), for example, referring to ships carrying timber on deck, recommends to limit the metacentric height to maximum 3% of the ship breadth (Part B, paragraph 3.7.5). Operational experience indicates that excessive initial stability should be avoided because it results in large accelerations in rolling and can cause huge stresses in lashings. Norby (1962) quotes researches carried on by Kempf, in Germany, in the 1930s. Kempf defined a non-dimensional rolling factor, $T\sqrt{g/B}$, and, on the basis of extensive statistics found that:

- for values of Kempf's factor under 8 the ship motions are **stiff**;
- for values between 8 and 14 the roll is **comfortable**;
- for factor values above 14 the motions are **tender**.

When the motions become too tender the ship master will worry because the metacentric height may be too low.

The exact value of the radius of gyration, i_m, can be calculated from Eq. (6.24) and requires the knowledge of all masses and their positions. This knowledge is not always available, certainly not in the first phases of ship design. Therefore, it is usual to assume that the radius of gyration, i_m, is proportional to the ship breadth, B,

$$i_m = aB$$

Let us define

$$c = 2a = \frac{2i_m}{B}$$

Substituting into Eq. (6.27) we obtain

$$T_0 = \frac{\pi c B}{\sqrt{g\widehat{GM}}} \tag{6.28}$$

As $\pi \approx \sqrt{g}$, we can rewrite Eq. (6.28) as

$$T_0 \approx \frac{cB}{\sqrt{\widehat{GM}}} \tag{6.29}$$

Rose (1952) quotes the following c values: large cargo and passenger vessels, 0.85; small cargo and passenger vessels, 0.77; loaded ore carriers, 0.81; tugs, 0.76; wide barges, 0.79. These values are based on old-type vessels. More recently, Costaguta (1981) recommends to take $i_m = B/3$ for merchant ships, and $c = 0.8$–0.9 for round-bilge, motor yachts. Some shipyards use $i_m = 0.35B$.

For actual ships, i_m can be obtained experimentally by measuring the roll period. When i_m is known, Eq. (6.27) can be used to control the metacentric height by measuring the roll period.

This can be done automatically and online with the help of modern technology.
Wendel (1960b) describes an instrument that did the job many years ago. The use of the roll
period as a stability indicator is discussed, for example, by Norby (1962) and Jons (1987).

Normally, the roll period is measured in the still water of a harbour, and the ship is tied by the
stern and by the aft to minimize other motions than roll. When measuring the roll period in a
seaway it is necessary to distinguish between the ships own period and the period of encounter
with the waves (see Jons, 1987, and Chapter 9).

6.9 Loads that Adversely Affect Stability

6.9.1 Loads Displaced Transversely

In Figure 6.8 we consider that a mass, m, belonging to the ship displacement, Δ, is moved
transversely a distance d. A heeling moment appears and its value, for any heeling angle, ϕ, is
$dm \cos \phi$. As a result, the ship centre of gravity, G, moves to a new position, G_1, the distance
$\overline{GG_1}$ being equal to

$$\overline{GG_1} = \frac{dm}{\Delta} \tag{6.30}$$

and the righting arm is reduced to an **effective** value

$$\overline{GZ_{eff}} = \overline{GZ} - \frac{dm}{\Delta} \cos \phi \tag{6.31}$$

The effect is the same as if the centre of gravity, G, moved to a higher position, G_{eff}. During
roll the ship inclines also to the other side, as in Figure 6.9. Then the **effective righting arm**,
$\overline{GZ_{eff}}$, increases and the ship behaves as if the centre of gravity moved to a lower position, G_{eff}.

Figure 6.8 The destabilizing effect of a mass moved transversely

Figure 6.9   The effect of a mass moved transversely when the ship rolls to the other side

6.9.2   Hanging Loads

In Figure 6.10 we consider a mass m suspended at the end of a rope of length h. When an external moment causes the ship to heel by an angle ϕ, the hanging mass moves transversely a distance $h \tan \phi$, and the ship centre of gravity moves in the same direction a distance

$$\overline{GG_1} = \frac{hm}{\Delta} \tan \phi \tag{6.32}$$

In Figure 6.11 we see that the righting arm is reduced from $\overline{GZ}$ to $\overline{G_1 Z_1} = \overline{GZ_{eff}}$. The effect is the same as if the centre of gravity, G, moved to a higher position, G_V, given by

$$\overline{GG_v} = \frac{\overline{GG_1}}{\tan \phi} = \frac{hm}{\Delta} \tag{6.33}$$

As a result, we use for initial-stability calculations a corrected, or **effective metacentric height**

$$\overline{GM_{eff}} = \overline{GM} - \frac{hm}{\Delta} \tag{6.34}$$

The destabilizing effect appears immediately after raising the load sufficiently to let it move freely. Looking at Eq. (6.34) we see that the metacentric height is reduced by the same amount that would result from raising the load by a distance h. In other words, we can consider that the mass acts in the hanging point.

6.9.3   Free Surfaces of Liquids

Liquids with free surfaces are a very common kind of moving load. Any engine-propelled vessel needs fuel and lubricating-oil tanks. Tanks are needed for carrying fresh water. The cargo can be liquid; then tanks occupy a large part of the vessel. Tanks cannot be filled to

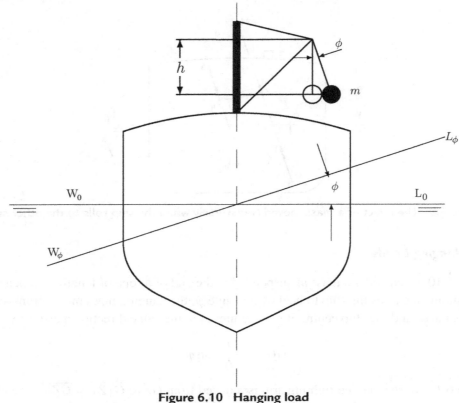

Figure 6.10   Hanging load

the top. Liquids can have large thermal expansion coefficients and space must be provided to accommodate for their expansion, otherwise unbearable pressure forces may develop. In conclusion, almost all vessels carry liquids that can move to a certain extent endangering thus the ship stability. A partially-filled tank is known as a **slack tank**.

Figure 6.12a shows a tank containing a liquid whose surface is free to move within a large range of heeling angles without touching the tank top or bottom. Let us consider that the liquid volume behaves like a ship hull and consider the free surface a waterplane. Then, the centre of gravity of the liquid is the buoyancy centre of the liquid hull. Therefore, we use for it the notation b_0. While the ship heels, the centre of gravity of the liquid moves along the curve of the centre of the buoyancy, "around" the metacentre, m. The horizontal distance between the initial position, b_0, and the inclined position, b_ϕ, is

$$\overline{b_0 m} \tan \phi$$

If v is the volume occupied by the liquid, i_B the moment of inertia of the liquid surface with respect to the barycentric axis (of the free surface) parallel to the axis of heeling, and ρ the

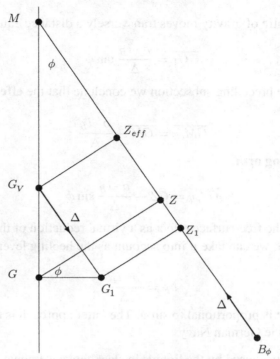

Figure 6.11   Effective metacentric height

Figure 6.12   The free-surface effect

liquid density, the heeling moment produced by the inclination of the liquid surface is

$$M_l = \rho v \frac{i_B}{v} \tan \phi = \rho i_B \tan \phi$$

where M_l has the dimensions of mass times length

As a result, the ship centre of gravity moves transversely a distance equal to

$$\overline{GG_1} = \frac{\rho \cdot i_B}{\Delta} \tan \phi \qquad (6.35)$$

By comparison with the preceding subsection we conclude that the **effective metacentric height** is

$$\overline{GM_{eff}} = \overline{GM} - \frac{\rho \cdot i_B}{\Delta} \qquad (6.36)$$

and the **effective righting arm**,

$$\overline{GZ_{eff}} = \overline{GZ} - \frac{\rho \cdot i_B}{\Delta} \sin \phi \qquad (6.37)$$

Instead of considering the free-surface effect as a virtual reduction of the metacentric height and of the righting lever, we can take it into account as the heeling lever of free movable liquids. Its value is

$$\ell_F = \frac{\rho \cdot i_B}{\Delta} \qquad (6.38)$$

and the respective curve is proportional to $\sin \phi$. The latter approach is that adopted in the stability regulations of the German Navy.

The reduction of stability caused by the liquids in slack tanks is known as **free-surface effect**. Two of its features must be emphasized:

- the mass of the liquid plays no role, only the moment of inertia of the free surface appears in equations;
- the effect does not depend on the position of the tank.

In general, ships have more than one tank, and different tanks can contain different liquids. The destabilizing effects of all tanks must be summed up when calculating the effective metacentric height

$$\overline{GM_{eff}} = \overline{GM} - \frac{\sum_{k=1}^{n} \rho_k \cdot i_{Bk}}{\Delta} \qquad (6.39)$$

and the effective righting arm,

$$\overline{GZ_{eff}} = \overline{GZ} - \frac{\sum_{k=1}^{n} \rho_k \cdot i_{Bk}}{\Delta} \sin \phi \qquad (6.40)$$

where n is the total number of tanks.

Often the liquid surface is not free to behave as in Figure 6.12a and its shape changes when it reaches the tank top or bottom. Then, we cannot use the equations shown above. The same happens when the heeling angle is large and the forms of the tank such that the shape of the free surface changes in a way that cannot be neglected. In such cases the exact trajectory of the centre of gravity must be calculated. As shown in Figure 6.12b, the resulting heeling moment is

$$M_\ell = W(\xi \cos\phi + \zeta \sin\phi) \tag{6.41}$$

where W is the liquid mass, ξ the horizontal distance and ζ the vertical distance travelled by the centre of gravity.

Some books and articles on Naval Architecture contain tables and curves that allow the calculation of the free-surface effect for various tank proportions. Present-day computer programmes can calculate exactly and quickly the position of the centre of gravity for any heel angle. For example, one can describe the tank form as a hull surface and run the option for cross-curves calculations. Therefore, correction tables and curves are not included in this book.

The free-surface effect can endanger the ship, or even lead to a negative metacentric height. Therefore, it is necessary to reduce the free-surface effect. The usual way to do this is to subdivide tanks by **longitudinal bulkheads**, such as shown in Figure 6.13. If the left-hand figure would refer to a parallelepipedic hull, the moment of inertia of the liquid surface in each tank would be $1/2^3 = 1/8$ that of the undivided tank. Having two tanks, the total moment of inertia, and the corresponding free-surface effect, are reduced in the ratio 1/4. An usual arrangement in tankers is shown in Figure 6.13b.

Some materials that are not really liquid can behave like liquids. Writes Price (1980), "Whole fish when carried in bulk in a vessel's hold behave like liquid," and should be considered as such in stability calculations.

We end this section by noting that transverse bulkheads do not reduce the free-surface effect of slack tanks.

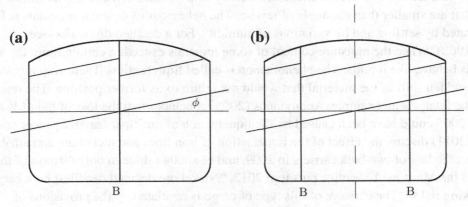

Longitudinal bulkheads are marked by 'B'

Figure 6.13 Reducing the free-surface effect

6.9.4 Shifting Loads

Shifting loads, also called **sliding loads**, such as grain, coal, and sand are a very dangerous type of moving loads. Arndt (1968) lists 31 incidents due to sliding loads, 13 of them leading to sinking, one to abandoning the ship. Those accidents occurred between July 1954 and November 1966. More cases are cited in the literature of specialty. To quote just one example, we mention that in 1998 shifting of granular sulfur caused the sinking of *Arcadia Pride*. Unlike liquid loads, materials considered in this subsection do not move continuously during the ship roll. Shifting loads stay in place until a certain roll angle is reached and then they slide suddenly.

The sides of a mass of **granular** materials, like those cited above, are inclined. The angle between the side and the horizontal is called **angle of repose** and is an important characteristic of the material. The angle of repose of most grain loads ranges between 20° and 22°, but for barley it reaches 46° (see Price, 1980). The angles of repose of ores range between 34° for copper from Norway, and 60° for copper from Peru.

Let the angle of repose be ρ_R. During roll, the mass of the granular material stays in place until the heel angle exceeds the angle of repose, that is $\phi > \rho_R$. Then, the granular load slides suddenly and its centre of gravity moves horizontally a distance, ξ, and up a distance ζ. By analogy with Figure 6.12b we can calculate a reduction of the metacentric height equal to

$$\overline{GG_V} = \frac{m_L(\xi \cos \phi + \zeta \sin \phi)}{\Delta}$$

While the ship rolls back, the load does not move until its angle exceeds the angle of repose. Wendel (1960b) describes this process and shows how the reduction of metacentric height can be represented by a loop that reminds the phenomenon of hysteresis known mainly from the theory of magnetism. The accelerations induced by ship motions can cause load shifting at angles that are smaller than the angle of repose. The behaviour of granular materials is further complicated by settling and by variations of humidity. For a detailed discussion see Arndt (1968). When the moisture content of some minerals exceeds a certain limit, those materials behave like liquids. The phenomenon is called **liquefaction**. Then, ship motions can cause a sudden shift of the material that would not return to its former position. The resulting list can be fatal. As an example, Anonymous (2005) concludes that the loss of the *M.V. Hui Long* in 2005 could have been caused by the liquefaction of fluorspar cargo. Spencer and Tilsley (2011) discuss the effect of the liquefaction of iron fines and nickel ore and attribute to this effect the loss of two bulk carriers in 2009, and of another three in only 40 days of the year 2010. As the Maritime Executive puts it in 2012, "Nickel ore deemed deadliest bulk cargo on ship sinking risks." The carriage of this type of cargo is regulated by the provisions of *IMSBC*—the International Maritime Solid Bulk Cargoes Code—that beginning from 1 January 2011 replaces the *BC* code. Cargoes that may liquefy are classified as Group A.

Figure 6.14 Moving loads as a case of positive feedback

6.9.5 Moving Loads as a Case of Positive Feedback

In all cases of moving loads we can assume that an external moment, m_h, caused the ship to heel by an angle ϕ. Consequently, the load moved to the same side producing another heeling moment, m_a, that is added to the external moment. This process is illustrated in Figure 6.14. Control engineers will recognize in this process an example of **positive feedback**. Following Birbănescu-Biran (1979) we can, indeed, use simple block-diagram techniques to retrieve some of the relationships found above. A simplified development follows; a more rigorous one can be found in the cited reference. Readers familiar with the elements of Control Engineering can understand this subsection without difficulty; other readers may skip it. However, making a little effort to understand the block diagram in Figure 6.14 can provide more insight into the moving-load effect.

In Figure 6.14, $G(s)$ is the **ship transfer function**, and $H(s)$, the moving-load transfer function. In the forward branch of the ship-load system, the **Laplace transform** of the heel angle, $\Phi(s)$, is related to the Laplace transform of the effective heeling moment, $M_e(s)$, by

$$\Phi(s) = G(s)M_e(s) \tag{6.42}$$

The Laplace transform of the additional heeling moment, $M_a(s)$, induced by the moving load, is related to the Laplace transform of the heel angle by

$$M_a(s) = H(s)\Phi(s) \tag{6.43}$$

Substituting $M_e(s)$ in Eq. (6.42) by the sum of the moments $M(s)$ and $M_a(s)$ yields

$$\Phi(s) = G(s)(M(s) + H(s)\Phi(s)) \tag{6.44}$$

Finally, the transfer function of the ship-load system is given by

$$\frac{\Phi(s)}{M(s)} = \frac{G(s)}{1 - G(s)H(s)} \tag{6.45}$$

To find the transfer function of the ship we refer to Eq. (6.25) to which we add a heeling moment, M_e, in the right-hand side

$$\frac{d^2\phi}{dt^2} + \frac{g\overline{GM}}{i_m^2}\phi = \frac{g}{i_m^2} \cdot \frac{M_e}{\Delta} \qquad (6.46)$$

Applying the Laplace transform, with zero initial conditions and rearranging, we obtain the ship transfer function

$$\frac{\Phi(s)}{M_e(s)} = \frac{\frac{g}{i_m^2\Delta}}{s^2 + \frac{g}{i_m^2}\overline{GM}} \qquad (6.47)$$

Substitution of the above transfer function into Eq. (6.45) yields

$$\frac{\Phi(s)}{M(s)} = \frac{\frac{g}{i_m^2\Delta}}{s^2 + \frac{g}{i_m^2}\left(\overline{GM} - \frac{H(s)}{\Delta}\right)} \qquad (6.48)$$

The factor

$$\left(\overline{GM} - \frac{H(s)}{\Delta}\right)$$

is the **effective metacentric height**.

From the preceding subsections it can be found that the transfer function of a hanging load is $H(s) = mh$, and the transfer function of a free liquid surface is $H(s) = \rho \cdot i_B$. Equation (6.48) yields the condition for bounded response:

$$\overline{GM} - \frac{H(s)}{\Delta} > 0$$

Indeed, if this condition is fulfilled, $\phi(t)$ is a sinusoidal function of time with bounded amplitude. If the condition is not fulfilled, the heel angle is given by a hyperbolic sine, a function whose amplitude is not bounded. We retrieved thus, by other means, the famous condition of initial stability. A diagram such as that in Figure 6.14 can be the basis of a SIMULINK© programme for simulating the roll of a ship with moving loads aboard.

6.10 The Stability of Grounded or Docked Ships

6.10.1 Grounding on the Whole Length of the Keel

Figure 6.15 shows a ship grounded on the whole length of the keel. If local tide lowers the sea level, at a certain draught the ship will loose stability and capsize. To plan the necessary actions, the ship master must know how much time remains until reaching the critical draught. A similar situation occurs when a ship is laid in a floating dock. While ballast water is pumped out of the dock, the draught of the ship decreases. Props must be fully in place before the critical draught is reached.

Figure 6.15 Ship grounded on the whole keel length

In Figure 6.15 we consider that the draught, T, descended below the value T_0 corresponding to the ship displacement mass, Δ. Then, the ship weight is supported partly by the buoyancy force $g\rho\nabla_T$ and partly by the reaction, R:

$$g\rho\nabla_T + R = g\Delta \tag{6.49}$$

where ∇_T is the submerged volume at the actual draught, T. The ship heels and for a small angle, ϕ, the condition of stability is

$$g\rho\nabla_T \overline{KM}\sin\phi > g\Delta\overline{KG}\sin\phi \tag{6.50}$$

or

$$\overline{KM} > \frac{\Delta\overline{KG}}{\rho} \cdot \frac{1}{\nabla_T} \tag{6.51}$$

Simplifying we obtain

$$\overline{KM} > \frac{\nabla}{\nabla_T} \cdot \overline{KG} \tag{6.52}$$

where ∇ is the displacement volume corresponding to the ship mass, Δ. As an example, Figure 6.16 shows the curves $\overline{KM}$ and $\nabla\overline{KG}/\nabla_T$ as functions of draught, that is local depth, T, for the ship *Lido 9*. The critical draught in this case is 1.53 m.

6.10.2 Grounding on One Point of the Keel

Figure 6.17 shows a ship grounded on one point of the keel; let this point be P_0. We draw a horizontal line through P_0; let P_1 be its intersection with the transverse plane passing through the centre of gravity, G, and P_3, the intersection with the transverse plane passing through the centre of buoyancy, B, and the metacentre, M. Taking moments about the line $P_0 P_3$ we write

$$\rho\nabla_T \overline{P_3M}\sin\phi > \Delta\overline{P_1G}\sin\phi \tag{6.53}$$

Figure 6.16 Finding the critical draught of a ship grounded on the whole keel length

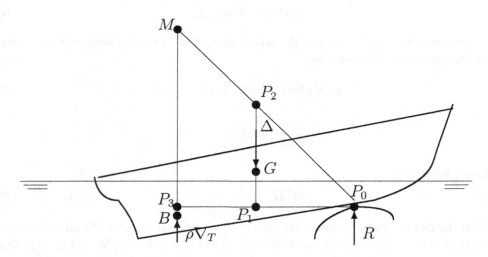

Figure 6.17 Ship grounded on one point of the keel

or

$$\overline{P_3 M} > \frac{\nabla}{\nabla_T} \cdot \overline{P_1 G} \tag{6.54}$$

The similarity of the triangles $P_0 M P_3$ and $P_0 P_2 P_1$ lets us write

$$\frac{\overline{P_3 M}}{\overline{P_1 P_2}} = \frac{\overline{P_3 P_0}}{\overline{P_1 P_0}} \tag{6.55}$$

Taking moments of forces about the point P_0 gives

$$\rho \nabla_T \overline{P_3 P0} = \Delta \overline{P_1 P0} \tag{6.56}$$

Combining Eqs. (6.54)–(6.56) yields the condition

$$\overline{P_1 P_2} > \overline{P_1 G} \tag{6.57}$$

In other words, the point P_2 plays the role of metacentre. From Figure 6.17 and Eq. (6.57) we see that pulling the ship to the left increases the distance $\overline{G P_2}$, while pulling the ship to the right reduces it.

6.11 Negative Metacentric Height

The metacentric height, $\overline{GM}$, can become negative if the centre of gravity is too high, or if the influence of moving loads is important. Even with a negative, initial metacentric height, ships with certain forms can still find a position of stable equilibrium at an angle of heel that does not endanger them immediately. An example is shown in Figure 6.18 where the $\overline{GZ}$ curve is

Small cargo ship, $\Delta = 1964$ t, KG = 4.75 m, δ KG = 0.75 m

Figure 6.18 The stability curve of a ship that can float with negative, initial metacentric height

Figure 6.19   Stability with negative initial metacentric height

based on the data of a small cargo ship built in 1958. The solid line represents the righting-arm curve in ballast, departure condition. Let us assume that for some reason the centre of gravity, G, moves upwards a distance $\delta \overline{KG} = 0.75$ m. The dot-point line curve represents the quantity $\delta \overline{KG} \sin \phi$ that must be subtracted from the initial righting-arm curve. The two curves intersect at approximately 10° and 55°; $\overline{GZ}$ is zero there. The resulting righting-arm is shown in Figure 6.19. The ship finds a position of stable equilibrium at $\phi_1 \approx 10°$; she sails permanently heeled at this angle called **angle of loll**. Looking again at Figure 6.18 we see that the first intersection of the two curves is possible because the first part of the $\overline{GZ}$ curve lies above the tangent in the origin. It can be shown that the corresponding metacentric evolute has ascending branches at $\phi = 0$.

In Figure 6.19 we can see that, if a ship sailing with a positive angle of loll receives a wave, a small gust of wind, or some other perturbation coming from the starboard, she will incline to the port side and stay there at a negative angle of loll, $\phi_2 = -\phi_1$. In a seaway, such a ship can oscillate between ϕ_1 and ϕ_2. This kind of abrupt oscillation, different from a continuous roll, is characteristic for negative metacentric heights.

An angle of loll can be corrected only by lowering the centre of gravity, not by moving loads transversely, or by filling ballast tanks on the higher side. Hervieu (1985) proves this in two

Figure 6.20 Stability with negative initial metacentric height

ways, first by considering the metacentric evolute, next by examining the righting-arm curve. We adopt here the second approach.

We first assume that the ship master tries to correct the loll by moving a mass $m = 2$ t. As the breadth of the ship is 11.9 m, we can assume that the mass is moved a distance $d = 6$ m toward port. The correcting' arm, $dm \cos\phi/\Delta$, is shown as a dotted line in Figure 6.20. Subtracting this correcting arm from the initial righting-arm curve, we obtain the dashed line. The starboard angle of loll, ϕ_3, is smaller than the initial angle, ϕ_1, but the port side angle of loll increases from ϕ_2 to ϕ_4. Also, we see that the area A under the $\overline{GZ}$ curve is somewhat reduced.

Next, we assume that, unsatisfied by the first result, the ship master moves more masses toward port, until $m = 4.25$ t. Figure 6.21 shows now the limit situation in which the correcting-arm curve is tangent to the initial $\overline{GZ}$ curve. The starboard angle of loll, ϕ_3, is smaller than in the previous case, but still not zero. On the other hand, the port side angle of loll, ϕ_4, is sensibly larger than the uncorrected one, and the area A is smaller.

Finally, we consider in Figure 6.22 a very grave case with a still higher centre of gravity ($\overline{KG} = 5.55$ m) and assume that the ship master decides to move more masses until $m = 6.5$ t. There is no position of equilibrium at starboard and the ship can find one only with the port

Figure 6.21 Correcting an angle of loll

side down, at an angle of loll ϕ_4 sensibly larger than the initial angle, ϕ_2. The area A under the righting-arm curve is small and a not-too-large moment tending to incline the vessel toward port can cause capsizing.

Spencer and Tilsley (2011) briefly describe the case of a bulk carrier that loaded 10 000 t of iron fines and developed a starboard list of more than 23° during the voyage. After being directed to an anchorage, the master tried to correct the list by filling port side ballast tanks. The ship developed a large port side list and eventually sank.

Ships whose righting-arm curves do not present inflexions like that shown in Figure 6.18 cannot find an angle of loll. The reader is invited to examine such a case in Exercise 6.7.

Once, it was not unusual to see that a ship carrying timber on deck sailed out of harbour with an angle of loll. The IMO code of 2008 specifies ships carrying timber deck cargo should have an initial metacentric height not lower than 0.15 m. Further, at all times during a voyage, the effective metacentric height GM_{eff} should not be less than 0.10 m taking into account absorption of water, ice accretion, variation of consumables, free-surface effect in tanks, and water trapped between logs.

Figure 6.22   Correcting an angle of loll

6.12 Wall-Sided Floating Bodies with Negative Metacentric Height

In Section 2.8.1 we defined as *wall-sided* those floating bodies whose hull surface is a cylinder with generators perpendicular to the waterplane. The assumption of a wall-sided hull can be a good approximation for many vessels, at small heel angles. In this section we are going to see that this approximation leads to a very interesting result in the case of an initial negative metacentric height.

Within the range in which a vessel can be considered as wall-sided, Eqs. (2.29) and (2.30) yield the following coordinates of the centre of buoyancy

$$y_B = \overline{BM} \tan \phi \tag{6.58}$$
$$z_B = \frac{1}{2} \overline{BM} \tan^2 \phi$$

Referring to Figure 5.2, we project the coordinates of the centres of buoyancy and gravity on the heeled waterplane $W_\phi L_\phi$ and obtain

$$\overline{GZ} = y_B \cos \phi + z_B \sin \phi - \overline{BG} \sin \phi \tag{6.59}$$

$$= \sin\phi \left(\overline{BM} - \overline{BG} + \frac{\overline{BM}}{2} \tan^2\phi \right) \tag{6.60}$$

$$= \sin\phi \left(\overline{GM} + \frac{\overline{BM}}{2} \tan^2\phi \right)$$

This is the equation that gives the righting arm of wall-sided vessels. As seen in Figure 6.19 and from Eq. (6.59), $\overline{GZ}$ is zero at $\phi = 0$, but also at the heel angle ϕ_1 for which the term between parentheses is zero

$$\tan\phi_1 = \pm\sqrt{-\frac{2\overline{GM}}{\overline{BM}}} \tag{6.61}$$

Soldà (1964) used this result and a geometric reasoning for the metacentric evolute of a wall-sided ship to show that at the angle of loll ϕ_1 the metacentric height is larger than the absolute value of the metacentric height in upright condition. We shall give here an analytic proof following Rawson and Tupper (2002) and a personal communication of Lawrence Doctors. In fact, the metacentric height at the angle of loll is given by

$$\overline{GM}\big|_{\phi=\phi_1} = \frac{d\overline{GZ}}{d\phi}\bigg|_{\phi=\phi_1}$$

$$= \cos\phi_1 \left(\overline{GM} + \frac{\overline{BM}}{2} \tan^2\phi_1 \right) \tag{6.62}$$

$$+ \frac{1}{2} \sin\phi_1 \overline{BM} \cdot \frac{2\tan\phi_1}{\cos^2\phi_1}$$

As the quantity between parentheses is zero at the angle of loll, and using Eq. (6.61) we obtain

$$\overline{GM}\big|_{\phi=\phi_1} = -\frac{2\overline{GM}}{\cos\phi_1} \tag{6.63}$$

Following the presentation of Soldà in a meeting of ATMA, one of the participants explained that Barillon had found this result even earlier, but he did not publish it.

6.13 The Limitations of Simple Models

In Sections 6.3 and 6.4 we assumed that the water reaction to the heeling force acts at half-draught. This assumption is obviously arbitrary, but practice has proven it acceptable. A better evaluation would require an amount of calculations unacceptable in practice. To find the exact location of the centre of pressure it is necessary to take into account the exact hull-surface form. Moreover, the position of the centre of pressure can change with heel. In practice stability calculations must be carried on for each change in load, in many cases by ship masters and mates. Under such circumstances computing resources are limited and one must be satisfied with an approximation of the centre of pressure consistent with other approximations assumed in the model. A documented discussion on the point of application of water reaction can be found in Wegner (1965). At this point it may be helpful to remind that the models developed in this chapter may be rough approximations of the reality, but they stand at the basis of national and international regulations that are compulsory. Stability regulations correspond to the notion of **codes of practice** as known in other engineering fields. All codes of practice accept simplifying assumptions that enable calculations with a reasonable amount of time and computing resources. Another situation occurs in research where more exact models must be assumed, powerful computer and experimenting resources are available, and more time is allowed.

Equation (6.2) developed in Section 6.3 yields a heeling arm equal to zero at the heel angle 90°. Such a result is obviously wrong as any vessel presents a sail area exposed to the wind even when lying on the side. Figure 1.103 in Henschke (1957) illustrates well this point. At small angles the results based on the curve proportional to $\cos^2 \phi$ differ little from those obtained with other approximations (see Chapters 8 and 10) and, therefore, they are acceptable for large vessels that do not heel much under wind, such as the capital ships of the US Navy. Smaller vessels tend to heel more under wind and then curves based on the $\cos^2 \phi$ assumption may become quite unrealistic.

The models developed in this chapter are based on further simplifications. In real life the water opposes the motions of a ship with forces that depend on the amplitude, the speed, and the acceleration of motion. Assuming negligible roll velocity and acceleration, our models take into account only the moment that depends on the amplitude of heeling, that is the righting moment.

The moment that depends on the heeling speed, $\dot{\phi}$, is called **damping moment**. Damping causes **energy dissipation**. If a system that includes damping is displaced from its equilibrium position and then it is left to oscillate freely, the amplitude of oscillations will decrease with time and eventually will die out. The damping of the roll motion is mainly due to the generation of waves, but viscous effects may increase it and become important for certain bilge forms or if the vessel is fitted with bilge keels or a large keel.

The moments proportional to heel acceleration belong to a category of forces and moments called **added masses** because they can be collected together with the mass moment of inertia of the ship.

The evaluation of damping and added masses requires special computer programmes or model experiments. Neglecting damping and added masses leads to overestimation of dynamic heeling angles and this is on the **safe side**. Therefore, no stability regulation takes explicitly into account the effects of damping or added masses, but some regulations, like the IMO code, consider indirectly their influence by using different parameters for ships fitted with sharp bilges, bilge keels, or deep keels.

Cardo et al. (1978), for example, discuss stability considering non-linear roll equations that include damping and added masses. Using advanced mathematical criteria, the authors reach the same qualitative results as those obtained in Section 6.7. An outline of the linear theory of ship motions is given in Chapter 12.

Last, but not least, we neglected until now the influence of waves, and we leave the discussion of this subject for Chapters 9, 10, and 12.

6.14 Other Modes of Capsizing

Capsizing can be defined as the sudden transition of a floating body from a position of equilibrium to another position of equilibrium. Depending on the ship forms and loading, her second position can be on the side or with the keel up. If in the new position water can enter in large quantities, the ship will eventually sink. Often the process is so fast that many lives are lost. Sometimes no survivor remains to tell the story.

In Chapter 2 we saw that a floating body can capsize if the metacentric height is negative. In this chapter we learnt that a vessel can capsize if the righting arm is too small in comparison with the heeling arm, or if the area under the righting-arm curve is too small in comparison with the area under the heeling-arm curve. In Chapter 9 we shall see that a ship can capsize because of the variation of the metacentric height and of the righting arm in waves that travel in the same direction as the ship (head or following seas) or at some angle with her. That dangerous phenomenon is called **parametric resonance** or **Mathieu effect**. What happens if the waves are parallel to the ship? Arndt (1960a) explains that a ship cannot capsize in regular, parallel waves. Adds Arndt, "From practice we know cases in which captains put the ship parallel to the wave crests in order to reduce the effect of storms, neither in experiments could anyone cause until now a model to capsize in lateral, regular waves." Otherwise seems to happen with freak, or breaking waves of great steepness whose impact on the ship side can be high enough to overturn the ship. Thus, for example, Morrall (1980) investigates the loss of the large stern trawler *Gaul*, and Dahle and Kjærland (1980) study the capsizing of the Norwegian research vessel *Helland-Hansen*. These studies support the hypothesis that the discussed disasters were due to high breaking waves.

It seems that the process of capsizing because of freak or breaking waves is not yet well understood and the methods proposed for its prediction are probabilistic (see Dahle and Myrhaug, 1996, Myrhaug and Dahle, 1994). de Kat (1990) studies numerical models for the simulation of capsizing and Grochowalski (1989) describes a research on ship models. Probabilistic and simulation studies are beyond the scope of this book.

Another mode of capsizing is **broaching-to**; it is a dynamic phenomenon due to the loss of control in severe following or quartering seas. The ship enters into a forced turning that cannot be corrected by the rudder, heels, and capsizes. Broaching-to is studied by Nicholson (1975), Spyrou (1995, 1996a,b).

It has been claimed that capsizing results from a combination of several factors. An example can be found in Hua (1996) who studies the capsize of the ferry *Herald of Free Enterprise* as a result of the interaction between heeling and turning motion, while great quantities of water were present on one deck.

6.15 Summary

The statical stability of ships is checked by comparing the righting-arm curve with the curves of heeling arms. A heeling arm is calculated by dividing a heeling moment by the ship displacement force. In general, a heeling-arm curve intersects the righting-arm curve in two points that correspond to positions of statical equilibrium. The equilibrium is stable only in the first position; there the slope of the righting-arm curve is larger than that of the heeling arm.

Heeling moments are caused by wind, by the centrifugal force developed in turning, by the crowding of passengers on one side of the ship, by towing, or by the tension in a cable that links two vessels during a replenishment-at-sea operation.

The wind heeling arm is proportional to the square of the wind velocity and depends on the area of the lateral projection of the above-sea ship surface. We call that area 'sail area.' Assuming that the wind velocity is constant over the whole sail area, the wind heeling arm is proportional to the sail area. This assumption is acceptable for quick calculations. In reality, the wind speed increases with height above the sea level and this 'wind gradient' is taken into account in more exact calculations.

The heeling arm in turning is proportional to the square of the ship speed in turning, and inversely proportional to the radius of the turning circle. When the heeling moment appears or increases suddenly we must check the dynamical stability of the vessel. This situation can be caused by a gust of wind or by losing a mass on one side of the ship. The area under the righting arm up to the maximum angle reached momentarily by the ship is equal to the area under the heeling arm up to that point. The process depends on the angle of roll at which the sudden moment is applied. For a gust of wind, for example, the situation is worse if the ship was heeled to the windward side, than if the ship was caught by the gust with the lee side down. If the area available under the righting arm is smaller than the area under the heeling arm the ship is lost.

The period of unresisted roll is proportional to the square root of the metacentric height. This imposes an upper limit on the $\overline{GM}$ value. If the roll period is too short the roll motion is stiff; it is unpleasant for passengers and crew and may be dangerous for equipment and cargo. If the motion is too tender, it may indicate a dangerously low metacentric height.

A load displaced transversely reduces the stability when heeling to the same side as the load. Moving loads too decrease the stability. Thus, a load suspended so that it can move freely produces a virtual reduction of the metacentric height as if the load were moved to the point of suspension. A very common type of moving loads is liquids whose surfaces are free to move inside tanks or on the deck. The reduction of stability is proportional to the moment of inertia of the free surface about a barycentric axis of the free surface that is parallel to the axis of ship inclination. The effect does not depend on the mass of the liquid (as long as the liquid surface does not reach the tank top or bottom) or the position of the tank. The usual way of reducing the free-surface effect is to subdivide the tanks by longitudinal bulkheads. Two other methods are to empty the tank or to fill it. In the latter case the effect of the thermal expansion of the liquid should be considered. Granular materials constitute another category of moving loads. Such loads stay in place until the heel angle exceeds a value characteristic for the material. This value is called angle of repose. The variation of stability reduction due to sliding loads follows a hysteresis loop. The effect of moving loads is a case of positive feedback.

If a ship is grounded in a region where the water level is descending, at a certain draught it can lose stability. The same happens with a ship on dock. The calculation of the critical draught is rather simple.

A ship with negative metacentric height can find a position of stable equilibrium, without capsizing, if the first part of the righting-arm curve lies above the tangent in the origin. This fixed angle of heel is called **angle of loll**. There are two angles of loll and they are symmetric about the origin. Under moderate perturbations the ship can heel suddenly from one angle of loll to the other. This motion is different from a continuous roll and is characteristic for negative metacentric height. The angle of loll cannot be corrected by moving masses transversely; such an action can endanger the ship. Angles of loll should be corrected only by lowering the centre of gravity.

6.16 Examples

Example 6.1 (Wind pressure). Let us calculate the pressure corresponding to a wind speed of 70 knots. This is the value specified by the German Navy for evaluating the intact stability of vessels operating in open seas that are not exposed to tropical storms. Assuming an aerodynamic resistance coefficient equal to 1.2, and an air density equal to 1.27 kg m^{-3} we obtain

$$p_W = c_W \frac{\rho}{2} V^2 = 1.2 \times \frac{1.27}{2} \times (0.5144 \times 70)^2 \frac{\text{kg}}{\text{m}^3} \left(\frac{\text{ms}^{-1}}{\text{knot}} \text{knot} \right)^2$$

$$= 987.99 \text{ kgm}^{-1}\text{s}^{-2}$$

Figure 6.23 Calculation of sail area and its centroid

Table 6.1 Ship *Lido 9*—sail area for $T = 1.85$ m

Area Component	Dimensions (m)	Area (m)	Centroid (m²)	Moment (m³)
1	0.6 × 2/2	0.60	1.33	0.80
2	2 × 16.4	32.80	1.00	32.80
3	0.8 × 2.4/2	0.96	1.60	1.54
4	2 × 11	22.00	3.00	66.00
5	1 × 3	3.00	4.50	13.50
Total		59.36	1.93	114.63

Rounding off yields 1 kN m^{-2}, or, using the SI term, 1 kPa. The conversion factor, 0.5144, results from the definition of the **knot** as **nautical mile per hour**. Substituting SI units we divide 1852 m by 3600 s and obtain 1852/3600 = 0.5144 m s^{-1}/knot.

Example 6.2 (Calculating a wind heeling arm). Figure 6.23 is a simplified sketch of the sail area of the ship *Lido 9* with the waterline corresponding to a draught of 1.85 m. To simplify calculations, the area is subdivided into five simple geometrical forms, namely rectangles and triangles. The calculations are carried on in the spreadsheet shown in Table 6.1.

Example 6.3 (The statical stability curves of HMS *Captain*). In the night between 6 and 7 September 1870 a British fleet was sailing off Cape Finisterre. The fleet was hit by a strong gale and one of the ships, HMS *Captain* capsized, but all other ships survived. The righting arms of HMS *Captain* are given in Reed (1872) and Attwood and Pengelly (1960), while the latter book contains also the righting arms of HMS *Monarch*, a ship that was part of the same fleet and survived. The statical stability curves of the two ships are compared in Figure 6.24. The slopes in the origin of the curves show that both ships had practically the same initial metacentric height. The angle of vanishing stability of HMS *Monarch* was much larger than that of HMS *Captain*. The same was true for the areas under the righting-arm curves. The difference between those qualities was due mainly to a substantial difference in the freeboards.

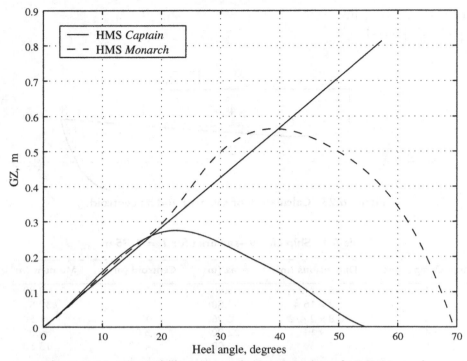

Figure 6.24 The stability curves of HMS *Captain* and HMS *Monarch*

Visual inspection of Figure 6.24 explains why HMS *Monarch* could survive the gust of wind that led to the capsizing of HMS *Captain*.

Example 6.4 (The roll of a 22.09 m fishing vessel).   Table 6.2 shows a part of the data of a fishing vessel studied by Pérez and Sanguinetti (2006).

The coefficient used by Rose is

$$c = \frac{2i_m}{B} = 6$$

Table 6.2   Data of 22.09 m fishing vessel

Waterline length (L_{WL})	22.09 m
Breadth (B)	6.86 m
Displacement mass (Δ)	170.30 t
Transverse radius of gyration (i_m)	2.06 m
Metacentric height ($\overline{GM}$)	0.37 m

Table 6.3 Small cargo ship—partial cross-curves values

Heel angle (°)	l_k (m)
10	0.918
20	1.833
30	2.717
45	3.847
60	4.653
75	5.007
90	4.994

The natural period of roll is given by

$$T = 2\pi \frac{i_m}{\sqrt{gGM}} = 2\pi \frac{2.06}{\sqrt{9.81 \times 0.37}} = 6.79 \text{ s}$$

The paper presents time stories of the roll amplitude. The period that can be measured in the graphs corresponds to that shown above. Further we calculate

$$T\sqrt{\frac{g}{B}} = 6.79\sqrt{\frac{9.81}{6.86}} = 8.12$$

According to Kempf the ship is comfortable.

Example 6.5 (Hanging, displaced load).   Exercises 6.1 and 6.6, and Tables 6.3 and 6.6 contain data of a small cargo ship. Let us assume that the loading condition corresponds to the volume of displacement $\nabla = 2549 \text{ m}^3$ and that a derrick on board the vessel lifts a mass $m = 10$ t. While this mass is hanging at the end of a cable of length $h = 8$ m, the derrick moves it by a transversal distance $d = 8$ m. As the displacement volume is calculated from moulded dimensions, we approximate the displacement mass by

$$\Delta = 1.03\nabla = 1.3 \times 2549 = 2625.47 \text{ t}$$

The righting arm is plotted as a solid line in Figure 6.25. The initial metacentric height is

$$\overline{GM} = \overline{KM} - \overline{KG} = 5.16 - 5 = 0.16 \text{ m}$$

With this value we plot in Figure 6.25 the tangent to the righting-arm curve. The lever arm of the free-surface effect is given as $f_s = 0.04$ m. To this we must add the virtual metacentric-height reduction due to the hanging load. The resulting effective metacentric height is

$$\overline{GM}_{eff} = \overline{GM} - f_s - \frac{mh}{\Delta} = 0.160 - 0.04 - \frac{10 \times 8}{2625.47} = 0.09 \text{ m}$$

Figure 6.25   Finding the heel angle with a displaced, hanging load

The effective righting arms are calculated from

$$\overline{GZ}_{eff} = \overline{GZ} - \left(f_s - \frac{mh}{\Delta} \right) \sin \phi$$

and are plotted in Figure 6.25, together with the tangent in origin, as dash-dot lines. To end the exercise let us find the angle of heel caused by the displaced, hanging load. Frequently this angle is found by equating the heeling moment to the righting moment

$$md \cos \phi_{st} = \Delta \overline{GM}_{eff} \sin \phi_{st} \qquad (6.64)$$

The result is

$$\phi_{st} = \arctan \left(\frac{md}{\Delta \overline{GM}_{eff}} \right) = \arctan \left(\frac{10 \times 8}{2625.47 \times 0.09} \right) = 18.8^\circ \qquad (6.65)$$

This is a very large list. However, if we plot the curve of the heeling arm (as a dashed line) over the other curves in Figure 6.25 we see that the intersection of the heeling-arm curve with the righting-arm curve corresponds to an angle of 8.8°. It can be immediately seen that the heel angle yielded by Eq. (6.65) corresponds to the intersection of the heeling-arm curve with the tangent in origin to the righting-arm curve. The lesson to be learned is that Eq. (6.65) gives an acceptable approximation for small angles of heel only. When the righting-arm curve lies entirely under the tangent in origin, the intersection of the heeling- and righting-arm curves may occur at an angle larger than that yielded by Eq. (6.65). We considered here the statical stability in still water only. In waves the hanging load can lead to dangerous situations.

The downloadable file `HangingOnCargo.m` contains a MATLAB solution of this problem. Running the programme produces also an output file whose contents are partly shown below.

Heel arm deg	Righting arm m	Effective righting arm m	Heeling arm m
0.0	0.000	0.000	0.030
2.5	0.006	0.003	0.030
...			
87.5	0.020	-0.051	0.001
90.0	-0.006	-0.076	0.000

6.17 Exercises

Exercise 6.1 (Stability in turning). Table 6.3 shows part of the cross-curves values of the small cargo ship exemplified in Section 6.11. The other data needed in this problem are the displacement volume, $\nabla = 2549$ m^3, the vertical centre of gravity, $\overline{KG} = 5$ m, the ship length, $L_{pp} = 75.5$ m, and the ship speed, $V = 16$ knots. Using the formulae given in Section 6.4 calculate the heeling arm in turning. Plot the heeling-arm curve over the righting arm and find the heel angle in turning. Next, consider a free-surface correction equal to $l_f = 0.04$ m, draw the corrected righting-arm curve, $\overline{GZ_{eff}}$, and see if the angle of heel is affected. Use the tangent in origin when drawing the righting-arm curve.

Exercise 6.2 (Dynamical stability). The organizers of a boat race must throw a buoy from the starboard of a boat. The boat is rolling. Would you advise the organizers to throw the buoy while the starboard is down, or when the port side is down?

Table 6.4 Data of 21.44 m fishing vessel

Length between perpendiculars (L_{pp})	21.44 m
Breadth (B)	2.49 m
Draught (T)	2.49 m
Displacement mass (Δ)	162.60 t
Metacentric height ($\overline{GM}$)	0.48 m
Transverse radius of gyration (i_m)	2.62 m

Table 6.5   Data of 247 m ore carrier

Length between perpendiculars (L_{pp})	247.000 m
Breadth (B)	40.600 m
Draught (T)	16.000 m
Displacement mass (Δ)	135950.000 t
Metacentric height ($\overline{GM}$)	4.130 m
Transverse radius of gyration (i_m)	0.22B m

Exercise 6.3 (Roll of a 21.44 m fishing vessel).   The data in Table 6.4 are taken from Santos Neves et al. (2002).

Your tasks are:

1. to calculate the coefficient of Rose, $c = 2i_m/B$;
2. to calculate the roll period, T;
3. following Kempf check whether this vessel is stiff, comfortable, or tender.

Exercise 6.4 (Roll of a 247 m ore carrier).   The data in Table 6.5 are given in Sartori and Podenzana-Bonvino (1981).

Your tasks are:

1. to calculate the transverse radius of inertia, i_m;
2. to calculate the roll period, T;
3. following Kempf check whether this vessel is stiff, comfortable, or tender.

Exercise 6.5 (Small cargo ship with hanging load).   Table 6.3 shows part of the cross-curves values of the small cargo ship exemplified in Section 6.11. Other data for this exercise are the displacement volume, $\nabla = 3786.4$ m^3, and the vertical centre of gravity, $\overline{KG} = 4.78$ m. The

Table 6.6   Small cargo ship—partial hydrostatic data

Draught, T (m)	∇ (m^3)	$\overline{KM}$ (m)	Draught, T (m)	∇ (m^3)	$\overline{KM}$ (m)
2.00	993	6.75	4.32	2549	5.16
2.20	1118	6.39	4.40	2609	5.16
2.40	1243	6.09	4.60	2757	5.16
2.60	1377	5.83	4.80	2901	5.17
2.80	1504	5.63	5.00	3057	5.18
3.00	1640	5.48	5.20	3210	5.20
3.20	1776	5.37	5.40	3352	5.23
3.40	1907	5.28	5.60	3507	5.27
3.60	2045	5.24	5.80	3653	5.31
3.80	2189	5.20	5.96	3786	5.34
4.00	2322	5.18	6.00	3811	5.36
4.20	2471	5.17	6.20	3972	5.42

lever arm of the free-surface effect is $f_s = 0.03$ m. Let us assume that a derrick on board of the vessel lifts a mass $m = 15$ t. While this mass is hanging at the end of a cable of length $h = 10$ m, the derrick moves it by a transversal distance $d = 10$ m. Your tasks are:

1. plot the $\overline{GZ}$ curve and the tangent in its origin;
2. calculate the corrections for the free-surface effect and the hanging load and plot the curve of the effective righting arm, $\overline{GZ}_{eff}$, and the tangent in origin;
3. find the heeling angle by Eq. (6.65) and compare it to the angle measured on the curves. Do you see any difference between this situation and the one analysed in Example 6.5?

Exercise 6.6 (Critical draught of grounded ship).   Table 6.6 contains part of the hydrostatic data of the small cargo ship exemplified in the analysis of the angle of loll (Section 6.11).

1. The docking condition of the ship is characterized by the displacement volume $\nabla = 1562.8$ m³, and $\overline{KG} = 5.34$ m. Find the critical draught at which props must be in place.
2. The data of the ship carrying a cargo of oranges and close to her destination (fuel tanks at minimum filling) are the displacement volume $\nabla = 2979.4$ m³, and $\overline{KG} = 4.92$ m. Find the critical draught if the ship is grounded on the whole length of the keel.

Exercise 6.7 (Negative metacentric height).   Using the data in Table 5.1 show that the vessel *Lido 9* cannot find an angle of loll if the metacentric height is negative.

CHAPTER 7

Weight and Trim Calculations

Chapter Outline

7.1 Introduction

All models of stability require the knowledge of the displacement mass, Δ, and of the height of the centre of gravity (vertical centre of gravity), $\overline{KG}$. Stevin's law (see Section 2.3.2) states that the ship trim is determined by the longitudinal position of the centre of gravity, LCG. The three quantities, Δ, $\overline{KG}$, and LCG are calculated by summing up the masses of all ship components and their moments about a horizontal and a transverse plane. The centre of gravity of a ship in upright condition is situated in the plane of port-to-starboard symmetry of the ship (centreline plane); therefore, the coordinate of the centre of gravity about this plane is zero. However, individual mass ship components may not be symmetrical about the centreline plane and it is necessary to calculate their moments about that plane and ensure that the transverse coordinate (y-coordinate) of the ship's centre of gravity is zero. It is usual to call the latter coordinate **transverse centre of gravity** and note it by **TCG**. Thus, we have a consistent notation for the triple of coordinates LCG, VCG, TCG. Systematic calculations of displacements and centres of gravity are known as **weight calculations** and they are the subject of the first part of this chapter. Recent literature and standards deal with masses rather than weights. We follow this trend in our book, but use the term *weight calculations* because it is rooted in tradition.

Another subject of this chapter is the calculation of the trim and of the forward and aft draughts. As mentioned in the previous chapter, the trim affects the ship stability. Also, a ship trimmed at a large angle can look unpleasant to the eye. Above all, the trim determines the forward and aft draughts and thus affects certain ship functions. For example, the aft draught must be large enough to ensure sufficient propeller submergence and avoid cavitation.

Frequently weight calculations are based on approximate or insufficient data. The sources of uncertainty are explained in this chapter when introducing the notions of **reserve** and **margin** of displacement and of $\overline{KG}$. Because of these uncertainties, statutory regulations require an experimental validation of the coordinates of the centre of gravity, and of the corresponding metacentric height, $\overline{GM}$, for all new buildings or for vessels that underwent alterations that can influence their stability. This validation is carried on in the **inclining experiment**, also known in some shipyards as **stability test**. In this chapter we deal with the general requirements of the inclining experiment. Statutory requirements are mentioned in Chapters 8 and 10.

7.2 Weight Calculations

7.2.1 Weight Groups

A vessel is composed of hundreds, sometimes thousands of mass items. To systematize calculations it is necessary to organize them into **weight groups**. The first subdivision is into two main sets: **lightship** and **deadweight**. The lightship (less frequently known as **lightweight**) is the mass of the empty ship; it is composed of the hull, the outfit, and the machinery masses, including the liquids in the machinery and various systems, but not those in tanks or storage spaces. The deadweight is the sum of the masses of crew, cargo, and passengers, fuel, lubricating oil, provisions, water, stores, and spare parts. The usual acronym for deadweight is DWT. In simpler terms, the deadweight is the weight that the ship "carries."

One should make a distinction between the term *lightship* used as above, and its homophone that designs a ship provided with a strong light and used to mark a position.

In the first stages of ship design, known as **preliminary design**, the lightship masses and their centres of gravity are estimated by empirical equations, based on statistics of similar ships, or are derived from the masses of a given **parent ship**. This subject is treated in books on **ship design** such as Kiss (1980), Schneekluth (1980), Schneekluth and Bertram (1998), and Watson (1998). For merchant ships, the lightship groups are the hull, the outfit, and the machinery. The classification of warship weight groups may be somewhat different. Thus, the classification system of the US Navy, SWBS, distinguishes the following main weight groups: hull structure, propulsion plant, electric plant, command and surveillance, auxiliary systems, outfit and furnishings, armament.

As the design progresses by successive iterations, the weight estimations are refined by subdividing the weight groups into subgroups, the subgroups into lower-level subgroups, and so on. Thus, the hull mass is subdivided into hull and superstructure, then the hull into bottom, sides, decks, bulkheads, etc. The machinery components are first subdivided into main, or propulsion machinery, and auxiliary machinery. In the final stages it is possible to calculate the masses and centres of gravity of individual items from detailed drawings or from data provided by equipment suppliers.

The procedure described above requires a classification of the various weight groups, subgroups, and so on that ensures that no item is forgotten and that no item belongs to two groups. Readers who like mathematics may say that the weight groups shall be **disjoint**. Such readers can also see that such a classification system can be described by a **tree graph** (see Birbănescu-Biran, 1988). Several authorities and organizations engaged in ship design and construction have developed their own classification systems. An example of classification system for merchant ships is shown in Kiss (1980). As mentioned above, the classification system adopted by the US Navy is known as SWBS, an acronym for **Ship Work Breakdown Structure**.

The main deadweight item is the **cargo**; it is pre-specified by the owner. The number of crew members depends on the functions to be carried aboard; frequently a minimum is prescribed by regulations. The masses of fuel, lubricating oil, and water result from the required ship speed and range, two characteristics specified by the owner.

To compensate for the uncertainties in weight estimation in the first design stages, Naval Architects introduce a weight item called **reserve**, or **weight margin**. Some regulations consider also a $\overline{KG}$ margin; that is the calculated height of the centre of gravity, $\overline{KG}$, is increased by a certain amount ensuring that stability calculations fall on the safe side. As the ship design progresses, the uncertainties are reduced and so must be the weight reserve and the $\overline{KG}$ margin.

When the detailed ship project is delivered for construction, all weight and centres of gravity are supposed to be exact; however, a 'building' weight reserve and a $\overline{KG}$ margin are still included in weight calculations. By doing so designers take into account acceptable tolerances in plate, profile, and pipe thicknesses, tolerances in metal densities, and changes in the catalogs of suppliers.

Even when the ship is delivered to the owner, weight calculations still include 'commissioning' margins that take into account future equipment additions, trapping of water in places from where it cannot be pumped out, and weight increase due to rust and paint. Certain codes of practice, such as the stability regulations of the US Navy and those of the German Federal Navy, impose well-defined margin values.

7.2.2 Weight Calculations

Once the ship is built and in service, the lightship displacement and its centre of gravity are taken in calculations as constants. For each possible **loading case**, that is for each combination of cargo and other deadweight items, the masses of those items and their moments are added to those of the lightship. The calculations yield the displacement and the coordinates of the centre of gravity of the loading case under consideration. To give an example, we return to the data of the small cargo ship considered in Chapter 6. Figure 7.1 shows the calculations corresponding to the load case **homogeneous cargo, departure**. By **departure condition** we mean the ship leaving the port, with all the fuel, lubricating oil and provisions.

The table in Figure 7.1 was calculated in MS Excel. Alternatively, the calculations can be performed in MATLAB. Then, the weight data can be stored in a matrix, for example in the format

$$m_i \ kg_i \ lcg_i$$

where m_i is the mass of the ith weight item, kg_i, its vertical centre of gravity, and lcg_i, its longitudinal centre of gravity. An example of calculations for the loading case considered in Table 7.1 is

```
Wdata = [
   1247.66 5.93 32.04
      3.60    9.60    11.00
      5.00    7.30     3.50
    177.21    1.56    30.88
      4.50    4.65     8.45
    103.09    4.61    27.19
    993.94    4.35    42.62
     90.00    6.08    38.66 ];
format bank, format compact
Displ = sum(Wdata (:, 1))
Displ = 2625.00
KG    = Wdata(:, 1)'*Wdata(:, 2)/Displ
KG    = 5.00
LCG   = Wdata(:, 1)'*Wdata(:, 3)/Displ
LCG   = 35.88
```

Unless all calculations are carried on by a computer programme, the results of weight calculations are used as described below.

1. The mean draught, T_m, corresponding to the calculated displacement, is read in the hydrostatic curves.

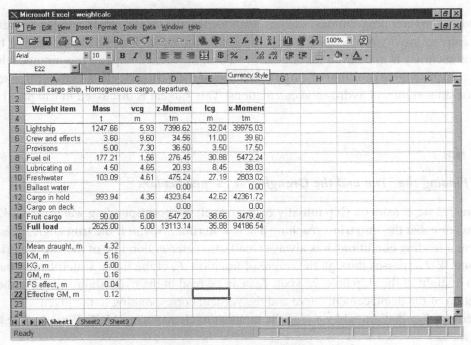

Figure 7.1   A spreadsheet of weight calculations

2. The trimming moment is calculated as

$$M_{\text{trim}} = \Delta(LCG - LCB) \tag{7.1}$$

where the *LCB* value corresponding to T_m is found in the hydrostatic curves. The moment to change trim, *MCT*, corresponding to T_m, is read from the hydrostatic curves and the trim is calculated as shown in Section 7.3. If the trim is small one can go to the next step, otherwise it is advisable to continue the calculations using the Bonjean curves or to resort to a computer programme.

3. The height of the metacentre above BL, $\overline{KM}$, corresponding to T_m, is read in the hydrostatic curves.

4. The metacentric height is calculated as

$$\overline{GM} = \overline{KM} - \overline{KG}$$

5. The free-surface effects of the tanks filled with liquids are added up and their sum is subtracted from the metacentric height to find the effective metacentric height, $\overline{GM}_{\textit{eff}}$.

6. The righting levers, $\overline{GZ}$, are calculated, and the effective righting levers are obtained by subtracting the free-surface effect

$$\overline{GZ} = l_k - \overline{KG}\sin\phi$$
$$\overline{GZ}_{\textit{eff}} = \overline{GZ} - l_F\sin\phi \tag{7.2}$$

7. The data are used to plot the statical stability curve.

With older computer programmes, such as Archimedes, the displacement and the coordinates of the centre of gravity can be used as input to obtain the mean draught and the trim of the ship. The accuracy is good even for large trim values. In recent computer programmes the user has to input the degree of filling of cargo holds and of the various tanks and the computer carries on all weight and hydrostatic calculations. This subject is discussed in Chapter 13.

7.3 Trim

7.3.1 Finding the Trim and the Draughts at Perpendiculars

In Figure 7.2 we consider a ship initially on even keel; the corresponding waterline is $W_0 L_o$. Let us assume that the ship trims reaching a new waterline, $W_\theta L_\theta$. If the trim angle, θ, is small (for normal loading conditions it is always small), the intersection line of the two waterlines, $W_0 L_o$ and $W_\theta L_\theta$, passes through the centre of flotation, F, of the initial waterplane. The midship draught of the ship on even keel, T_m, can be read in the hydrostatic curves at the intersection of the displacement curve and the vertical corresponding to the given displacement. For that draught we read the moment to change trim, MCT. We calculate the trim, in m, as

$$\text{trim} = T_F - T_A = \frac{\Delta(LCG - LCB)}{MCT} \tag{7.3}$$

The trim angle is given by

$$\tan \theta = \frac{T_F - T_A}{L_{pp}} \tag{7.4}$$

Figure 7.2 Finding the forward and aft draughts

As the ship trims around a transversal axis passing through the centre of flotation F, we consider this point as fixed during trim and we assume that the draught of this point has the value T_m found above. From Figure 7.2 we see that

$$T_A = T_m - LCF \cdot \tan\theta = T_m - LCF \cdot \frac{\text{trim}}{L_{pp}} \qquad (7.5)$$

and

$$T_F = \text{trim} + T_A = T_m + \text{trim}\left(1 - \frac{LCF}{L_{pp}}\right) \qquad (7.6)$$

To give an example we consider again the loading case of the small cargo ship analysed in Section 7.2.2. In Tables 7.2 and 7.3 we find $T_m = 4.32$ m, $LCB = 0.291$ m, $LCF = -0.384$ m, and $MCT = 3223$ mtm^{-1}. We know that the length between perpendiculars is $L_{pp} = 75.40$ m. In the table LCB is measured from midship, positive forwards. As LCG is measured from AP, we calculate

$$75.40/2 + 0.291 = 37.99 \text{ m}$$

and the trim

$$\frac{\Delta(LCG - LCB)}{MCT} = \frac{2625(35.88 - 37.99)}{3223} = -1.72 \text{ m}$$

The ship is trimmed by the stern. In Table 7.3 LCF is measured from the midship, positive forward; the value measured from AP is

$$75.40/2 - 0.384 = 37.32 \text{ m}$$

and we calculate

$$T_A = 4.32 - 37.32\frac{-1.72}{75.4} = 5.17$$
$$T_F = -1.72 + 5.17 = 3.45$$

where the results are in m.

7.3.2 Equilibrium at Large Angles of Trim

For small angles of trim Stevin's law yields $LCB = LCG$ where both lengths are measured from the same origin. As Figure 7.3 shows, when the trim is large things are not so simple and the heights of the centres of buoyancy and gravity must be taken into account. In Figure 7.3 we assume again that both LCB and LCG are measured in the same system and from AP and write

$$LCG + (\overline{KG} - \overline{KB})\tan\theta = LCB \qquad (7.7)$$

The longitudinal centre of gravity, LCG, is always measured in a system fixed in the ship. Some computer programmes may measure LCB in a system fixed in space. Therefore, when using the output of a computer programme it is necessary to read carefully the definitions used by it.

Figure 7.3   Equilibrium at large trim

7.4 The Inclining Experiment

Because of the importance of this subject we give here the term in four foreign languages:

French	Expérience de stabilité
German	Krängungsversuch
Italian	Prova di stabilità
Spanish	Prueba de estabilidad

It is usual to carry on the inclining experiment a short time before the completion of the ship. The vessel must float in calm water and the work done while no wind is blowing. The number of persons aboard should be limited to that strictly necessary for the experiment; their masses and positions should be exactly recorded. Tank fillings and free surfaces in tanks should be well known. Free surfaces should not reach tank bottoms or ceilings for the expected heel and trim angles. All draught marks should be read, that is forward, at midship, at stern, both on starboard and on port side. Good practice requires to put a glass pipe before the draught mark and to read the draught value corresponding to the water level in the pipe. This procedure minimizes errors due to small waves. The water density should be read at several positions around the ship.

Figure 7.4 shows a common set-up for the inclining experiment. A plumb line with a bob B is hung at A. The bob is immersed in a water tank that serves as an oscillation damper. A mass p is displaced transversely a distance d. The resulting heel angle, assumed small, is given by

$$\tan \theta = \frac{pd}{\Delta \overline{GM}} \tag{7.8}$$

Figure 7.4 Set-up for the inclining experiment

The deflection of the plumb line is measured on a graduated batten $R_1 R_2$ and is used to calculate

$$\tan \theta = \frac{\overline{CD}}{\overline{AC}} \tag{7.9}$$

A recommended practice is to displace the mass once to starboard and measure $\tan \theta_S$, then to port and measure $\tan \theta_P$. The value to be substituted into Eq. (7.8) is

$$\tan \theta = \frac{\tan \theta_S + \tan \theta_P}{2}$$

It is recommended to repeat the set-up in Figure 7.4 at three stations along the ship. The masses used for inclining the vessel should be chosen so that the heel angles fall within that range in which Eq. (7.8) is applicable. Moore (1967) recommends angles of 1° for very large vessels, 1.5° for ships of 120 m length, and 2–3° for small vessels. Kastner (1989) cites German regulations that require heel angles ranging between 1° and 3.5°. Equation (7.8) can be used for the estimation of suitable masses.

According to Hansen (1985) the length of the plumb line should be chosen so that the length measured on the batten should be maximum 150–200 mm. Writes Hansen, "In general, long pendulums used on stiff ships and short pendulums used on tender ships result in about the same accuracy in measuring the ship list." Kastner (1989) studies the dynamics of a compound pendulum consisting of the ship and the plumb line. A long plumb line ensures a good resolution in reading the graduation on the batten. On the other hand, a long plumb line can yield a large dynamic response to small-amplitude ship motions and increase reading errors. Kastner concludes that a length of 1.146 m is sufficient. Today, the set-up shown in Figure 7.4 can be replaced by electronic instruments that measure the heel angle (inclinometers, gyroscopic platforms) whose output can be fed directly to an on-board computer.

A common way of checking the accuracy of the results consists in plotting the tangents of heel angles, $\arctan \theta$, against the heeling moments, pd. Equation (7.8) shows that the ideal plot should be a straight line. Years ago Naval Architects fitted by eye a straight line passing through the plotted points. Nowadays computers and many hand calculators yield easily a **least-squares fit**. Example 7.1 shows how to do it.

When analysing the results of the inclining experiment, Eq. (7.8) is rewritten as

$$\Delta \overline{GM} = \frac{pd}{\tan \theta}$$

The interpretation of the results of inclining experiments requires the knowledge of the displacement, Δ, and of the height of the metacentre above the baseline, $\overline{KM}$. If the trim is small, one can read the desired values in the hydrostatic curves, entering them with the measured mean draught, T_m, as input. Hansen (1985) quotes the limits imposed on the trim by the US Navy and the US Coast Guard. The recommended value for naval vessels is 0.67%, and for commercial ships 1% of the ship length. If the trim is not small one can use the Bonjean curves or a computer programme for hydrostatic calculations. When drawing the waterline on the Bonjean curves we must not forget that, in general, the forward and the aft draught marks are not placed in the transverse planes of the forward and aft perpendiculars. Therefore, the values read on the marks must be reduced to the FP and AP positions.

A computer programme for hydrostatic calculations can be used if the offsets of the ship are stored in the required input format. Then, it is sufficient to run the programme for the mean draught and the trim read during the inclining experiment.

The ship hull behaves like a beam that can deflect under bending moments. Bending moments arise from differences between the longitudinal distribution of masses and that of hydrostatic pressures. Deflections of the hull beam also can be caused by differences between thermal expansions of the deck and of the bottom. The deflection can be calculated as the difference between the average of forward and aft draught and the draught T_m measured at midship

$$d = T_M - \frac{T_F + T_A}{2} \tag{7.10}$$

Various authorities and authors publish formulae for calculating an **equivalent draught** that allows the calculation of the displacement of a deflected hull. For example, Hansen (1985) uses a rather complicated formula recommended by NAVSEA, a design authority of the US Navy. Ziha (2002) analyses the displacement change due to hull deflection and proposes ways of taking it into account. Hervieu (1985) simplifies the problem by assuming a parabolic elastic line (the deflected shape of the beam). Then, for a rectangular waterplane and vertical sides in the region of the actual waterline (wall-sided hull) the added or lost volume equals

$$\delta\Delta = \frac{2}{3}A_W d$$

where A_W is the waterplane area, and d is the deflection. In most cases the waterplane area is not rectangular, but still in a first approximation we can use as equivalent draught

$$T_{eq} = T_M + \frac{2}{3}d$$

The sign of d results from Eq. (7.10). The equivalent draught is used as input to hydrostatic curves.

We think that with present-day computers, and even hand calculators, it is possible to obtain with little effort and in a reasonable time more exact hydrostatic data. Moreover, assuming that the equivalent draught yields a good approximation of the displacement, what about the height of the metacentre above the baseline?

It is easy to calculate the hydrostatic data of a deflected hull by using the Bonjean curves. To do so one must simply draw a waterline passing through the three measured draughts, that is the forward, the midship and the aft draughts. The exact shape of the waterline is not known, but for small hull deflections that line cannot differ much from the shape taken by a draughting spline. Once the waterline is drawn, the Naval Architect has to read the Bonjean curves and use the readings as explained in Section 4.4. If a computer programme is available, and the ship offsets are stored in the required input format, one has to run the programme option for hydrostatic calculations in waves. The input wavelength is equal to twice the waterline length. The input wave height (trough-to-crest) to be considered is equal to twice the hull deflection. If $T_m > (T_F + T_A)/2$, a bending situation known as **sagging**, the wave crest shall be placed in the midship section. This case is exemplified in Figure 7.5. The upper figure, (a), shows what happens in reality. The lower figure, (b), shows the corresponding computer input. If $T_M < (T_F + T_A)/2$, a bending situation known as **hogging**, the wave trough shall be placed in the midship section. The midship draught and the trim measured during the experiment shall be those supplied as input.

Example 7.1 shows an analytic treatment of the results of the inclining experiment; it yields the product $\Delta\overline{GM}$ corresponding to the ship loading during the test. As described above, the displacement, Δ, is read in the hydrostatic curves, or is calculated from Bonjean curves or by a computer programme. Thus, one obtains the metacentric height, $\overline{GM}$, of the same ship loading. The height of the metacentre above the baseline, $\overline{KM}$, is obtained in the same way as

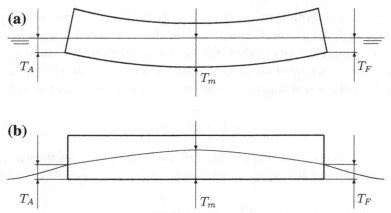

Figure 7.5 Deflected hull—sagging. (a) actual situation; (b) computer input

the displacement, that is from the hydrostatic curves, by integrating values read in the Bonjean curves, or by running a computer programme. The height of the centre of gravity above baseline is calculated as

$$\overline{KG} = \overline{KM} - \overline{GM}$$

For small trim angles we can assume that the x-coordinates of the centre of gravity and of the centre of buoyancy are equal, that is $LCG = LCB$; otherwise see Section 7.3.2. The longitudinal centre of buoyancy is obtained in the same way as the displacement. At this point the Naval Architect knows the displacement and the centre of gravity of the ship loaded as during the inclining experiment. To calculate the data of the lightship one must first subtract the masses and the moments of the items that do not belong to the lightship, but were aboard during the test. Such items are, for example, the masses used to incline the ship. Next, one has to add the masses and the moments of the items that belong to the lightship, but were not yet assembled at the time of the inclining experiment.

Sometimes the authorities that must approve the ship have their own inclining experiment regulations. Alternatively, the designer may be asked to abide by certain codes of practice that include provisions for the inclining experiment. Then, it is imperious to read those regulations before carrying on the work. An example of such regulations is the standard F1321-92 developed by ASTM (2001). Highlights of the relevant IMO requirements are given in Chapter 8.

7.5 Summary

Stability and trim calculations require the knowledge of the displacement and of the position of the centre of gravity. To calculate these quantities it is necessary to organize the ship masses into weight groups. The sum of the weight groups that do not change during operation is called lightship displacement; for merchant vessels it is the sum of hull, outfit, and machinery masses. The sum of the masses that are carried in operation according to the different loading cases is called deadweight; it includes the crew and its equipment, the cargo and passengers, the fuel, the lubricating oil, the fresh water, and the stores.

To find the displacement of a given loading case it is necessary to add the masses of the lightship and of the deadweight items carried on board in that case. To find the coordinates of the centre of gravity, LCG, and VCG $(\overline{KG})$, it is necessary to sum up the moments of the above masses with respect to a transverse plane for the first, a horizontal plane for the second. The calculations can be conveniently carried on in an electronic spreadsheet or by software such as MATLAB or MS Excel.

Once the displacement, Δ, is known, one can find the corresponding mean draught, T_m, by reading the hydrostatic curves. These curves also yield the values of the longitudinal centre of buoyancy, LCB, the longitudinal centre of flotation, LCF, and the moment to change trim by 1 m, MCT. If the trim is small it can be found from

$$T_F - T_A = \frac{\Delta(LCG - LCB)}{MCT}$$

For normal loading situations the trim is always small. Then, the trimmed waterline, $W_\theta L_\theta$, intersects the waterlines of the ship on even keel, $W_0 L_0$, along a line passing through the centre of flotation, F, of $W_0 L_0$. To obtain the forward draught, T_F, and the aft draught, T_A, it is necessary to add to, or subtract from the mean draught a part of the trim proportional to the distance of the respective perpendicular from the centre of flotation

$$T_A = T_m - \text{trim} \cdot \frac{LCF}{L_{pp}}, \text{ metres}$$

$$T_F = T_m + \text{trim} \left(1 - \frac{LCF}{L_{pp}, \text{ metres}}\right)$$

If the trim is large, the heights of the centres of buoyancy and flotation must be taken into account.

Because of uncertainties in the calculation of masses and centres of gravity, it is necessary to validate them experimentally. This is done in the inclining experiment, an operation to be carried on for new buildings and for ships that underwent substantial changes. The ship is brought in sheltered waters and when no wind is blowing. A known mass, p, is displaced transversely a known distance, d, and the tangent of the resulting heel angle, $\tan \theta$, is measured. The statistical analysis of several inclining tests yields the product

$$\Delta \overline{GM} = \frac{pd}{\tan \theta}$$

The displacement, Δ, is found as a function of the draughts measured during the experiment. If a hull deflection is measured it must be taken into account. The vertical centre of gravity is calculated as

$$\overline{KG} = \overline{KM} - \overline{GM}$$

If the trim is large the hydrostatic curves cannot be used. The Bonjean curves are helpful here, as is a computer programme. Both Bonjean curves and computer programmes can be used to calculate the effect of hull deflection.

7.6 Examples

Example 7.1 (Least-squares fit of the results of an inclining experiment).   The results of the inclining experiment presented here are taken from an example in Hansen (1985), but are converted into SI units. The data are plotted as points in Figure 7.6. At a first glance it seems reasonable to fit a straight line whose slope equals the mean of $pd/\tan\theta$ values. In this example some trials performed with very small pd values produced zero heel-angle tangents. Those cases must be discarded when averaging because they yield $pd/\tan\theta = \infty$. After eliminating the pairs corresponding to zero heel-angle tangents we calculate the mean slope and obtain 53679.638. The reader can easily verify that the line having this slope is far from being satisfactory. Available programmes for linear least-squares interpolation cannot be used

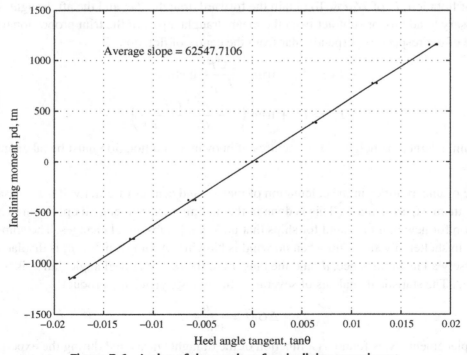

Figure 7.6   A plot of the results of an inclining experiment

Table 7.1 Results of inclining experiment

Inclining Moment (tm)	Heel Angle Tangent	Inclining Moment (tm)	Heel Angle Tangent
1156.9	0.0187	−1136.5	−0.0179
1156.9	0.0185	−1136.5	−0.0180
1156.9	0.0179	−1136.5	−0.0185
771.5	0.0126	−757.5	−0.0119
771.5	0.0126	−757.5	−0.0120
771.5	0.0121	−757.5	−0.0124
386.3	0.0065	−379.4	−0.0057
386.3	0.0064	−379.4	−0.0060
386.3	0.0062	−379.4	−0.0065
1.1	0.0004	−0.2	0.0000
1.1	0.0005	−0.2	0.0000
1.1	0.0000	−0.2	0.0006

because, in general, they fit a line having an equation of the form

$$y = c_1 x + c_2$$

Obviously, in our case the line must pass through the origin, that is $c_2 = 0$. Therefore, let us derive by ourselves a suitable procedure.

To simplify notations let x_i be the tangents of the measured heeling angles, and y_i the corresponding inclining moments. As said, we want to fit to the measured data a straight line passing through the origin

$$y = Mx \tag{7.11}$$

The error of the fitted point to the ith measured point is

$$y_i - Mx_i \tag{7.12}$$

Table 7.2 Small cargo ship—homogeneous cargo, arrival

Weight Item	Mass (t)	vcg (m)	lcg (m)
Lightship	1247.66	5.93	32.04
Crew and effects	3.60	9.60	11.00
Provisions	1.00	7.00	3.50
Fuel oil	27.74	2.17	23.15
Lubricating oil	3.49	0.62	17.08
Fresh water	8.70	1.61	9.75
Ballast water	248.87	0.55	39.62
Cargo in hold	993.94	4.35	42.62
Fruit cargo	90.00	6.08	38.66

Table 7.3 Small cargo ship—partial hydrostatic data, 2

Draught, T (m)	MCT (m)	LCB from Midship (m)	LCF from Midship (m)	Draught, T (m)	MCT (m)	LCB from Midship (m)	LCF from Midship (m)
2.00	2206	0.607	0.518	4.32	3223	0.291	−0.384
2.20	2296	0.600	0.460	4.40	3260	0.272	−0.430
2.40	2382	0.590	0.398	4.60	3336	0.225	−0.560
2.60	2470	0.575	0.330	4.80	3413	0.180	−0.698
2.80	2563	0.557	0.260	5.00	3485	0.131	−0.839
3.00	2645	0.537	0.190	5.20	3567	0.083	−0.960
3.20	2732	0.510	0.119	5.40	3639	0.033	−1.066
3.40	2824	0.480	0.041	5.60	3716	−0.018	−1.158
3.60	2906	0.442	−0.035	5.80	3793	−0.067	−1.231
3.80	2293	0.406	−0.017	5.96	3863	−0.108	−1.281
4.00	3085	0.360	−0.210	6.00	3880	−0.118	−1.293
4.20	3167	0.319	−0.314	6.20	3951	−0.167	−1.348

We want to minimize the sum of the squares of errors

$$e = \sum (y_i - Mx_i)^2 \tag{7.13}$$

To do this we differentiate e with respect to M and equal the derivative to zero

$$\sum x_i(y_i - Mx_i) = 0 \tag{7.14}$$

The solution is

$$M = \frac{\sum x_i y_i}{\sum x_i^2} \tag{7.15}$$

An example of a MATLAB script file that plots the data, calculates the slope, M, and plots the fitted line is

```
%INCLINING Analysis of Inclining Experiment
% Format of data is [ moment tangent ], initial units
    [ ft-tons - ]
incldata = [
    . . .
    . . . ];
% separate data
moment = incldata (:, 1); tangent = incldata (:, 2);
plot(tangent, moment, 'k.'), grid
ylabel('Inclining moment, pd, tm')
xlabel('Heel angle tangent, tan\theta')
hold on
tmin = min(tangent); tmax = max(tangent);
```

```
M    = sum(tangent.*moment)/sum(tangent.^2);
Mmin = M*tmin; Mmax = M*tmax;
plot( [ tmin tmax ], [ Mmin Mmax ], 'k-')
text(-0.015, 1100, [ 'Average slope = ' num2str(M) ])
hold off
```

Above, the user has to write the data of the inclining experiment in the matrix `incldata`. The MATLAB programme shown here can be easily transformed so that the user can input the name of a separate file that stores the `incldata` matrix.

7.7 Exercises

Exercise 7.1 (Small cargo ship homogeneous load, arrival). Using the data in Table 7.2 calculate the loading case **homogeneous cargo, arrival**, of the small cargo ship earlier encountered in this book. By **arrival** we mean the situation of the ship entering the port of destination with the fuel, the lubricating oil and the provisions consumed in great part.

Exercise 7.2 (Trim and draughts). Check that substituting in $T_F - T_A$ the expressions given by Eqs. (7.5) and (7.6) we obtain, indeed, the trim.

CHAPTER 8

Intact Stability Regulations I

Chapter Outline

8.1 Introduction

In the preceding chapters we presented the laws that govern the behaviour of floating bodies. We learnt how to find the parameters of a floating condition and how to check whether or not that condition is stable. The models we developed allow us to check the stability of a vessel under the influence of various heeling moments. At this point we may ask what is satisfactory stability, or, in simpler terms, how much stable a ship must be. Analyzing the data of vessels that behaved well, and especially the data of vessels that did not survive storms or other adverse conditions, various researchers and regulatory bodies prescribed criteria for deciding if the stability is satisfactory. In this chapter we present examples of such criteria. To use picturesque language, we may say that in Chapters 2 to 7 we described *laws of nature*, while in this chapter we present *man-made laws*. Laws of nature act independently of man's will and they always govern the phenomena to which they apply. Man-made laws, in our case stability regulations, have another meaning. Stability regulations prescribe criteria for approving ship designs, accepting new buildings, or allowing ships to sail out of harbour. If a certain ship fulfils the requirements of given regulations, it does not mean that the ship can survive all challenges, but her chances of survival are good because stability regulations are based on considerable experience and reasonable theoretic models. Conversely, if a certain ship does not fulfil certain regulations, she must not necessarily capsize, only the risks are higher and the owner has the right to reject the design, or the authority in charge has the right to prevent the ship from sailing out of harbour. Stability regulations are, in fact, **codes of practice** that provide reasonable safety margins. The codes are compulsory not only for designers and builders, but also for ship masters who must check if their vessels meet the requirements in a proposed loading condition.

The codes of stability presented in this chapter take into consideration only phenomena discussed in the preceding chapters. The stability regulations of the German Federal Navy are based on the analysis of a phenomenon discussed in the next chapter; therefore, we defer their presentation until Chapter 10. In addition, the International Maritime Organization, IMO, issued some international recommendations for ship masters that enable them to minimize or avoid the influence of other phenomena whose explication requires some knowledge of the theory of ship motions in waves. We briefly discuss this theory in Chapter 12 and, therefore, we place there also the citation of the respective recommendations.

For obvious reasons it is not possible to include in this book all existing stability regulations; we only choose a few representative examples. Neither is it possible to present all the provisions of any single regulation. We only want to draw the attention of the reader to the existence of such codes of practice, to show how the models developed in the previous chapters are applied, and to help the reader in understanding and using the regulations. Following these guidelines we present the most important provisions of the *IMO Intact Stability Code* (2008 IS Code), the main instrument for checking the stability of merchant ships, and the principal requirements of the stability regulations of the US (DDS-079), and UK (NES 109) Navies. In continuation we explain the criterion of a UK code for sailing ships because we consider that its approach and the reason behind it are an example of good engineering. We end with some provisions for other small boats.

Technological developments, experience accumulation, and especially major marine disasters can impose revisions of existing stability regulations. For all the reasons mentioned above, before checking the stability of a vessel according to given regulations, the Naval Architect must read in detail their newest, official version.

All stability regulations specify a number of loading conditions for which calculations must be carried on. Some regulations add a sentence like 'and any other condition that may be more dangerous.' It is the duty of the Naval Architect in charge of the project to identify such situations, if they exist, and check if the stability criteria are met for them. In some cases, all possible service load conditions cannot be foreseen in the design stage. For this reason, some international regulations like SOLAS (International Convention for the Safety of Life at Sea) and ILLC (International Load Lines Convention) require the preparation of operational guidance for masters that enables the checking of any load condition against applicable criteria. This guidance includes limiting, or allowable $\overline{KG}$ or $\overline{GM}$ curves.

8.2 The IMO Code of Intact Stability

The *InterGovernmental Maritime Consultative Organization* was established in 1948 and was initially known as *IMCO*. The name was changed in 1982 to **IMO—International Maritime Organization**. The purpose of IMO is the intergovernmental cooperation in the development of regulations regarding shipping, maritime safety, maritime security, navigation, and the prevention of marine pollution from ships. IMO is a specialized agency of the United Nations; at the time of writing it has 170 Member States and three Associate Members. The headquarters are in London. A critical sketch of IMO's history can be found in Francescutto (2007).

The main international instruments that address adequate buoyancy, subdivision (see Chapter 11 for this term) and stability are 'live' regulations that evolve with time and are made at IMO: the SOLAS convention, the ILLC and the IS code. Some provisions of the SOLAS convention are presented in Chapter 11, those of ILLC belong mainly to general ship design. In this chapter we refer to the IS code version issued in 2008 and entered into force the day of 1 July 2010 (see IMO, 2009). Part A of the code covers the **mandatory** criteria for 'ships and other marine vehicles of 24 m in length and above.' Part B of the code describes **recommendations** for particular sizes of certain types of ships and other marine vehicles not included in Part A, or recommendations intended to supplement the provisions of Part A for particular sizes or modes of operation. The 2008 IS Code applies to cargo ships and those carrying timber deck cargo, passenger ships, fishing vessels, special purpose ships, offshore supply vessels, mobile offshore drilling units (MODUs), pontoons, and containerships. Countries that adopted these regulations enforce them either directly, or by converting them into national ordinances.

8.2.1 General Mandatory Criteria for Passenger and Cargo Ships

Part A of the code establishes the general criteria to be applied to all loading conditions taking into account the free-surface effect. There are two main criteria:

- the righting-arm curve ($\overline{GZ}$ curve);
- the severe wind and rolling criterion, known more as the **weather criterion**.

The required properties of the $\overline{GZ}$ (in fact $\overline{GZ}_{eff}$) curve are:

1. The area under the righting-arm curve should not be less than 0.055 m rad up to 30°, and not less than 0.09 m rad up to 40° or the angle of downflooding if this angle is smaller than 40°. Additionally, the area under the righting-arm curve between 30° and 40°, or between 30° and the flooding angle, if this angle is less than 40°, should not be less than 0.03 m rad;
2. The righting arm, $\overline{GZ}$, shall be at least 0.2 m at an angle of heel equal to or greater than 30°;
3. The maximum righting arm should occur at an angle of heel not less than 25°. If this is not possible an equivalent criterion may be used with the approval of the Administration;
4. The initial metacentric height, $\overline{GM}_{eff}$, should not be less than 0.15 m.

The code uses frequently the terms **angle of flooding** and **downflooding**; they refer to the smallest angle of heel at which an opening in the hull, superstructure or deckhouse, that cannot be closed watertightly, submerges. The requirements of this criterion are inspired by Rahola's work published in 1939 and cited in Section 6.1 (see also Rahola, 1939). Example 8.1 illustrates its application. One caveat to this IMO criterion: Rahola's work was based on statistics over a sample of ships that suffered accidents linked with low stability, while the IMO criterion is based on statistics over Rahola's sample plus an additional number of 166 casualties reported to the Organization (IMO, 2009). IMO advises that the criterion may not be applicable to certain ships like those with large beam and small depth, i.e., $B/D \gg 2.5$. We can see here how criteria of prescriptive nature may have their limitations as hull forms evolve with time. Something to have in mind when applying regulations.

The second criterion, applicable to cargo and passenger ships, is the weather criterion; it tests the ability of a ship to withstand the combined effects of beam wind and rolling. We explain this criterion with the help of Figure 8.1. First, the code assumes that the ship is subjected to a steady-wind heeling arm

$$\ell_{w1} = \frac{PAZ}{1000g\Delta} \tag{8.1}$$

where $P = 504$ N m^{-2}, A is the projected lateral area of the ship and deck cargo above the waterline, in m^2, Z is the vertical distance from the centroid of A to the centre of the underwater lateral area, or approximately to half-draught, in m, Δ is the displacement mass, in t, and $g = 9.81$ m s^{-2}. Unlike the model developed in Section 6.3 (model used by the US Navy), IMO accepts the more severe assumption that the wind heeling arm does not decrease as the heel angle increases. The static angle caused by the wind arm l_{w1} is ϕ_0. It is assumed

Small cargo ship, $\Delta = 2625$ t, KG = 5 m, IMO weather criterion

Figure 8.1 The IMO weather criterion

that, due to wave action, the ship rolls from the angle ϕ_0 windward by an angle ϕ_1. The static angle of heel, ϕ_0, should not exceed 16° or 80% of the angle for deck immersion, which is less. Next, the code assumes that the ship is subjected to a gust of wind causing a heeling arm $\ell_{wl2} = 1.5\ell_{w1}$. The angle of roll is given by

$$\phi_1 = 109k X_1 X_2 \sqrt{rs} \tag{8.2}$$

where ϕ_1 is measured in degrees, X_1 is a factor given in Table 2.3.4-1. of the code, X_2 is a factor given in Table 2.3.4-2. of the code, and k a factor defined as follows:

- $k = 1.0$ for round-bilge ships;
- $k = 0.7$ for a ship with sharp bilges;
- k as given by Table 2.3.4-3. of the code for a ship having bilge keels, a bar keel or both.

As commented in Section 6.13, by using the factor k, the IMO code considers indirectly the effect of damping on stability. More specifically, it acknowledges that sharp bilges, bilge keels, and bar keels reduce the roll amplitude. However, the code explicitly says that the angle of roll of ships with active anti-rolling devices should be determined without taking into account the operation of those devices, unless the Administration is satisfied by a proof that the system will be effective even in the event of a sudden power shutdown.

By assuming that the ship is subjected to the wind gust while heeled windward from the static angle, the dynamical effect appears more severe, as explained in Section 6.6 and the lower plot of Figure 6.5.

The factor r is calculated from

$$r = 0.73 + 0.6\,\overline{OG}/T_m \tag{8.3}$$

where $\overline{OG}$ is the distance between the waterline and the centre of gravity, positive upwards. The factor s is given in Table 2.3.4-4 of the code, as a function of the roll period, T. The code states that in the absence of other information the following formula should be used for calculating the roll period, in seconds,

$$T = \frac{2CB}{\sqrt{\overline{GM}_{eff}}} \tag{8.4}$$

where

$$C = 0.373 + 0.023(B/T_m) - 0.043(L_{WL}/100) \tag{8.5}$$

The code assumes that the lever arm of the wind gust is

$$\ell_{w2} = 1.5\ell_{w1} \tag{8.6}$$

Plotting the curve of the arm ℓ_{w2} we distinguish the areas a and b. The area b is limited to the right at 50° or at the angle of flooding, whichever is the smaller. The area b should be equal to or greater than the area a. This provision refers to dynamical stability, as explained in Section 6.6.

When applying the criteria described above, the Naval Architect must use values corrected for the free-surface effect, that is $\overline{GM}_{eff}$ and $\overline{GZ}_{eff}$. An older version of the code gave an expression for calculating the free-surface effect. The expression was exact under limiting assumptions. Today, computer programmes for hydrostatic calculations yield values of the free-surface lever for any tank form described in the input, and for any heel angle.

The code specifies the loading cases for which stability calculations must be performed. For example, for cargo ships the criteria shall be checked for the following four conditions:

1. Full-load departure, with cargo homogeneously distributed throughout all cargo spaces.
2. Full-load arrival, with 10% stores and fuel.
3. Ballast departure, without cargo.
4. Ballast arrival, with 10% stores and fuel.

Similarly with what we mentioned for the righting-arm curve, there is one important caveat also for the weather criterion. All the tables used in the calculation of the factors X_1, X_2, k, and s are derived from ships characterized by

1. $B/T_m < 3.5$;
2. $-0.3 \leq \overline{KG}/T_m - 1 \leq 0.5$:
3. $T < 20$ s.

Although this criterion has been criticized by several authors who also exemplified their inadequacy for certain ships (Krueger, 2002; Francescutto, 2007), the tables and formulae remained as in the 1995 version. Therefore, IMO allows some deviations. Thus, for ships with parameters outside of the above limits the angle of roll, ϕ_1, can be determined from model experiments following the procedure described in MSC.1/Circ1200 (see IMO, 2009). The administration may accept this alternative, if deemed appropriate. The reason behind this is that especially in the case of cruise vessels and yachts the formula cannot be applied because of the evolution of submerged forms and the enlargement of the sail area. This is one of the limitations of regulations of prescriptive nature; they are simply checklists that allow one to establish whether a ship complies with a given equation or not. The safety of ships is more than this, as we will have an opportunity to see at the end of this chapter (Sections 8.7 and 8.8).

8.2.2 Special Mandatory Criteria for Passenger Ships

In addition to the general criteria described above, passenger ships shall meet two additional criteria. First, the angle of heel caused by the crowding of passengers to one side should not exceed 10°. For calculations, the mass of a passenger is assumed to be equal to 75 kg, and the centre of gravity of a standing passenger is assumed to lie 1 m above the deck, while that of a seated passenger is taken as 0.30 m above the seat. The second additional requirement refers to the angle of heel caused by the centrifugal force developed in turning. The heeling moment due to that force is calculated with the formula

$$M_T = 0.2 \frac{V_0^2}{L_{WL}} \Delta \left(\overline{KG} - \frac{T_m}{2} \right) \tag{8.7}$$

where M_T is measured in N m, V_0 is the service speed in m s^{-1}, L_{WL} the waterline length in m, Δ the displacement mass in t, $\overline{KG}$ the vertical centre of gravity in m, and T_m the mean draught in m. Again, the resulting angle shall not exceed 10°. The reason for limiting the angle of heel is that at larger values the passengers may panic. The application of this criterion is exemplified in Figure 8.2 and in Example 8.3. In fact, to check the stability in turning we compare the heeling arm due to the centrifugal force with the righting arm. Instead of multiplying with Δ and dividing by the same value, we would rather calculate directly the heeling arm as

$$l_T = 0.2 \frac{V_0^2}{L_{WL}} \frac{\left(\overline{KG} - \frac{T_m}{2} \right)}{g}$$

where g is the acceleration of gravity.

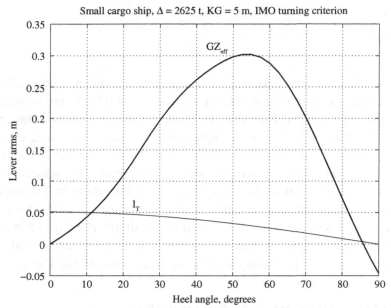

Figure 8.2 The IMO turning criterion

8.2.3 Special Mandatory Criteria for Cargo Ships Carrying Timber Deck Cargoes

Cargo ships carrying timber deck cargoes shall comply with the general mandatory requirements stated above unless the Administration is satisfied with the application of the following alternative provisions that hold if the timber deck cargo extends longitudinally between superstructures and transversally on the full deck breadth, except for a reasonable gunwale. Where there is no limiting superstructure at the aft, the cargo should extend at least to the after end of the aftermost hatch. For such ships the area under the righting-arm curve should not be less than 0.08 m rad up to $40°$ or the angle of flooding, whichever is smaller. The maximum value of the righting arm, $\overline{GZ}$, shall be at least 0.25 m. Finally, at all times during a voyage, the metacentric height shall not be less than 0.1 m, taking into account the absorbtion of water by the deck cargo and/or ice accretion on the exposed surfaces. The buoyancy of the deck cargo can be taken into account, assuming that the cargo can absorb a volume of water equal to 25% of its volume (see Chapter 11 for the notion of permeability). In such a case the Administration may consider it necessary to investigate the influence of different percentages of absorbtion and/or the height of the deck cargo.

8.2.4 Oil Tankers of 5000 t Deadweight and Above

An *oil tanker* is defined as a ship constructed or adapted primarily to carry oil in bulk in its cargo spaces. The definition includes combination carriers and any chemical tanker when it is carrying a cargo or partial cargo of oil in bulk. When in port, the initial metacentric height corrected for the free-surface effect shall not be less than 0.15 m, and when at sea:

1. The area under the righting-arm curve shall be not less than 0.055 m rad up to 30°, and not less than 0.09 m rad up to 40° or the angle of flooding, whichever is smaller. Additionally, the area under the righting-arm curve, between 30° and 40° or the angle of flooding, whichever is smaller, shall not be less than 0.03 m rad.
2. The righting lever shall be at least 0.20 m at an angle of heel equal to or greater than 30°.
3. The maximum righting arm shall occur at an angle of heel preferably exceeding 30° but not less than 25°.
4. The initial metacentric height, corrected for the free-surface effect, shall not be less than 0.1 m.

8.2.5 Cargo Ships Carrying Grain in Bulk

In Section 6.9.4 we study the destabilizing influence of shifting loads. One kind of such loads is grain. The intact stability of ships engaged in the carriage of grain shall comply with the requirements of the *International code for the safe carriage of grain in bulk* adopted in resolution MSC.23(59). The angle of heel due to the shift of grain shall not exceed 12°. The metacentric height shall be not less than 0.30 m. The code indicates how to approximate by a straight line the heeling arm due to the transverse shifting of grain and specifies a criterion for the area between the heeling- and the righting-arm curves.

8.2.6 High-Speed Craft

A vessel is defined as **high-speed craft**, shortly **HSC**, if it is capable to attain a maximum speed, in m s^{-1}, equal to

$$V = 3.7 \nabla^{0.1667}$$

where ∇ is the displacement volume in m^3. The code distinguishes between *hydrofoil craft* (whether fitted with surface piercing or submerged foils), *monohull craft*, and *multihull craft*. High-speed craft built on or after 1 January 1996, to which Chapter 10 of the 1974 SOLAS Convention applies, shall also comply with the stability requirements of the 1994 HSC Code. Any high-speed craft built according to SOLAS 1974 and subsequently affected by repairs, or modifications of major character, and high-speed craft constructed on or after 1 July 2002, shall comply with the stability requirements of the *2000 HSC Code*. The two HSC Codes derived from the *Code of safety for dynamically supported craft*, shortly DSC, adopted by IMO in November 1977. Following the rapid evolution of these vessels, the international community had to adjourn the codes. A **dynamically supported craft**, shortly DSC, as a type of HSC, is defined in the DSC Code as one of the following cases:

A vessel is a **dynamically supported craft (DSC)** in one of the following cases:

1. if, in one mode of operation, a significant part of the weight is supported by other than buoyancy forces;
2. if the craft is able to operate at Froude numbers, $F_n = V/\sqrt{gL}$, equal to or greater than 0.9.

The first category includes air-cushion vehicles and hydrofoil boats. Hydrofoil boats float, or sail, in the **hull-borne** or **displacement mode** if their weight is supported only by the buoyancy force predicted by Archimedes' principle. At higher speeds hydrodynamic forces develop on the foils and they balance an important part of the boat weight. Then, we say that the craft operates in the **foil-borne** mode.

Annex 6 of the 2000 HSC Code contains requirements for hydrofoil craft operating between two ports situated in different countries. The prescriptions for surface-piercing hydrofoil boats in hull-borne mode are described in Paragraph 41.1 of the code. The heeling moment in turning is calculated as

$$M_R = \frac{0.196 V_0^2 \Delta \overline{KG}}{L}$$

where V_0 is the speed in turning, in m s^{-1}, and M_R results in kN m. The formula is valid if the radius of the turning circle lies between $2L$ and $4L$. The resulting angles of inclination should not exceed 8°.

The wind heeling moment, in the displacement mode, in kN m, should be calculated as

$$M_V = 0.001 P_V A_V Z$$

and is considered constant within the whole heeling range. The area subjected to wind pressure, A_V, is called here **windage area**. The wind pressure, P_V, is a function of the wind speed, V_w, corresponding to the worst intended conditions and equals $P_V = 750(V_W/26)^2$, measured in N m^{-2}. The **windage area lever**, Z, is the distance between the waterline and the centroid of the windage area. A minimum capsizing moment, M_C, is calculated as shown in paragraph 1.1.5.1 of the code and as illustrated in Figure 8.3. The curve of the righting arm is extended to the left by a roll angle ϕ_z averaged from model or sea tests. In the absence of such data the angle is assumed equal to 15°. Then, a horizontal line is drawn so that the two grey areas shown in the figure are equal. The ordinate of this line defines the value M_C. According to the theory developed in Section 6.6 the ship capsizes if this moment is applied dynamically. The stability is considered sufficient if $M_C/M_V \geq 1$.

The code also prescribes criteria for the transient and foil-borne modes. Such criteria consider the forces developed on the foils, a subject that is not discussed in this book.

Figure 8.3 Defining the minimum capsizing moment of a dynamically supported craft (DSC)

8.2.7 Fishing Vessels

Part B, Chapter 2 of the IS Code is devoted to *Fishing Vessels*. The general mandatory criteria for intact stability given in Part A of the code apply to fishing vessels having a length of 24 m and over. An exception is the requirement that the initial metacentric height should not be less than 0.35 m for single-deck vessels. If the vessel has a complete superstructure, or the ship length is equal to or larger than 70 m, the metacentric height can be reduced with the agreement of the government under whose flag the ship sails, but it should not be less than 0.15 m. The weather criterion applies in full to ships of 45 m length and longer. For fishing vessels whose length ranges between 24 and 45 m the code prescribes a wind gradient such that the pressure ranges between 316 and 504 N m^{-2} for heights of 1 to 6 m above sea level. Decked vessels shorter than 30 m must have a minimum metacentric height calculated with a formula given in paragraph 2.1.5.1 of the code.

8.2.8 Mobile Offshore Drilling Units

A **mobile offshore drilling unit**, shortly MODU, is a ship capable of engaging in drilling operations for the exploration or exploitation of resources beneath the sea bed, such as liquid or gaseous hydrocarbons. Part B, Chapter 2.6 of the code applies to mobile drilling units whose keels were laid after 1 March 1991. The wind force is calculated by considering the shape factors of structural members exposed to the wind, and a height coefficient ranging between 1.0 and 1.8 for heights above the waterline varying from 0 to 256 m. The area under the righting-arm curve up to the second static angle, or the downflooding angle, whichever is smaller, should exceed by at least 40% the area under the wind arm. The code also describes an alternative intact stability criterion for two pontoon, column-stabilized semi-submersible units.

8.2.9 Containerships Greater than 100 m

For container ships greater than 100 m length it may be possible to apply criteria alternative to those stated in Part A of the code. However, since the alternative criteria were empirically developed with data of containerships less than 200 m in length, they should be applied with care to ships beyond the above limit. Part B, Section 2.3.2 of the code defines a form factor C depending on the main dimensions of the ship and the configuration of hatches (Figure 4.9-1 in the code). The minimum values of areas under the righting-arm curve are prescribed in the form a/C, where a is specified for several heel intervals.

8.2.10 Allowable $\overline{GM}$ or $\overline{KG}$ Curves

All conventional stability criteria studied up to now set limits to different parameters of the $\overline{GZ}$ curve. As discussed in Chapter 5, the $\overline{GZ}$ curve is obtained from the formula

$$\overline{GZ} = l_k - \overline{KG}\sin\phi$$

For a given hull geometry the $\overline{GZ}$ curve will depend only on the displacement, or, equivalently, the draught, and the vertical centre of gravity, $\overline{KG}$. The **limiting**, or **allowable** $\overline{KG}$ is the maximum $\overline{KG}$ value that complies with the applicable stability criteria, for a given displacement. This value is calculated by trying various values until the point is found where the result of comparison with the relevant criteria changes from "pass" to "fail." The plot of allowable $\overline{KG}$ values as function of displacement (or draught) is called **allowable**, or **limiting** $\overline{KG}$ **curve**. A possible curve is shown in Figure 8.4.

As discussed in Chapter 6, we should bear in mind that in each loading condition there will be some slack tanks that adversely affect the stability. The influence of the free-surface effect can be taken into account as a virtual elevation of the centre of gravity equal to the sum of the values calculated for each tank as in Eq. (6.38). Therefore, to evaluate the stability of a ship in a given loading condition, the master has to plot the point corresponding to the corrected $\overline{KG}$ value in a diagram such as shown in Figure 8.4.

Usually limiting $\overline{KG}$ curves are calculated for different trim values and plotted together in the same diagram to cover the full operational trim range. Examples of such curves are given in a figure of IMO (2008). Limiting curves provide an overall indication of the stability performance of a design. Therefore, they can be used to compare different hull forms and geometries including watertight superstructures.

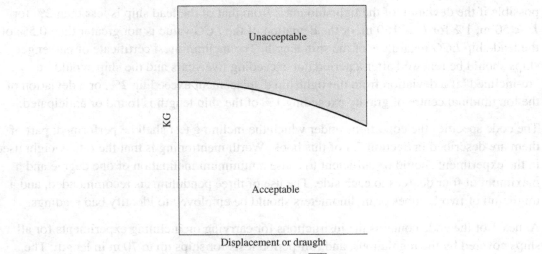

Figure 8.4 The limiting $\overline{KG}$ curve

8.2.11 Icing

Part B, Chapter 6 of the code bears the title "Ice considerations." For any ship operating in areas where ice accretion is likely to occur, adversely affecting the stability, corresponding weight allowances should be included in the analysis of loading conditions. In the case of fishing vessels and cargo ships carrying timber on deck, the allowance for additional weight should be made for the arrival condition. The code specifies clearly, in a chart, the geographical areas in which ice accretion can occur. Most important examples are the regions of Iceland, the Baltic Sea, the north of North America, the Bering and Okhotsk Seas, the Tartary Strait, and the seas south of 60° S. The following values, prescribed for fishing vessels, illustrate the severity of the problem. Stability calculations should be carried on assuming ice accretion (this is the term used in the code) with the surface densities.

- 30 kg m^{-2} on exposed weather decks and gangways;
- 7.5 kg m^{-2} for projected lateral areas on each side, above the waterplane.

8.2.12 Inclining and Rolling Tests

Part B, Chapter 8 of the code deals with the determination of lightship parameters. The term *lightship* is defined in Section 7.2.1. The IMO regulations specify that an inclining test shall be performed for any single passenger ship, regardless of her size, and for every single cargo ship, regardless of its length. The theoretical background of the inclining test is explained in Section 7.4. Exemption from the test is possible if the ship is one of a series of ships and basic stability data are available from the inclining test of a sister ship in the same series. This is possible if the deviation of the lightship mass from that of the lead ship is less than 2% for $L < 50$ m, 1% for $L > 160$ m, or the deviation of the *LCG* value is not greater than 0.5% of the lead-ship *LCG* regardless of the ship length. The inclining-test certificate of passenger ships should be renewed after a period not exceeding five years and the ship should be "re-inclined" if a deviation from the lightship displacement exceeding 2%, or a deviation of the longitudinal centre of gravity exceeding 1% of the ship length is found or anticipated.

The code specifies the conditions under which the inclining test shall be performed; part of them are described in Section 7.4 of this book. Worth mentioning is that the total weight used in the experiment should be sufficient to cause a minimum inclination of one degree and a maximum of four degrees to each side. The use of three pendulums is recommended, and a minimum of two U-tubes or inclinometers should be employed to identify bad readings.

Annex 1 of the code contains the instructions for carrying on inclining experiments for all ships covered by the regulations, and roll-period tests for ships up to 70 m in length. The relationship between the metacentric height, $\overline{GM}_O$, and the roll period, T, is given as

$$\overline{GM}_0 = \left(\frac{fB}{T}\right)^2$$

where B is the ship breadth.

An interesting part of the Annex refers to the plot of heel angle tangents against heeling moments; it explains the causes of deviations from a straight line, such as free surfaces of liquids, restrictions of movements, steady wind, or wind gust.

8.2.13 Stability Booklet

The IS code gives extensive guidance for preparing stability information that would allow control officers, mostly naval architects, to grant permission for the ship to sail, and masters to operate the ship in compliance with applicable requirements. This information must be compiled in a document called **stability booklet**. Among others, the stability booklet should contain hydrostatic curves or tables, cross-curves of stability calculated on a free trimming basis, loading restrictions, such as limiting $\overline{KG}$ or minimum $\overline{GM}$ curves, and the inclining test report. The code allows the use of board computers in which all the information necessary to carry on all checks is stored, but it notes that the computer may be only a supplement, not a substitute for the stability booklet. More details on such computers are discussed in Chapter 13.

8.3 The Regulations of the US Navy

In 1944 an American fleet was caught by a tropical storm in the Pacific Ocean. In a short time three destroyers capsized, a fourth one escaped because a funnel broke down under the force of the wind reducing thus the sail area. This disaster influenced the development of stability regulations for the US Navy. They were first published by Sarchin and Goldberg in 1962. These regulations were subsequently adopted by other navies.

The intact stability is checked under a wind whose speed depends on the service conditions. Thus, all vessels that must withstand tropical storms should be checked for winds of 100 knots. Ocean-going ships that can avoid the centre of tropical storms should be checked under a wind of 80 knots, while coastal vessels that can avoid the same dangers should be checked for winds of 60 knots. Coastal vessels that can be called to anchorage when expecting winds above Force 8, and all harbour vessels should be checked under the assumption of 60 knots winds.

We explain the weather criterion in Figure 8.5. The righting arm, $\overline{GZ}$, is actually the effective righting arm, $\overline{GZ}_{eff}$, calculated by taking into account the free-surface effect. The wind arm is obtained from the formula

$$l_V = \frac{0.017V_w^2 A\ell \cos^2 \phi}{1000\Delta} \tag{8.8}$$

where V_w is the wind velocity in knots, A the sail area in m^2, ℓ the distance between half-draught and the centroid of the sail area in m, and Δ the displacement in t. The first angle of static equilibrium is ϕ_{st1}. The criterion for static stability requires that the righting arm at this angle be not larger than 0.6 of the maximum righting arm. To check dynamical stability the regulations assume that the ship is subjected to a gust of wind while heeled 25° to the windward of ϕ_{st1}. We distinguish then the area a between the wind heeling arm and the

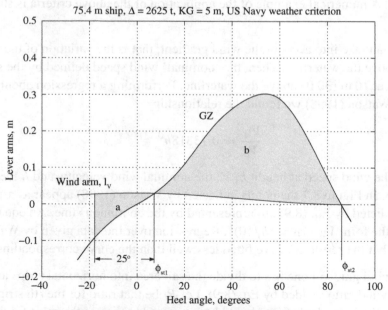

Figure 8.5 The US Navy weather criterion

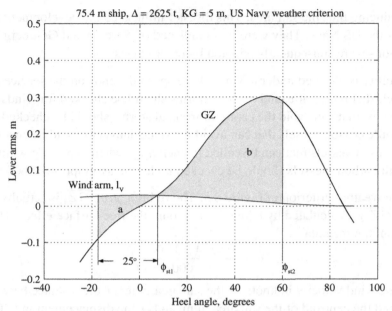

Figure 8.6   The US Navy weather criterion, downflooding angle 60°

righting-arm curves up to ϕ_{st1}, and the area b between the two curves, from the first static angle, ϕ_{st1}, up to the second static angle, ϕ_{st2} (see Figure 8.5), or up to the angle of downflooding, whichever is less (see Figure 8.6). The ratio of the area b to the area a should be at least 1.4. A numerical example of the application of the above criteria is shown in Example 8.4.

The designer can take into account the wind gradient, that is the variation of the wind speed with height above the waterline. Then, the 'nominal' wind speed defined by the service area is that measured at 10 m (30 ft) above the waterline. Performing a regression about new data presented by Watson (1998) we found the relationship

$$\frac{V_W}{V_0} = 0.73318 h^{0.13149} \tag{8.9}$$

where V_W is the wind speed at height h, V_0 the nominal wind velocity, and h the height above sea level, in m. In Figure 8.7 the points indicated by Watson (1998) appear as asterisks, while the values predicted by Eq. (8.9) are represented by the continuous line. An equation found in literature has the form $V_W/V_0 = (h/10)^b$. Regression over the data given by Watson yielded $b = 0.73318$, but the resulting curve fitted less well than the curve corresponding to Eq. (8.9).

To apply the wind gradient one has to divide the sail area into horizontal strips and apply in each strip the wind ratio yielded by Eq. (8.9). Let R_i be that ratio for the ith strip. The results for the individual strips should be integrated by one of the rules for numerical

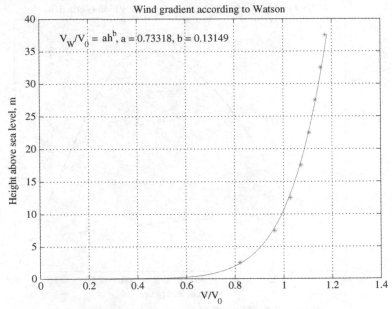

Figure 8.7 Wind gradient

integration. The coefficient in Eq. (8.8) should be modified to 0.0195 and then, the wind arm is given by

$$\ell_V = \frac{0.0195 V_0^2}{1000\Delta}\, h \left(\sum \alpha_i R_i^2 A_i \ell_i \right) \cos^2 \phi \tag{8.10}$$

where V_0 is the nominal wind speed, h is the common height of the horizontal strips, α_i is the trapezoidal multiplier, A_i is the area of the ith strip, and ℓ_i the vertical distance from half-draught to the centroid of the ith strip. It can be easily shown that

$$\ell_i = \frac{2i - 1}{2}h + \frac{T}{2} \tag{8.11}$$

To explain the criterion for stability in turning we use Figure 8.8. The heeling arm due to the centrifugal force is calculated from

$$\ell_{TC} = \frac{V^2(\overline{KG} - T/2)}{g\,R} \cos \phi \tag{8.12}$$

where V is the ship speed in m s^{-1} and R is the turning radius in metres. Ideally, R should be taken as one half of the **tactical diameter** measured from model or sea tests at full scale. Where this quantity is not known, an estimation must be made. In Section 6.4 we describe an empirical formula developed for this aim, in the following section, about the UK Navy, we give another approximate relationship. The stability is considered satisfactory if

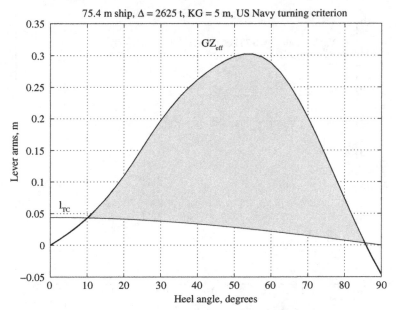

Figure 8.8   The US Navy turning criterion

1. the angle of heel does not exceed 15°;
2. the heeling arm at the angle of static equilibrium is not larger than 0.6 the maximum righting-arm value;
3. the grey area in the figure, called **reserve of dynamical stability**, is not less than 0.4 of the whole area under the positive righting-arm curve.

If the downflooding angle is smaller than the second static angle, the area representing the reserve of stability should be limited to the former value. An application of the above criteria is given in Example 8.5.

Another hazard considered in the regulations of the US Navy is the lifting of heavy weights over the side. The corresponding heeling arm is yielded by

$$l_W = \frac{wa}{\Delta} \cos \phi \tag{8.13}$$

where w is the lifted mass, a the transverse distance from the centreline to the boom end, and Δ the displacement mass including w. The criteria of stability are the same as those required for stability in turning.

The crowding of personnel to one side causes an effect similar to that of a heavy weight lifted transversely to one side. The heeling arm is yielded by Eq. (8.13), assuming that the personnel moved to one side as far as possible when five men crowd in one square metre. Again, the stability is considered sufficient if the requirements given for stability in turning are met.

8.4 The Regulations of the UK Navy

The stability standard of the Royal Navy evolved from the criteria published by Sarchin and Goldberg in 1962. The first British publication appeared in 1980 as NES 109. The currently valid version is Issue 4 (see MoD, 1999a). The document should be read in conjunction with the publication SSP 42 (MoD, 1999b). The British standard is issued by the Ministry of Defence, shortly MoD, and is applicable to vessels with a military role, to vessels designed to MoD standards but without a military role, and to auxiliary vessels. Vessels with a military role are exposed to enemy action or to similar dangers during peacetime exercises. We shall discuss here only the provisions related to such vessels. The standard NES 109 has two parts, the first dealing with conventional ships, the second with unconventional vessels. The second category includes:

1. monohull vessels of rigid construction having a speed in knots larger than $4\sqrt{L_{WL}}$, where the waterline length is measured in m;
2. multihull vessels;
3. dynamically supported vessels.

In this book we briefly discuss only the provisions for conventional vessels.

According to NES 109 the displacement and $\overline{KG}$ values used in stability calculations should include growth margins. For warships the weight growth margin should be 0.65% of the lightship displacement, for each year of service. The $\overline{KG}$ margin should be 0.45% of the lightship $\overline{KG}$, for each year of service.

The shape of the righting-arm curve should be such that:

*   the area under the curve, up to 30°, is not less than 0.08 m rad;
*   the area up to 40° is not less than 0.133 m rad;
*   the area between 30° and 40° is not less than 0.048 m rad;
*   the maximum $\overline{GZ}$ is not less than 0.3 m and should occur at an angle not smaller than 30°.

One can immediately see that all these requirements are considerably more severe than those prescribed by IMO 95 for merchant ships.

The stability under beam winds should be checked for the following wind speeds:

*   90 knots for ocean-going vessels;
*   70 knots for ocean-going or coastal vessels that can avoid extreme conditions;
*   50 knots for coastal vessels that can be called to anchorage to avoid winds over Force 8, and for harbour vessels.

These values are lower than those required by the US Navy and partially coincide with those specified by the German Navy. The angle of heel caused by the wind should not exceed 30°. The criterion for statical stability is the same as that of the US Navy, that is, the righting arm at the first static angle should not be greater than 0.6 the maximum righting arm. As in the American regulations, it is assumed that the ship rolls 25° windwards from the first static angle, and it is required that the reserve of stability should not be less than 1.4 times the area representing the wind heeling energy. Figure 1.3 in the UK regulations shows that the area representing the reserve of stability is limited at the right by the downflooding angle.

When checking stability in turning the corresponding ship speed should be 0.65 times the speed on a straight-line course. If no better data are available, it should be assumed that the radius of turning equals 2.5 times the length between perpendiculars. The angle of heel in turning should be less than 20°, a requirement less severe than that of the US Navy. The static criterion, regarding the value of the righting arm at the first static angle, and the dynamic criterion, regarding the reserve of stability, are the same as those of the US Navy.

To check stability when lifting a heavy mass over the side, the heeling arm should be calculated from

$$l_W = \frac{w(a\cos\phi + d\sin\phi)}{\Delta} \tag{8.14}$$

where a is the horizontal distance of the tip of the boom from the centreline, and d is the height of the point of suspension above the deck. Stability is considered sufficient if the following criteria are met:

1. the angle of heel is less than 15°;
2. the righting arm at the first static angle is less than half the maximum righting arm;
3. the reserve of stability is larger than half the total area under the righting-arm curve. The area representing the reserve of stability is limited at the right by the angle of downflooding.

It can be easily seen that criteria 2 and 3 are more stringent than those of the US Navy.

The NES 109 standard also specifies criteria for checking stability under icing. A thickness of 150 mm should be assumed for all horizontal decks, with an ice density equal to 950 kg m^{-3}. Only the effect on displacement and $\overline{KG}$ should be considered, and not the effect on the sail area.

8.5 A Criterion for Sail Vessels

The revival of the interest for large sailing vessels and several accidents justified new researches and the development of codes of stability for this category of ships. Thus, the UK Department of Transport sponsored a research carried on at the Wolfson Unit for Marine Transportation and Industrial Aerodynamics of the University of Southampton

(Deakin, 1991). The result of the research is the code of stability described in this section. A more recent research is presented by Cleary et al. (1966). The authors compare the stability criteria for sailing ships adopted by the US Coast Guard, the Wolfson Unit, the Germanischer Lloyd, the Bureau Veritas, the Ateliers and Chantiers du Havre, and Dr. Ing. Alimento of the University of Genoa. These criteria are illustrated by applying them to one ship, the US Coast Guard training barque *Eagle*, formerly *Horst Wessel*, built in 1936 in Germany.

In this section we describe the intact stability criteria of 'The code of practice for safety of large commercial sailing and motor vessels' issued by the UK Maritime and Coastguard Agency (Maritime, 1997). The code 'applies to vessels in commercial use for sport or pleasure $\cdots$ that are 24 m in load line length and over $\cdots$ and that do not carry cargo and do not carry more than 12 passengers.' For shorter sailing vessels, the UK Marine Safety Agency published another code, namely 'The safety of small commercial sailing vessels.'

The research carried on at the Wolfson Unit yielded a number of interesting results:

1. Form coefficients of sail rigs vary considerably and are difficult to predict. We mean here the coefficient c in

$$p = \frac{1}{2} c \rho v^2$$

   where p is the pressure, ρ the air density, and V the speed of the wind component perpendicular to the sail.
2. The wind-arm curve behaves like $\cos^{1.3} \phi$.
3. Wind gusts do not build up instantly, as conservatively assumed (see Section 6.6). The wind speed of gusts due to atmospheric turbulence are unlikely to exceed 1.4 times the hourly mean, have rise times of 10–20 s and durations of less than a minute. Other gusts, due to other atmospheric phenomena, are known as **squalls** and they can be much more dangerous. Because the rise-up times of significant gusts are usually larger than the natural roll periods of sailing vessels, ships do not respond as described in Section 6.6, but have time to find equilibrium positions close to the intersection of the gust-arm curve and the righting-arm curve.
4. Sails considerably increase the damping of the roll motion, limiting the response to a wind gust and enhancing the effect described above. Thus, the heel angle caused by a wind gust is smaller than that predicted by the balance of areas representing wind energy and righting-arm work (Section 6.6).

Based on the above conclusions, the criterion of intact stability adopted by the UK Maritime and Coastguard Agency does not consider the sail rig and the wind moment developed on it. The code simply provides the skipper with a means for appreciating the maximum allowable heel angle under a steady wind, if wind gusts are expected. Sailing at the recommended angle will avoid the submergence under gusts of openings that could lead to ship loss.

The code defines the downflooding angle as the angle at which openings having an 'aggregate area' whose value in metres is greater than $\Delta/1500$, submerge. The displacement, Δ, is measured in t. Deakin (1991) explains that under his assumptions the mass of water flowing through the above openings during 5 min equals the ship displacement. No ship is expected to float after a flooding of this extent, and 5 min are considered a maximum reasonable time of survival. For those who wish to understand Deakin's reasoning we remind that the flow through an orifice is proportional to the orifice area multiplied by the fluid speed

$$Q = a \cdot c_V \cdot \sqrt{2gh}$$

where a is the orifice area, c_V a discharge coefficient always smaller than 1, g the acceleration of gravity, and h the level of water above the orifice. The authors of the code assume $c_V = 1$ and $h = 1$ m. We calculate

$$Q = \frac{\Delta}{1500} \times 1 \times \sqrt{2 \times 9 \times 1} = 0.003\Delta \text{ m}^3 \text{ s}^{-1}$$

It follows that in sea water 5.5 min are required for a mass of water equal to the displacement mass.

We use Figure 8.9 to describe the criterion for intact stability. The righting-arm curve is marked GZ; it is based on the data of an actual training yacht. At the downflooding angle we measure the value of the righting arm, $\overline{GZ}_f$. We assume here the downflooding angle $\phi_f = 60°$. We calculate a gust-wind lever in upright condition

$$WLO = \frac{\overline{GZ}_f}{\cos^{1.3} \phi_f}$$

The dashed line curve represents the gust arm. Under the assumption that the gust speed is 1.4 times the speed of the steady wind, the pressure due to steady wind is one half that of the gust, and so is the corresponding heeling arm. Therefore, we draw the 'derived curve' as the dash-dot line beginning at $WLO/2$ and proportional to $\cos^{1.3} \phi$. This curve intercepts the GZ curve at the angle of **steady heel**, here a bit larger than 40°.

The code requires that:

1. the $\overline{GZ}$ curve should have a positive range not shorter than 90°;
2. if the downflooding angle is larger than 60°, ϕ_f should be taken as 60°;
3. the angle of steady heel should not be less than 15°.

Figure 8.9   Intact stability criterion for sail ships

8.6 A Code of Practice for Small Workboats and Pilot Boats

The regulations presented in this section (see Maritime, 1998) apply to small UK commercial sea vessels of up to 24 m load line length and that carry cargo and/or not more than 12 passengers. The regulations also apply to service or pilot vessels of the same size. By 'load line length' the code means either 96% of the total waterline length on a waterline at 85% depth, or the length from the fore side of the stern to the axis of the rudder stock on the above waterline.

The lightship displacement to be used in calculations should include a margin for growth equal to 5% of the lightship displacement. The x-coordinate of the centre of gravity of this margin shall equal LCG, and the z-coordinate shall equal either the height of the centre of the weather deck amidships or the lightship $\overline{KG}$, whichever is the higher. Curves of statical stability shall be calculated for the following loading cases:

- loaded departure, 100% consumables;
- loaded arrival, 10% consumables;
- other anticipated service conditions, including possible lifting appliances.

The stability is considered sufficient if the following two criteria are met in addition to the $\overline{GZ}$−curve criteria Section 8.2.1.

1. The maximum of the righting-arm curve should occur at an angle of heel not smaller than 25°.

2. The effective, initial metacentric height, $\overline{GM}_{eff}$, should not be less than 0.35 m.

If a multihull vessel does not meet the above stability criteria, the vessel shall meet the following alternative criteria.

1. If the maximum of the righting-arm curve occurs at 15°, the area under the curve shall not be less than 0.085 m rad. If the maximum occurs at 30°, the area shall not be less than 0.055 m rad.
2. If the maximum of the righting-arm curve occurs at an angle ϕ_{GZmax} situated between 15° and 30°, the area under the curve shall not be less than

$$A = 0.055 + 0.002(30^0 - \phi_{GZmax})\tag{8.15}$$

where A is measured in m rad.
3. The area under the righting-arm curve between 30° and 40°, or between 30° and the angle of downflooding, if this angle is less than 40°, shall not be less than 0.03 m rad.
4. The righting arm shall not be less than 0.2 m at 30°.
5. The maximum righting arm shall occur at an angle not smaller than 15°.
6. The initial metacentric height shall not be less than 0.35 m.

The intact stability of new vessels of less than 15 m length that carry a combined load of passengers and cargo of less than 1000 kg is checked in an inclining experiment. The passengers, the crew without the skipper, and the cargo are transferred to one side of the ship, while the skipper may be assumed to stay at the steering position. Under these conditions the angle of heel shall not exceed 7°. For vessels with a watertight weather deck the freeboard shall be not less than 75 mm at any point. For open boats the freeboard to the top of the gunwale shall not be less than 250 mm at any point.

8.7 Understanding the Limits of Rules and Regulations

Today, the maritime industry and shipping are not specially risk occupations, except shipbreaking ashore and the fishing industry in which deaths occur mostly in small artisanal fishing boats, in developing countries (ILO, 2003; FAO, 2005). However, there are more than 200 ship casualties per year (IMO, 2012) with special incidence on cargo ships and big fishing vessels (17 in 2011) and all those ships complied with the IMO regulations. So, as we can see, despite the fact that ships comply with international regulations, maritime accidents happen. Can we say that the ships were safe? Can we say that the international rules are safe regardless of those losses? The IS Code is solid and robust, the best and finest instrument on intact stability up to date, but the safety of a ship against capsizing depends mainly on four factors (Kobyliński and Kastner, 2003; Kobyliński, 2007) that involve: environmental conditions, ship loading, ship handling, and the human factor. These factors can cause a sequence of events that leads to the loss of the ship. Therefore, the safety of ships depends not only on their design, but also on their operation and maintenance.

As we saw in Chapter 1, the foundations of the science of stability were laid down in the 18th century by Bouguer, Euler and others, but just for a few exceptions (see Ferreiro, 2010), they little influenced the way ships were designed during their time. Those great men, however, developed certain simple rules on how to design stable ships and attention was focused on ensuring adequate metacentric height and righting-arm curve properties (Rahola, 1939). Since then, every single development in the field of ship stability has been design oriented, or, in other words, meant to be applied during the design stage. All present-day instruments regarding intact stability are prescriptive and deterministic; they are based on physical models of stability failure considering phenomena in a deterministic manner. These are check-list-like criteria: the ship must comply with this and that value of some stability parameter. As pointed out by Kobyliński (2007), this way of operating has many advantages: written in simple language, the prescriptions can be easily understood and applied. However, such prescriptions have also growing disadvantages: they are difficult to apply to new and innovative designs (e.g., HSC, cruise liners, x-bows) that, in many cases, render the regulations obsolete or impossible to apply, and restrain the designer from introducing new and novel solutions. What is the answer to that? As a first step, we can put in place performance-based criteria. As explained by Hoppe (2005), a 'Goal-based regulation' does not specify the means of achieving compliance but sets goals that allow alternative ways of achieving compliance.' The author gives an example of prescriptive regulation, 'You shall install a 1 m high rail at the edge of a cliff,' compared with the goal-based rule 'People shall be prevented from falling over the edge of a cliff.'

According to Kobyliński (2007), the opposite of prescriptive regulations, however, is the **risk-based** approach in which there is no need to comply with a certain value, but, talking in probabilistic terms, one assesses the risk involved and decides whether it may be accepted or not. The IMO documentation describes the risk-based approach as a *goal-oriented* approach. We say about a design that it is risk-based if it is supported by risk assessment or if the design basis resulted from a risk assessment. This procedure should ensure safety performance and cost effectiveness. The concept was introduced at IMO during the 89th session of the Maritime Safety Committee in 2011. Details on the risk approach can be found, for example, in IMO (1997) and Kobyliński (2008).

8.8 Future IMO Developments

The 2008 IS Code is a living working document and it is intended to be amended or supplemented in the future, but always taking into account the discussion between prescriptive versus goal-based approach. The intention is to develop performance-based criteria not only deterministic. The IMO is already laying the way ahead with the foundation of "second-generation," intact stability criteria (Belenky et al., 2011; Francescutto and Umeda, 2010). In 2007 the IMO agreed upon a plan of action for the development of new criteria that would rely upon a more realistic approach and whose aim will be to provide methods for the direct assessment of the stability of certain ships that could not be properly addressed by the existing criteria. The second-generation criteria should be developed in five stages called in a new terminology *tiers*. Hoppe (2005) defines them as follows:

Tier I Goals.
Tier II Functional requirements.
Tier III Verification of compliance criteria.
Tier IV Technical procedures and guidelines, classification rules and industry standards.
Tier V Codes of practice and safety and quality systems for shipbuilding, ship operation, maintenance. training and manning.

The first tier sets the goals that would ensure the safe and environmentally-friendly operation of the ship during a given life span. The second tier will establish requirements that will enable the achievements of the goals. The third tier will provide the instruments for proving that the detailed requirements developed within tier IV will comply with the provisions that result in tiers I and II. In tier IV IMO, national administrations and *classification societies* will issue detailed requirements that met the goals of tier I and the functional requirements of tier II. The fifth tier refers to industry standards, codes of practice, and quality systems. A flow diagram of these tiers is presented in IMO (2004), the proposal made by The Bahamas, Greece, and IACS (International Association of Classification Societies). Briefly we can say that *classification societies* are national, but non-governmental organizations that issue rules for the design and construction of ships and certify the compliance with these rules. The subject belongs to books on ship design and construction.

8.9 Summary

The IMO Code on Intact Stability applies to ships and other marine vehicles of 24 m length and above. The most important change in comparison with the original version adopted in 1993 is that Part A, the basic criteria, is made compulsory via reference in the SOLAS convention. One new feature is that some ships may comply with the weather criterion through an alternative method if the model tests for them were carried out. The metacentric height of passenger and cargo ships should be at least 0.15 m, and the areas under the righting-arm curve, between certain heel angles, should not be less than the values indicated in the document. Passenger vessels should not heel in turning more than 10°. In addition, passenger and cargo ships should meet a weather criterion in which it is assumed that the vessel is subjected to a wind arm that is constant throughout the heeling range. The heeling arm of wind gusts is assumed equal to 1.5 times the heeling arm of the steady wind. If a wind gust appears while the ship is heeled windwards by an angle prescribed by the code, the area representing the reserve of buoyancy should not be less than the area representing the heel energy. The former area is limited to the right by the angle of downflooding or by 50°, whichever is less.

The IMO code contains special requirements for ships carrying timber on deck, for fishing vessels, for mobile offshore drilling units, for dynamically supported craft, and for containerships larger than 100 m. The code also contains recommendations for inclining and for rolling tests.

The existing codes are prescriptive and based on deterministic theoretical models. In the last years, however, IMO and many researchers work for the development of goal-based and risk-based rules that take into consideration also probabilistic models. Such rules could suit better new types of ships and will facilitate the introduction of innovative solutions.

The stability regulations of the US Navy prescribe criteria for statical and dynamical stability under wind, in turning, under passenger crowding on one side, and when lifting heavy weights over the side. The static criterion requires that the righting arm at the first static angle should not exceed 60 % of the maximum righting arm. When checking dynamical stability under wind, it is assumed that the ship rolled 25° windwards from the first static angle. Then, the area representing the reserve of stability should be at least 1.4 times the area representing the heeling energy. When checking stability in turning, or under crowding or when lifting heavy weights, the angle of heel should not exceed 15° and the reserve of stability should not be less than 40 % of the total area under the righting-arm curve.

The stability regulations of the UK Navy are derived from those of the US Navy. In addition to static and dynamic criteria such as those mentioned above, the UK standard includes requirements concerning the areas under the righting-arm curve. The minimum values are higher than those prescribed by IMO for merchant ships. While the wind speeds specified by the UK standard are lower than those in the US regulations, the stability criteria are more severe.

A quite different criterion is prescribed in the code for large sailing vessels issued by the UK Ministry of Transport. As research proved that wind-pressure coefficients of sail rigs cannot be predicted, the code does not take into account the sail configuration, and the heeling moments developed on it. The document presents a simple method for finding a heel angle under steady wind, such that the heel angle caused by a gust of wind would be smaller than the angle leading to downflooding and ship loss. The steady heel angle should not exceed 15°, and the range of positive heeling arms should not be less than 90°.

An additional regulation mentioned in this chapter is a code for small workboats issued in the UK.

8.10 Examples

Example 8.1 (Application of the IMO general requirements for cargo and passenger ships).   Let us check if the small cargo ship used in Section 7.2.2 meets the IMO general requirements. We assume the same loading condition as in that section. The vessel was built five decades before the publication of the 2008 version of the IMO code for intact stability; therefore, it is not surprising if several criteria are not met. Table 8.1 contains the calculation of righting-arm levers and areas under the righting-arm curve. Figure 8.1 shows the corresponding statical stability curve. The areas under the righting-arm curve are obtained by means of the algorithm described in Section 3.4. The analysis of the results leads to the following conclusions.

1.   The area under the $\overline{GZ}_{eff}$ curve, up to 30°, is 0.043 m rad, less than the required 0.055. The area up to 40° equals 0.084 m rad, less than the required 0.09 m rad. The area between 30° and 40° equals 0.04 m rad, more than the required 0.03 m rad.
2.   The righting-arm lever equals 0.2 m at 30°; it meets the requirement at limit.
3.   The maximum righting arm occurs at an angle exceeding the required 30°.
4.   The initial effective metacentric height is 0.12 m, less than the required 0.15 m.

Example 8.2 (Application of the IMO weather criterion for cargo and passenger ships). We continue the preceding example and illustrate the application of the weather criterion to

Table 8.1   Small cargo ship—the IMO general requirements

Heel Angle (°)	ℓ_p (m)	$(\overline{KG} + \ell_F) \sin \phi$ (m)	$\overline{GZ}_{eff}$ (m)	Area Under Righting Arm (m^2)
0.0	0.000	0.000	0.000	0.000
5.0	0.459	0.439	0.019	0.001
10.0	0.918	0.875	0.043	0.004
15.0	1.377	1.304	0.072	0.009
20.0	1.833	1.724	0.109	0.017
25.0	2.283	2.130	0.153	0.028
30.0	2.717	2.520	0.197	0.043
35.0	3.124	2.891	0.233	0.062
40.0	3.501	3.240	0.262	0.084
45.0	3.847	3.564	0.283	0.107
50.0	4.159	3.861	0.298	0.133
55.0	4.431	4.129	0.302	0.159
60.0	4.653	4.365	0.288	0.185
65.0	4.821	4.568	0.253	0.208
70.0	4.937	4.736	0.201	0.228
75.0	5.007	4.868	0.139	0.243
80.0	5.036	4.963	0.073	0.252
85.0	5.030	5.021	0.009	0.256
90.0	4.994	5.040	-0.046	0.254

the same ship, in the same loading condition. The main dimensions are $L = 75.4$, $B = 11.9$, $T_m = 4.32$, and the height of the centre of gravity is $\overline{KG} = 5$, all measured in metres. The sail area is $A = 175$ m^2, the height of its centroid above half-draught $Z = 4.19$ m, and the wind pressure $P = 504$ N m^{-2}. The calculations presented here are performed in MATLAB keeping the full precision of the software, but we display the results rounded off to the first two or three digits. To keep the precision we define at the beginning the constants, for example $L = 75.4$, and then call them by name, for example L.

The wind heeling arm is calculated as

$$l_{w1} = \frac{PAZ}{1000g\Delta} = 0.014 \text{ m}$$

The lever of the wind gust is

$$l_{w2} = 1.5 l_{w1} = 0.022 \text{ m}$$

We assume that the bilge keels are 15 m long and 0.4 m deep; their total area is

$$A_k = 2 \times 15 \times 0.4 = 12 \text{ m}^2$$

To enter Table 2.3.4-3 of the code we calculate

$$\frac{A_k \times 100}{L \times B} = 1.337$$

Interpolating over the table we obtain $k = 0.963$. To find X_1 we calculate $B/T_m = 2.755$ and interpolating over Table 2.3.4-1 we obtain $X_1 = 0.94$. To enter Table 2.3.4-2 we calculate the block coefficient

$$C_B = 2635/(1.03 \times L \times B \times Tm) = 0.66$$

Interpolation yields $X_2 = 0.975$. The height of centre of gravity above waterline is

$$\overline{OG} = KG - T_m = 0.68$$

In continuation we calculate

$$r = 0.73 + 0.6 \times \overline{OG}/T_m = 0.824$$

To find the roll period we first calculate the coefficient

$$C = 0.373 + 0.023 \times (B/T_m) - 0.043 \times (L/100) = 0.404$$

With $\overline{GM}_{eff} = 0.12$ m the formula prescribed by the code yields the roll period

$$T = \frac{2CB}{\sqrt{GM_{eff}}} = 27.752 \text{ s}$$

With this value we enter Table 2.3.4-4 and retrieve $s = 0.035$. Then, the angle of roll windwards from the angle of statical stability, under the wind arm l_{w1}, is

$$\phi_1 = 109k\,X_1X_2\sqrt{rs} = 16.34°$$

Visual inspection of Figure 8.1 shows that the weather criterion is met. This fact is explained by the low sail area of the ship.

Example 8.3 (The IMO turning criterion). To illustrate the IMO criterion for stability in turning we use the data of the same small cargo ship that appeared above. Cargo ships are not required to meet this criterion, but we can assume, for our purposes, that the ship carries more than 12 passengers.

The ship length is $L = 75.4$ m, the mean draught $T_m = 4.32$ m, the ship speed $V_0 = 16$ knots, and the vertical centre of gravity $\overline{KG} = 5.0$ m. The speed in m s^{-1} is

$$V_0 = 16 \times 0.5144 = 8.23 \text{ m s}^{-1}$$

and the heel arm due to the centrifugal force is

$$l_T = 0.2 \cdot \frac{V_0^2}{L} \cdot \frac{(\overline{KG} - T_m/2)}{g} = 0.051 \text{ m}$$

Figure 8.2 shows the resulting statical stability curve. We see that the heel angle is slightly larger than $11°$.

Example 8.4 (The weather criterion of the US Navy). To allow comparisons between various codes of stability we use again the data of the small cargo ship that appeared in the previous examples. We initiate the calculations by defining the wind speed, $V_W = 80$ knots, the sail area $A = 175$ m^2, the height of its centroid above half-draught, $\ell = 4.19$ m, and the displacement, $\Delta = 2625$ t. The corresponding stability curve is shown in Figure 8.5. The wind heeling arm is given by

$$l_V = \frac{0.017V_w^2A\ell\cos^2\phi}{1000\Delta} = 0.03\cos^2\phi, \text{ m}$$

At the intersection of the righting arm and the wind-arm curves we find the first static angle, $\phi_{st1} \approx 7.5°$, and the righting arm at that angle equals 0.03 m. Rolling $25°$ windwards from the first static angle the ship reaches $-17.5°$. The second static angle is $\phi_{st2} = 85.7°$. The ratio of the $\overline{GZ}$ value at the first static angle to the maximum $\overline{GZ}$ is 0.03/0.302, that is close to 0.1 and smaller than the maximum admissible 0.6. The area b equals 0.235 m rad, and the area a equals 0.024 m rad. The ratio of the area b to the area a is nearly 10, much larger than the minimum admissible 1.4. We conclude that the vessel meets the criteria of the US Navy.

Example 8.5 (The turning criterion of the US Navy). We continue the calculations using the data of the same ship as above. We assume the speed of 16 knots, and the vertical centre of gravity, $\overline{KG} = 5$ m, as in Example 8.3. In the absence of other recommendations we consider, as in NES 109, that the speed in turning is 0.65 times the speed on a straight-line course. that is

$$V_0 = 0.65 \times 16 \times 0.5144 = 5.35 \text{ m s}^{-1}$$

Also, we assume that the radius of the turning circle equals 2.5 times the waterline length

$$R = 2.5 \times 75/4 = 188.5 \text{ m}$$

Then, the heeling arm in turning is given by

$$l_{TC} = \frac{V_0^2(\overline{KG} - T/2)}{gR} \cos\phi = 0.044 \cos\phi \text{ m}$$

Drawing the curves as in Figure 8.8 we find that the first static angle is $\phi_{st1} = 10.3°$, and the corresponding righting arm equals 0.044 m. The ratio of this arm to the maximum righting arm is $0.044/0.302 = 0.15$, less than the maximum admissible 0.6. The reserve of dynamical stability, that is the grey area in Figure 8.5, equals 0.205 m rad, while the total area under the positive righting-arm curve is 0.256 m rad. The ratio of the two areas equals 0.8, the double of the minimum admissible 0.4. We conclude that the ship meets the criteria of the US Navy.

8.11 Exercises

Exercise 8.1 (IMO general requirements). Let us refer to Example 8.1. Find the $\overline{KG}$ value for which the general requirement 4 is fulfilled. Check if with this value the first general requirement is also met.

Exercise 8.2 (The IMO turning criterion). Return to Example 8.3 and find the limit speed for which the turning criterion is met.

Exercise 8.3 (The IMO turning criterion). Return to the example in Exercise 8.1 and check if with the vertical centre of gravity, $\overline{KG}$, found in Section 7.2.2 the turning criterion is met.

Exercise 8.4 (The US-Navy turning criterion). Return to Example 8.4 and redo the calculations assuming a wind speed of 100 knots.

Exercise 8.5 (The code for small vessels). Check that for $\phi_{GZmax} = 15°$ and $30°$ Eq. (8.15) yields the values specified in criterion 1 for small vessels (Section 8.6).

CHAPTER 9

Stability in Waves

Chapter Outline

9.1 Introduction

Up to this chapter we assumed that the sea surface is plane. Actually, such a situation never occurs in nature, not even in the sheltered waters of a harbour. Waves always exist, even if very small. Can waves influence ship stability? And if yes, how? Arndt and Roden (1958) and Wendel (1965) cite French engineers that discussed this question at the end of the 19th century (J. Pollard and A. Dudebout, 1892, *Théorie du Navire*, Vol. III, Paris). In the 1920s Doyère explained how waves influence stability and proposed a method to calculate that influence. After 1950 the study of this subject was prompted by the sinking of a few ships that, according to what was known at that time, were considered stable.

At a first glance *beam seas*—that is waves whose crests are parallel to the ship—seem to be the most dangerous. In fact, waves parallel to the ship (beam waves) cause large angles of heel; loads can get loose and endanger stability. However, it can be shown that the resultant of the weight force and of the centrifugal force developed in waves is perpendicular to the wave surface. Therefore, a correctly-loaded vessel will never capsize in parallel waves, unless hit by large breaking waves or subject to some resonance phenomenon. Ships **can capsize** in *head seas*—that is waves travelling against the ship—and especially in *following seas*—that is waves travelling in the same direction with the ship. This is the lesson learnt after the sinking of the ship *Irene Oldendorff* in the night between 30 and 31 December 1951. Kurt Wendel analysed the case and reached the conclusion that the disaster was due to the variation of the righting arm in waves. Divers that checked the wreck found it intact, an observation that confirmed Wendel's hypothesis. Another disaster was that of *Pamir*. Again, the calculation of the righting arm in waves surprised the researchers (Arndt, 1962).

Kerwin (1955) analysed a simple model of the variation of $\overline{GM}$ in waves and its influence on ship stability. His investigations included experiments carried on at Delft and he reports difficulties that we attribute to the equipment available at that time.

To confirm the results of their calculations, researchers from Hamburg carried on model tests in a towing tank (Arndt and Roden, 1958) and with self-propelled models on a nearby lake (Wendel, 1965). *Post-mortem* analysis of other marine disasters showed that the righting arm was severely reduced when the ship was on the wave crest. Sometimes it was even negative.

Pauling (1961) discussed 'The transverse stability of a ship in a longitudinal seaway.'

Storch (1978) analysed the sinking of 13 king-crab boats. In one case he discovered that the righting arm on wave crest must have been negative, and in two others, greatly reduced.

Lindemann and Skomedal (1983) report a ship disaster they attribute to the reduction of the stability in waves. On 1 October 1980 the RO/RO (*roll-on/roll-off*) ship *Finneagle* was close to the Orkney Islands and sailing in *following seas*, that is with waves travelling in the same direction as the ship. All of a sudden three large roll cycles caused the ship to heel up to 40°. It is assumed that this large angle caused a container to break loose. Trimethylphosphate leaked from the container and reacted with the acid of a car battery. Because of the resulting fire the ship had to be abandoned.

Chantrel (1984) studied the large-amplitude motions of an offshore supply buoy and attributed them to the variation of properties in waves leading to the phenomenon of **parametric resonance** explained in this chapter. Interesting experimental and theoretical studies into the phenomenon of parametric resonance of trimaran models were performed at the University College of London, within the framework of Master's courses supervised by D.C. Fellows (Zucker, 2000).

Investigations of parametric resonance in the roll motion of fishing vessels are described by Santos Neves et al. (2002), Juana de et al. (2005), and Pérez and Sanguinetti (2006).

More recent incidents have shown that large roll angles due to parametric resonance can develop not only in the case of small ships, but also in that of large container ships and cause loss of cargo. Due to their forms, container, RoRo, and RoPax ships are susceptible to develop what is called *parametric roll*. This results in large angles of roll that endanger the containers carried on deck. France et al. (2001) analyze an accident that occurred in the north Pacific to a container ship of 262 m length between perpendiculars. A severe storm caused angles of roll of 35° to 40°. Out of nearly 1300 containers, one third was lost and another third suffered damage. Pérez Rojas and Belenky (2005) review papers on this subject presented at the *STAB 2003* conference in 2003. Russel (2011) analyzes an accident that occurred in 2006 to a ship having a length between perpendiculars equal to 197.10 m. Angles of 25–30° caused the loss of 27 containers and the collapse of another 28.

The frequency of incidents increased the interest for studies in parametric roll to such an extent that possibly all conferences related to ship theory held in the last decade include papers dealing with parametric roll. The subject was found attractive for academic diploma and graduate theses (see, for example, de Juana, 2004; Erkin, 2006; Kleiman and Gottlieb, 2011) and very interesting also for mathematicians (see, for example, Archer et al., 2009; Sheikh, 2008; Holden, 2011). The resulting number of papers, reports, and theses is so large that it is practically possible to cite only a few of them.

The onset of parametric resonance is quick, there are no signs predicting its approach. The phenomenon can be amplified by coupling with the pitch and other ship motions. IMO and several classification societies initiated studies of *parametric roll* and developed guidelines for avoiding it. Studies are carried on with the aim of developing *second-generation intact stability criteria* that cover modes of capsizing identified in the last years (Belenky et al., 2011).

The influence of waves on ship stability can be modelled by a linear differential equation with periodic coefficients known as the **Mathieu equation**. Under certain conditions, known as **parametric resonance**, the response of a system governed by a Mathieu equation can be unstable, that is, grow beyond any limits. For a ship, unstable response means capsizing. This is a new mode of ship capsizing; the first we learnt are due to **insufficient metacentric height** and to **insufficient area under the righting-arm curve**. This chapter contains a practical introduction to the subjects of parametric excitation and resonance known also as **Mathieu effect**.

9.2 The Influence of Waves on Ship Stability

In this section we explain why the metacentric height varies when a wave travels along the ship. We illustrate the discussion with data calculated for a 29 m fast patrol boat (further denoted as FPB) whose offsets are described by Talib and Poddar (1980). For hulls like the one chosen here the influence of waves is particularly visible. Figure 9.1 shows an outline of the boat and the location of three stations numbered 36, 9, and 18. This is the original numbering in the cited reference. The shapes of those sections are shown in Figure 9.2. We calculated the hydrostatic data of the vessel for the draught 2.5 m, by means of the same ARCHIMEDES programme that Talib and Podder used. The waterline corresponding to the above draught appears as a solid line in Figures 9.1 and 9.2. Let us see what happens in waves.

Figure 9.1   Wave profiles on a fast patrol boat outline—S = still water, T = wave trough, C = wave crest

Figure 9.2   Wave profiles on FPB transverse sections—S = still water, T = wave trough, C = wave crest

Calculations and experiments show that the maximum influence of longitudinal waves on ship stability occurs when the wavelength is approximately equal to that of the ship waterline. Consequently, we choose a wavelength

$$\lambda = L_{pp} = 27.3 \text{ m}$$

The wave height prescribed by the German Navy is

$$H = \frac{\lambda}{10 + 0.05\lambda} = \frac{27.3}{10 + 0.05 \times 27.3} = 2.402 \text{ m}$$

The dot-dot lines in Figures 9.1 and 9.2 represent the waterline corresponding to the situation in which the **wave crest** is in the midship section plane. We say that the ship is on **wave crest**. In Figure 9.2 we see that in the midship section the waterline lies above the still-water line. The breadth of the waterline almost does not change in that section. In sections 36 and 18 the waterline descends below the still-water position. In section 18 the breadth decreases. This effect occurs in a large part of the forebody. In the calculation of the metacentric radius, $\overline{BM}$, breadths enter at the third power (at constant displacement!). Therefore, the overall result is a decrease of the metacentric radius.

The dash-dash lines in Figures 9.1 and 9.2 represent the situation in which the position of the wave relative to the ship changed by half a wavelength. The **trough** of the wave reached now the midship section and we say that the ship is in a **wave trough**. In Figure 9.2 we see that the breadth of the waterline increased significantly in the plane of station 18, decreased insignificantly in the midship section, and increased slightly in the plane of station 36. The overall effect is an increase of the metacentric radius.

Figure 9.3   The influence of waves on *KM*

A quantitative illustration of the effect of waves on stability appears in Figure 9.3. For some time the common belief was that the minimum metacentric radius occurs when the ship is on a wave crest. It appeared, however, that for forms like those of the FPB the minimum occurs when the wave crest is approximately $0.3L_{pp}$ astern of the midship section. Calculations carried by us for various ship forms showed that the relationships can change. Figure 9.3 shows, indeed, that for draughts under 1.6 m $\overline{KM}$ is larger on wave crest than in wave trough. Similar conclusions can be reached for the righting-arm curves in waves. For example, the righting arm in wave trough can be the largest in a certain heeling-angle range, and ceases to be so outside that range. The reader is invited to use the data in Exercise 9.1 and check the effect of waves on the righting arm of another vessel, named *Ship No. 83074* by Poulsen (1980).

More explanations of the effect of waves on righting arms can be found in Wendel (1958), Arndt (1962), and Abicht (1971). Detailed stability calculations in waves, for a training ship, are described by Arndt et al. (1960), and results for a cargo vessel with $C_B = 0.63$, are presented by Arndt (1964). A few results of calculations and model tests for RoRo ships can be found in Sjöholm and Kjellberg (1985).

To develop a simple model of the influence of waves we assume that the wave is a periodic function of time with period T. Then, also $\overline{GM}$ is a periodic function with period T. We write

$$\overline{GM(t)} = \overline{GM_0} + \delta\overline{GM(t)}$$

where

$$\delta\overline{GM(t)} = \delta\overline{GM(t+T)}$$

for any t. In Section 6.7 we developed a simple model of the free rolling motion. To include the variation of the metacentric height in waves we can rewrite the roll equation as

$$\ddot{\phi} + \frac{g}{i^2}(\overline{GM_0} + \delta\overline{GM})\phi = 0$$

Going one step further we assume that the wave is harmonic (*regular wave*) and the $\overline{GM}$ variation also harmonic, so that the free rolling motion can be modelled by

$$\ddot{\phi} + \frac{g}{i^2}(\overline{GM_0} + \delta\overline{GM}\cos\omega_e t)\phi = 0 \tag{9.1}$$

This is a **Mathieu equation**; those of its properties that interest us are described in Section 9.4.

9.3 The Influence of New Ship Forms

The amplitude of the $\overline{GM}$ variations in waves depends on the ship forms. The forms of certain, relatively recent types of ships greatly amplifies these changes. Krüger et al. (2004) identify in this category container, RoRo- and RoPax ships. *RoRo* is an abbreviation for *Roll-on/roll-off* and it designs ships carrying vehicles that are driving in and out the ship on their own wheels. *RoPax* is an abbreviation for ferries that combine RoRo facilities with accommodations for passengers. The trend in the design of ships belonging to the above three types is to achieve hydrodynamically efficient forms for the submerged hull and large volumes and/or deck areas above the water. The resulting forms greatly enhance the variation of the waterplane moment of inertias during the wave passage along the ship. To explain this effect, Krüger et al. (2004) compare the lines of a RoPax vessel with those of a conventional cargo ship. We are going to explain these facts in Figures 9.4 and 9.5. Thus, the first of these figures schematically shows the stern of a typical container ship. To provide more area for containers, the ship ends in the aft with a bread transom stern. Also, the load waterline in still water is

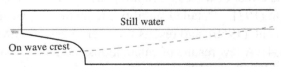

Figure 9.4 New longitudinal RoPax lines compared with conventional cargo lines, 1

Figure 9.5 New transversal RoPax lines compared with conventional cargo lines, 2

only a little above the end of the stern. Therefore, the moments of inertia of the waterplane in still water is very large. On the other hand, when the ship finds herself on wave crest, a large part of the stern emerges completely and the moments drop considerably.

Let us look now to Figure 9.5; it shows typical lines in the forebodies of a modern container ship and of a traditional cargo ship. To ensure a fine shape under the water and a large deck area, the lines of modern container ships tend to have a large flare in the forebody. It can be easily seen that this also leads to large fluctuations of the moments of inertia when the waterline height varies as the wave passes along the ship. No wonder then why ships having these forms experience difficulties of a kind and magnitude not known before. We have here a good example of how the desire for increased commercial efficiency conflicts with safety, hence the need for restrictive regulations.

We mentioned that the lines of RoRo ships also lead to large fluctuations of the righting moment. A corresponding analysis can be found in Hua and Wang (2001). Hass (2002) compares the performances of several ship forms.

9.4 The Mathieu Effect—Parametric Resonance

9.4.1 The Mathieu Equation—Stability

A general form of a differential equation with periodic coefficients is *Hill's equation*:

$$\ddot{x} + h(t)x = 0$$

where $h(t) = h(t + T)$. In the particular case in which the periodic function is a cosine we have the *Mathieu equation*; it is frequently written as

$$\ddot{\phi} + (\delta + \epsilon \cos 2t)\phi = 0 \tag{9.2}$$

This equation was studied by Mathieu (Émile-Léonard, French, 1835–1900) in 1868 when he investigated the vibrational modes of a membrane with an elliptical boundary. Floquet (Gaston, French, 1847–1920) developed in 1883 an interesting theory of linear differential equations with periodic coefficients. Since then many other researchers approached the subject; a historical summary of their work can be found in McLachlan (1947).

A rigorous discussion of the Mathieu equation is beyond the scope of this book; for more details the reader is referred to specialized books, such as Arscott (1964), Cartmell (1990), Grimshaw (1990), or McLachlan (1947). A comprehensive bibliography on 'parametrically excited systems' and a good theoretic treatment are given by Nayfeh and Mook (1995). For our purposes it is sufficient to explain the conditions under which the equation has stable solutions. By 'stable' we understand that the response, ϕ, is *bounded*. Correspondingly, 'unstable' means that the response grows beyond any boundaries. For a ship whose rolling motion is governed by the Mathieu equation, unstable response simply means that the ship capsizes. The reader may be familiar with the condition of stability of an ordinary, linear differential equation with constant coefficients: A system is stable if all the poles of the transfer function have negative real parts (Dorf and Bishop, 2011). This is not the condition of stability of the Mathieu equation; the behaviour of its solutions depends on the parameters ϵ and δ. This behaviour can be explained with the aid of Figure 9.6. In this figure, sometimes known as *Strutt diagram*, but attributed by McLachlan (1947) to Ince, the horizontal axis represents the parameter δ, and the vertical axis, the parameter ϵ. The δ, ϵ plane is divided into two kinds of regions. For δ, ϵ combinations that fall in the grey areas, the solutions of the Mathieu equation are stable. The δ, ϵ points in white regions and on the boundary curves correspond to unstable solutions. The diagram is symmetric about the δ axis; for our purposes it is sufficient to show only half of it.

The theory reveals the following properties of the Strutt-Ince diagram.

- The lines separating stable from unstable regions intercept the δ axis in points for which

$$\delta = \frac{n^2}{4}, \quad n = 0, 1, 2, 3, \ldots$$

- As δ grows larger, so do the stable regions.
- As ϵ grows, the stable regions become smaller. Remember, ϵ is the "disturbance."

Cesari (1971) considers the equation

$$\ddot{x} + (\sigma^2 + \epsilon \cos \omega t)x = 0 \tag{9.3}$$

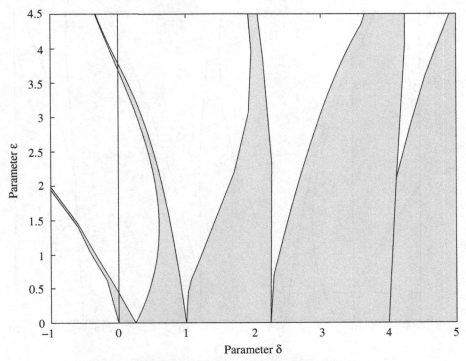

Figure 9.6   The Strutt-Ince diagram, the δ-ϵ plane

The natural frequency of the 'undisturbed' equation – that is for $\epsilon = 0$ - is $\sigma/2\pi$, while the frequency of the periodic disturbance is $\omega/2\pi$. With the transformation

$$\omega t = 2t_1 \tag{9.4}$$

we calculate

$$\dot{x} = \frac{dx}{dt_1}\frac{dt_1}{dt} = \frac{\omega}{2}\frac{dx}{dt_1}$$

$$\ddot{x} = \frac{d\dot{x}}{dt_1}\frac{dt_1}{dt} = \frac{\omega^2}{4}\frac{d^2x}{dt_1^2} \tag{9.5}$$

Substituting Eqs. (9.4) and (9.5) into Eq. (9.3) yields an equation in the standard form

$$\ddot{x} + (\delta_1 + \epsilon_1 \cos 2t_1)x = 0$$

where

$$\delta_1 = \frac{4\sigma^2}{\omega^2}, \quad \epsilon_1 = \frac{4\epsilon}{\omega^2}$$

The general aspect of the δ_1-ϵ_1 plane is shown in Figure 9.7. Visual inspection shows us that for small ϵ values the danger of falling into an unstable region is greater in the

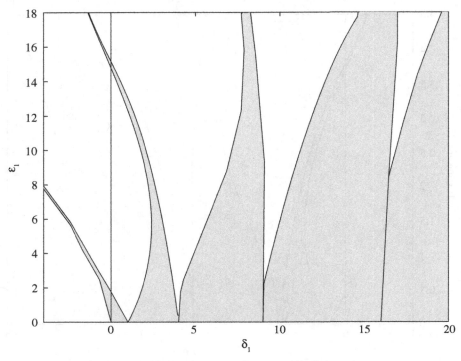

Figure 9.7 The Strutt-Ince diagram, the δ_1-ϵ_1 plane

neighbourhoods of $\delta_1 = 1^2, 2^2, 3^2, \ldots$ This means that for small ϵ **parametric resonance** occurs at circular frequencies $\omega = 2\sigma/n^2$, where $n = 1, 2, 3, \ldots$ The first dangerous situation is met when $\omega = 2\sigma$. We reach the important conclusion that *the danger of parametric resonance is greatest when the frequency of the perturbation equals twice the natural frequency of the undisturbed system.* This statement is rephrased in terms of ship-stability parameters in Example 9.1 where σ becomes the natural roll circular frequency, ω_0, of the ship, and ω becomes ω_E, the frequency of encounter, that is the frequency with which the ship encounters the waves. This theoretical conclusion was confirmed by basin tests.

Surprising as it may seem, the phenomenon of parametric excitation is well known. The main character in Molière's *Le bourgeois gentilhomme* has been writing prose for many years without being aware of it. Similarly, readers are certainly familiar with parametric excitation since their childhood. Here are, indeed, three well-known examples.

The motion of a pendulum is stable. However, if the point from which the pendulum hangs is moved up and down periodically, with a suitable amplitude and frequency, the pendulum can be caused to overturn.

Figure 9.8   Two familiar uses of parametric excitation

Try to "invert" a pendulum so that its mass is concentrated above the centre of oscillation. The pendulum will fall. Still, at circus we see clowns that keep a long rod clasped in their hand, as shown in Figure 9.8a. The rod can be *stabilized* by moving the hand up and down with a suitable amplitude and frequency.

A third, familiar example of parametric excitation is that of a swing. To increase the amplitude of motion the person on the swing kneels close to the extreme positions and stands up in the middle position (Figure 9.8b). Thus, the distance between the hanging point and the centre of gravity of the person varies periodically. The swing behaves like a *pendulum with varying length*.

More examples of parametrically excited systems can be found in Den Hartog (1956). That author also studies a case in which the periodic function is a *rectangular ripple* whose analytic treatment is relatively simple and allows the derivation of an explicit condition of stability.

9.4.2 The Mathieu Equation—Simulations

In this section we show how to simulate the behaviour of the Mathieu equation and give four examples that illustrate the conclusions reached in the preceding subsection. To solve numerically the Mathieu equation we define

$$\phi_1 = \phi, \quad \phi_2 = \dot{\phi}_1$$

and replace Eq. (9.2) by the first-order system

$$\dot{\phi}_1 = \phi_2$$
$$\dot{\phi}_2 = -(\delta + \epsilon \cos 2t)\phi_1 \tag{9.6}$$

The following MATLAB function, written on a file `mathieu.m`, calculates the derivatives in Eq. (9.6):

```
%MATHIEU    Derivatives of Mathieu equation.
        function    dphi = mathieu(t1, phi, d1, e1)
dphi = [phi(2); -(d1 + e1*cos(2*t1))*phi(1)];
```

We write a second MATLAB function, `mathisim.m`, that calls the function `mathieu`:

```
function ms = mathisim(omega_0, epsilon, omega_e, tf)

%MATHISIM Simulates the Mathieu equation

clf                              % clean window

d1 = 4*omega_0^2/omega_e^2;
e1 = 4*epsilon/omega_e^2;
w0 = [ 0.1; 0.0 ];               % initial conditions;
ts = [ 0; tf ];                  % time span
hmathieu = @mathieu;
[ t1, phi ] = ode45(hmathieu, ts, w0, [], d1, e1);

t = 2*t1/omega_e;
subplot(2, 2, 1), plot(t, phi (:, 1)), grid
ns = num2str(omega_0);
nd = num2str(d1);
ne = num2str(e1);
no = num2str(omega_e);
title('Time domain'), ylabel('\phi')
subplot(2, 2, 3), plot (t, phi(:, 2)), grid
xlabel('t'), ylabel('\phi"')
subplot(2, 2, 2), plot (:, 1), phi(:, 2));   % phase plot
grid
title('Phase plane'), xlabel('\phi'), ylabel('\phi"')
hold on
plot(phi(1, 1), phi(1, 2), 'r*', 'LineWidth', 1.5)
subplot(2, 2, 4), axis off
text(0.1, 0.66, ['\omega_0 = ' ns ', \delta_1 = ' nd ])
text(0.1, 0.33, ['\epsilon_1 = ' ne ', \omega_e = ' no ])
text(0.1, 0.00, ' * - starting point')
hold off
```

Figures 9.9–9.12 show results of simulations carried on by means of the function `mathisim`. Figure 9.9 corresponds to the parameters

$$\sigma = 4, \quad \epsilon_1 = 0, \quad \omega = \pi/4$$

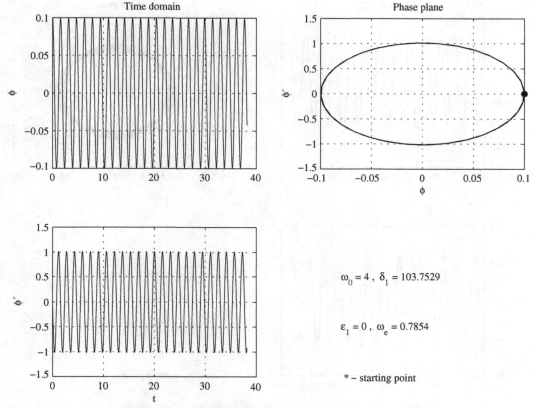

Figure 9.9 Simulation of Mathieu equation; sinusoidal response

In this case we deal with the well-known equation

$$\ddot{\phi} + \delta\phi = 0$$

whose solution is a sinusoid with circular frequency $\sqrt{\delta}$:

$$\phi = C_1 \sin\left(\sqrt{\delta}t + C_2\right)$$

The constants C_1, C_2 can be found from the initial conditions of the problem. The first derivative, $\dot{\phi}$, shown in the second subplot, is also a sinusoid:

$$\dot{\phi} = C_1\sqrt{\delta}\cos\left(\sqrt{\delta}t + C_2\right)$$

The third subplot is the *phase plane* of the motion. The curve is an ellipse. Indeed, simple calculations show us that

$$\frac{\phi^2}{C_1^2} + \frac{\dot{\phi}^2}{\left(\sqrt{\delta}C_2\right)^2} = 1$$

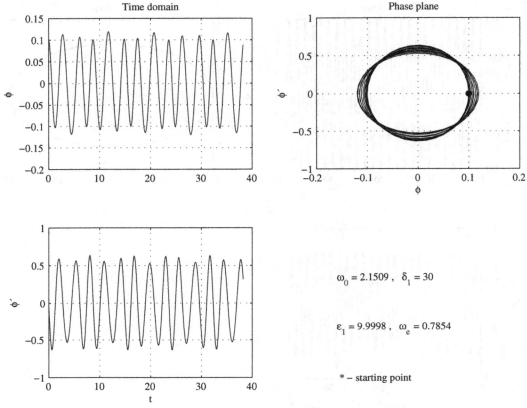

Figure 9.10   Simulation of Mathieu equation; stable response

The run parameters that generate Figure 9.10 are

$$\sigma = 2.1509, \quad \epsilon 1 = 10, \quad \omega = \pi/4$$

These values define in Figures 9.6 and 9.7 a point in a stable region. As the simulation shows, the solution is bounded, periodic, but not sinusoidal.

The run parameters that generate Figure 9.11 are

$$\sigma = \pi/4, \quad \epsilon_1 = 103.7529, \quad \omega = \pi/4$$

These values define in Figures 9.6 and 9.7 a point in an unstable region. As the simulation shows, the solution is unbounded. This can be best seen in the phase plane where the start of the curve is marked by the word 'start.'

The run parameters for Figure 9.12 are

$$\sigma = 2, \quad \epsilon_1 = 0.05, \quad \omega = 4$$

Figure 9.11 Simulation of Mathieu equation; unstable response

These values define in the Strutt diagram a point in an unstable region, very close to a boundary curve. As the simulation shows, the solution is periodic and steadily growing. This can be best seen in the phase plane where the start of the curve is marked by an 'asterisk'. The case shown in this figure corresponds to the most dangerous condition of parametric resonance, $\omega = 2\sigma$.

9.4.3 Frequency of Encounter

When judging ship stability, the frequency to be used in the Mathieu equation is the number of waves 'seen' by the ship in one time unit. This is the **frequency of encounter**, ω_E; to calculate it we use Figure 9.13. Let v be the ship speed, c, the **wave celerity**, that is the speed of the wave, λ, the wavelength, ω_w, the wave circular frequency, and α, the angle between ship speed and wave celerity. By convention, $\alpha = 180°$ in *head seas* and $0°$ in *following seas*. The relative speed between ship and wave is

$$c - v \cos \alpha$$

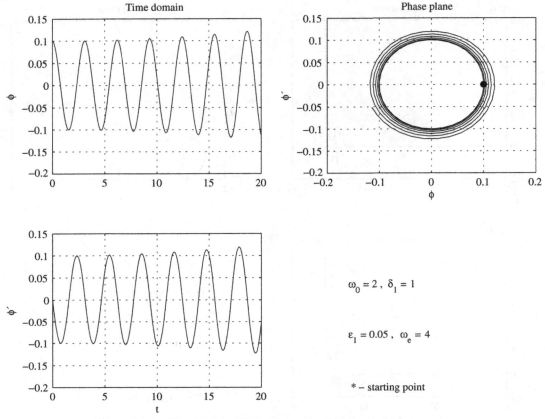

Figure 9.12 Simulation of Mathieu equation; unstable response

Figure 9.13 Calculating the frequency of encounter

The ship encounters wave crests (or wave troughs) at time intervals equal to

$$T_E = \frac{\lambda}{c - v \cos \alpha}$$

This is the **period of encounter**. By definition, the wave circular frequency is

$$\omega_w = \frac{2\pi}{T_w}$$

where T_w is the *wave period*. Similarly, the *circular frequency of encounter* is defined by

$$\omega_E = \frac{2\pi}{T_E}$$

In wave theory (see Section 12.2) it is shown that the relationship between wave length and wave circular frequency, in water of infinite depth, is

$$\lambda = \frac{2\pi g}{\omega_w^2}$$

Putting all together we obtain

$$\omega_E = \omega_w - \frac{\omega_w^2}{g} v \cos \alpha \tag{9.7}$$

In calculations of ship behaviour in waves the frequency of encounter plays the role of exciting, or driving frequency. The research has shown that the effects of parametric resonance can be amplified if the frequency of the pitch motion (see Chapter 12) equals the frequency of encounter, or, in other words, it is the double of the roll frequency (Umeda and Peters, 2002).

9.5 Pure Loss of Stability

In Section 9.2 we have learned that the righting arm on wave crest can be severely reduced and even become negative. If the speed of the ship relative to that of waves is such that the reduction of stability persists over a certain time, the ship can capsize. This phenomenon is called **pure loss of stability**; it can be enhanced by other effects, including ship motions. Belenky and Bassler (2010), Bulian (2010), and Belenky et al. (2011) have investigated methods for predicting the *vulnerability* to this mode of capsizing. The approach is mainly probabilistic.

9.6 The Activities of IMO and of Professional Societies

The IMO Subcommittee on Stability and Load Lines and on Fishing Vessels is dealing also with the phenomenon of parametric roll. The various proposals submitted there can be read on Internet looking for *SLF*. See, for example, the SLF48/4/4 proposal (Anonymous, 2005). IMO (2007) is a *Revised guide to the master for avoiding dangerous situations in adverse weather and sea conditions.*The American classification society ABS issued a guide (see ABS, 2008) whose justification can be found in more detail in Shin et al. (2004) SNAME, The Society of Naval Architects and Marine Engineers, established an Ad Hoc Panel #13 whose task is to 'identify sea and vessel characteristics that initiate the coupling of pitch and extreme rolling' and to propose an amendment to IMO's Maritime Safety Circ. 707, 'Guidance to the master for avoiding dangerous situations in following and quartering seas.' Details can be seen on SNAME's site http://www.sname.org. A simple criterion for the onset of parametric roll is proposed in ITTC (2005). As it refers to terms explained in more detail in Chapter 12, we defer its presentation until there.

9.7 Summary

Longitudinal and quartering waves influence the stability of ships and other floating bodies. The moment of inertia of the waterline surface in waves differs from that of the waterplane in still water and, consequently, so do the metacentric height and the righting arms. The way in which those quantities vary depends on the ship form; however, it can be said that in many cases the righting moment in wave trough is larger than in still water, while on wave crest it is smaller. If the wave is periodic, also the variation of the righting arm is periodic. Then, into the equation of rolling developed in Chapter 6 we must add to the coefficient of the roll angle a term that is a periodic function of time:

$$\ddot{\phi} + \frac{g}{i^2}(\overline{GM_0} + \delta\overline{GM}\cos\omega_e t)\phi = 0$$

For small heel angles the above equation can be reduced to the canonical form of the Mathieu equation

$$\ddot{\phi} + (\delta + \epsilon\cos 2t)\phi = 0$$

The condition of stability is not the same as for a linear differential equation with constant coefficients. In other words, the condition of positive metacentric height, $\overline{GM} > 0$, is no longer sufficient. The theory of differential equations with periodic coefficients shows that the plane of the parameters δ and ϵ can be subdivided into regions so that if in one of them the solution of the Mathieu equation is stable, in the adjacent regions it is not. This means that for certain pairs $[\delta, \epsilon]$ the solution is unstable and we say that parametric resonance occurs. The partition of the $\delta - \epsilon$ plane into stable and unstable regions can be best visualized in the Strutt-Ince diagram. Thus, it can be easily discovered that even for small ϵ values a particularly dangerous situation arises when the frequency of the periodic coefficient is twice the natural frequency of the system without periodic excitation.

Parametric excitation occurs in several systems we are familiar with. Thus, the amplitude of oscillation of a swing can be increased by periodically changing the position of the centre of gravity of the person on the swing. As another example, a conventional pendulum is usually stable, but it can be forced to overturn if the point of hanging is moved up and down with appropriate frequency and amplitude. Conversely, an inverted pendulum, although inherently unstable, can be stabilized by applying a suitable periodic motion to its centre of oscillation.

Ships have capsized although they fulfilled the criteria of stability commonly accepted at the time of the disaster. *Post-mortem* analysis of some cases pinpointed the Mathieu effect as the cause of capsizing. The surprising discovery was that the righting arm could be negative on wave crest. If this effect persists for a certain time, it may lead to a mode of failure called *pure loss of stability*.

The analysis of the Mathieu effect confirms a fact well known to experienced seafarers: following seas are more dangerous than head seas. In fact, when the direction of the waves is the same as that of the ship, the relative velocity is small and the time interval in which the stability is reduced is longer. Then, there is more time to develop large heeling angles. Still worse, in following seas the effect of reduced stability can be enhanced by waves flowing over the deck. The latter effect will increase the height of the centre of gravity because it means an extra weight loaded high up on the ship. It also adds a free-surface effect.

We considered in this chapter only the roll motion. In reality it is coupled to other motions. The full analysis is based on notions introduced in Chapter 12 and it is beyond the scope of this book.

9.8 Examples

Example 9.1 (Parametric resonance in ship stability). In this example we are going to explain the significance of the parameters δ and ϵ for ship stability. In Chapter 6 we developed the equation of free roll

$$\ddot{\phi} + \frac{g\overline{GM}}{i^2} = 0 \tag{9.8}$$

The natural, circular roll frequency is

$$\omega_0 = \frac{\left[g\overline{GM}\right]^{1/2}}{i} \tag{9.9}$$

Let us assume that the wave produces a periodic variation of the metacentric height equal to

$$\delta\overline{GM}\cos\omega_E$$

where ω_E is the circular frequency of encounter. With this assumption and with the notation introduced by Eq. (9.9) we rewrite Eq. (9.8) as

$$\ddot{\phi} + \left(\omega_0^2 + \frac{g\delta\overline{GM}}{i^2}\cos\omega_E t\right)\phi = 0 \tag{9.10}$$

Following Cesari (1971) we use the substitution $\omega_E t = 2t_2$ and proceeding like in Section 9.4.1 we transform Eq. (9.10) to

$$\frac{\omega_E^2}{4} \cdot \frac{d^2\phi}{dt_1^2} + \left(\omega_o^2 + \frac{g\delta\overline{GM}}{i^2} \cos 2t_1 \right) \phi = 0 \qquad (9.11)$$

Substituting Eq. (9.9) we obtain

$$\frac{d^2\phi}{dt_1^2} + \left\{ 4\left(\frac{\omega_o}{\omega_E}\right)^2 + 4\frac{g\delta\overline{GM}}{\overline{GM}}\left(\frac{\omega_o}{\omega_E}\right)^2 \cos 2t_1 \right\} \phi = 0 \qquad (9.12)$$

Equation (9.12) can be brought to the standard Mathieu form with

$$\delta1 = 4\left(\frac{\omega_o}{\omega_E}\right)^2, \quad \epsilon_1 = 4\frac{g\delta\overline{GM}}{\overline{GM}}\left(\frac{\omega_o}{\omega_E}\right)^2 \qquad (9.13)$$

We know that the most dangerous situation occurs at $\delta_1 = 1$, that is for $\omega_E = 2\omega_0$. This result is used by Ünsalan (2006) in the analysis of a marine disaster.

Example 9.2 (Sail ship in longitudinal waves).   The righting-arm curve in still water shown in Figure 9.14 was calculated for an actual training yacht. We assume that the righting-arm

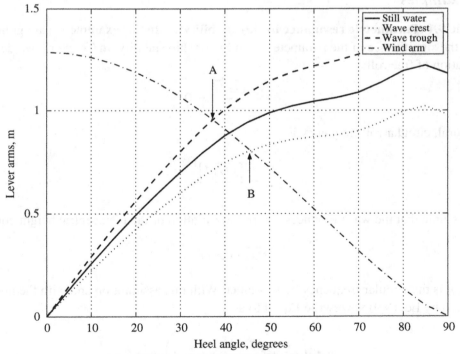

Figure 9.14 Sail ship in longitudinal waves

curves on wave crest and in wave trough, and the wind heeling arm are as shown in the figure. It is obvious that while advancing in waves the yacht will roll between points A and B. Thus, the Mathieu effect induces roll in head or following seas, a behaviour that is not predicted by the conventional roll equation. Readers involved in yachting may have experienced the phenomenon.

9.9 Exercises

Exercise 9.1 (*Ship 83074*, levers of stability in seaway).

Table 9.1 shows the cross-curves of stability of the *Ship No. 83074* for a displacement volume equal to 20 000 m^3. Plot in the same graph the curves for still water, in wave trough and on wave crest.

Table 9.1   Levers of stability of *Ship 83074*, 20 000 m^3

Heel Angle (°)	Wave Trough (m)	Still Water (m)	Wave Crest (m)
0	0.000	0.000	0.000
10	2.617	2.312	2.309
20	4.985	4.606	4.635
30	6.912	6.759	6.892
45	9.095	9.361	9.235
60	9.734	10.447	10.073
75	10.783	10.425	9.917

CHAPTER 10

Intact Stability Regulations II

Chapter Outline

10.1 Introduction

In this section we describe a few highlights of the stability regulations of the German Navy, as an example of philosophy different from those illustrated in Chapter 8. Some engineering data specified in these regulations can be useful in engineering practice, certainly in university exercises. Our text is based on the BV 1030-1 version issued by the end of the year 2001. From personal communications we are aware that these regulations are now under revision. As we suggested for other regulations, for checks of stability that must be submitted for approval it is highly recommended to inquire about the latest, complete edition of *BV 1030-1* and consult it for updatings and specific details.

10.2 The Regulations of the German Navy

Kurt Wendel wrote in 1961 the first draft of stability regulations for the German Federal Navy. Wendel issued in 1964 a new edition known as *BV 103*. An early detailed explanation of the regulations and their background is due to Arndt (1965). His paper was soon translated into English by the British Ship Research Association and appeared as *BSRA Translation No. 5052.* An updated version of the regulations was published in 1969 and since then they were known during many years as *BV 1033.* As pointed out by Brandl (1981), the German regulations were adopted by the Dutch Royal Navy (see, for example, Harpen, 1971) and they also served in the design of some ships built in Germany for several foreign navies.

In the preceding chapter we mentioned experiments performed by German researchers before the publication of the regulations. The authors continued to experiment after the implementation of *BV 1033* and thus confirmed the validity of the requirements and showed that the German regulations and the regulations of the US Navy confer to a large extent equivalent safety against capsizing. For details we refer the reader to Brandl (1981) and Arndt et al. (1982).

10.2.1 Categories of Service

Stability requirements vary according to the intended use of the ships. The regulations of the German Federal Navy classify vessels into four categories, as explained below.

Group A.   There are no limitations to the area of operation of ships belonging to this category. Calculations for Group A should be carried on for a wind speed equal to 90 knots.

Group B.   This category includes ships that can avoid winds whose speed exceeds 70 knots. The wind speed to be considered for this group is 70 knots.

Group C.   The category consists of coastal vessels that can reach a harbour if a storm warning is received. Stability calculations shall be based on a wind speed of 50 knots.

Group D.   It consists of ships that do not experience wind speeds above 40 knots and significant wave heights above 1.5 m. The wind speed to be considered is 40 knots.

Further categories are E and F and regard ships, boats, and floating equipment that operate under more stringent restrictions.

10.2.2 Loading Conditions

The BV 1030-1 regulations require the verification of stability in a number of loading conditions. We shall exemplify here only three of them. The detailed description of the loading cases involves the term *empty ship, operationally ready for operation.* By this the regulations mean the ship with fuel, feed water, and lubricating oil in machines, piping, weapons, and other systems, if necessary also with fixed ballast.

Loading case 0—Empty ship

The weight groups to be included are the empty ship ready for operation, crew, and personal effects.

Loading case 1—Limit displacement

The items to be included are the empty ship ready for operation, crew, and personal effects, 50% consumables, 33% provisions, 10% fresh water, or 50% if a fresh water generator with capacity of minimum 30 l per head and day is on board, 10% fuel, 10% aviation fuel, 50% lubricating oil, 100% foam, 33% ammunitions, where launching tubes and weapons are charged, and the rest of the ammunition is in the corresponding storing places, aircraft, transported loads, ballast water, if necessary for stability.

Loading case 2—Operational displacement

This case corresponds to the design displacement and includes 100% of all items.

The BV 1030-1 regulations require that certain displacement and $\overline{KG}$ margins shall be taken into account. The reasons for these margins may present interest to any Naval Architect and we reproduce them below adding our own comments.

Design margin. Uncertainties due to approximate design methods and to future changes in the catalogs of suppliers. For instance, add 3% to the estimated $\overline{KG}$-value.

Building margin. Tolerances of supplied materials, changes during the detail design.

Maintenance margin. Weight increase due to corrosion (oxygen intake), painting, additional equipment.

Retrofit margin. Retrofitting of equipment and systems.

10.2.3 Waves

The previous version of the regulations specified that the stability on waves should be checked in **trochoidal waves**. This wave form was used also for other naval-architectural calculations, mainly those of *longitudinal bending*. The trochoidal wave theory is the oldest among wave theories. The new version of the regulations, BV 1030-1, assumes *sinusoidal* waves, such as predicted by the theory described in Chapter 12 of this book. The characteristics specified by the regulations are:

> wave length equal to ship length, that is,
> $$\lambda = L$$
> wave height $H = \lambda/(10 + 0.05\lambda)$.

The height shall be rounded up to one decimal place. The relationship between wave length and height is based on statistics and probabilistic considerations.

10.2.4 Righting Arms

The cross-curves of stability shall be calculated in still water and in waves and denoted as follows:

$\overline{GZ}_s$ in still water;
$\overline{GZ}_c$ for ship on crest;
$\overline{GZ}_T$ for ship in wave trough;
$\overline{GZ}_w$ for *seaway*, that is the mean of the on-crest and in-trough values.

Usually ships are not symmetric about a transverse plane (notable exceptions are Viking ships and some ferries). Therefore, during heeling the centre of buoyancy travels in the longitudinal direction causing trim changes. According to the old regulations this effect had to be considered in the calculation of cross-curves. The present BV 1030-1 regulations specify that cross-curves shall be calculated for fixed trim. From private correspondence, however, we know that this subject is under discussion and the trend is to specify in the future that calculations shall be performed for various initial trim values that will cover the expected domain of operation, while allowing for free trimming, or, in other words, with trim compensation. The data in Table 9.1 and in Example 10.2 are calculated in this way.

When calculating the cross-curves the volumes of closed superstructures, deckhouses, and hatchways are to be taken into account.

10.2.5 Free Liquid Surfaces

The German regulations consider the influence of free liquid surfaces as a heeling arm, rather than a quantity to be deducted from metacentric height and righting arms. The first formula to be used is

$$\ell_{F1} = \frac{\sum_{j=1}^n \rho_j i_j}{\Delta} \sin \phi \tag{10.1}$$

where, as shown in Chapter 5, n is the number of tanks or other spaces containing free liquid surfaces, ρ_j, the density of the liquid in the jth tank, and i_j, the moment of inertia of the free liquid surface, in the same tank, with respect to a baricentric axis parallel to the centreline. As convened, Δ is the mass displacement.

If ℓ_{F1} calculated with Formula (10.1) exceeds 0.03 m at 30°, an exact calculation of the free-surface effect is required. The formula to be used is

$$\ell_{F2} = \frac{1}{\Delta} \sum_{j=1}^n p_j b_j \tag{10.2}$$

where p_j is the mass of the liquid in the jth tank and b_j, the actual transverse displacement of the centre of gravity of the liquid at the heel angle considered. Obviously, calculations with Formula (10.2) should be repeated for enough heel angles to allow a satisfactory plot of the ℓ_{F2} curve. The appendix of this chapter contains the table of liquid densities (Table 10.4) to be used with the BV regulations.

10.2.6 Wind Heeling Arm

The wind heeling arm is calculated from the formula

$$\ell_V = \frac{A_v(z_A - 0.5T_m)}{g\Delta} p_w(0.25 + 0.75\cos^3\phi) \tag{10.3}$$

where

A_v is the sail area in m^2

z_A the height coordinate of the sail area centroid, in m, measured from the same line as the mean draught

T_m the mean draught, in m

p_w the wind pressure, in kN/m^2

$g\Delta$ the ship displacement in kN

The wind pressure is taken from Table 10.1, which contains rounded off values.

The sail area, A_v, is the lateral projection of the ship outline above the sea surface. The BV 1030-1 regulations allow for the multiplication of area elements by aerodynamic coefficients, C_w, that take into account their shape. The area of circular elements should be multiplied by 0.6, while for plane surfaces $C_w = 1$, for free-standing, flat lattice structures $C_w = 1.7$, and for round lattice structures $C_w = 1.3$.

Arndt (1965) attributes Formula (10.3) to Kinoshita and Okada who published it in the proceedings of a symposium held at Wageningen in 1957. The above equation yields non-zero values at 90° of heel; therefore, as pointed out by Arndt, it gives realistic values in the heel range 60–90°.

Table 10.1 Wind pressures

Knots	m/s	Beaufort	Pressure, kN/m^2(kPa)
90	46	14	1.5
70	36	12	1.0
50	26	10	0.5
40	21	8	0.3
20	10	5	0.1

10.2.7 The Wind Criterion

Compliance with the wind criterion should be checked in still water with a wind of 40 knots, and in seaway, on wave crest, or in wave trough with the wind specified in Section 10.2.1. With reference to Figure 10.1 let us explain how to proceed after drawing the righting-arm curve:

1. Plot the heeling arm, ℓ_{F1} or ℓ_{F2}, due to free liquid surfaces.
2. Draw the curve of the wind arm, ℓ_V, by measuring from the ℓ_F curve upward.
3. Find the intersection of the $\ell_F + \ell_V$ curve with the curve of the righting arm, $\overline{GZ}$; it yields the angle of static equilibrium, ϕ_{ST}.
4. Look at a reference angle, ϕ_{REF}, defined by

$$\phi_{REF} = \begin{cases} 35° & \text{if } \phi_{ST} \leq 15° \\ 5° + 2 \cdot \phi_{ST} & \text{otherwise} \end{cases} \tag{10.4}$$

5. At the reference angle, ϕ_{REF}, measure the difference between the righting arm, $\overline{GZ}$, and the heeling arm, $\ell_F + \ell_V$. This difference, h_{RES}, called **residual arm**, shall not be less than the value yielded by

$$h_{RES} = \begin{cases} 0.1 & \text{if } \phi_{ST} \leq 15° \\ 0.01 \cdot \phi_{ST} - 0.05 & \text{otherwise} \end{cases} \tag{10.5}$$

Figure 10.1 Statical stability curve of the example Maestral, according to BV1030-1

The explicit display of the free liquid surface effect as a heeling arm makes it possible to compare its influence to that of the wind and take correcting measures, if necessary. For example, a too large surface effect, compared to the wind arm, can mean that it is desirable to subdivide some tanks.

The heel angle caused by winds up to 50 knots shall not exceed 15°, but for 70 knots it may be 20°, and for 90 knots 25°.

The reader may have observed that the regulations assume a wind blowing perpendicularly on the centreline plane, while the waves run longitudinally. Arndt et al. (1982) write:

> This combination is accounting for the fact that even strong winds may change their direction in short time only, whereas the waves are proceeding in the direction in which they were excited. Waves and winds from different directions can be observed especially near storm centres ...

Figure 10.1 was plotted with the help of the function described in Example 10.1. Example 10.2 details the data used in the above-mentioned figure. Both examples can provide a better insight into the techniques of the German regulations.

10.2.8 Stability in Turning

For a known radius of turning the heeling arm due to the centrifugal force shall be calculated with the formula

$$\ell_{TC} = \frac{V_D^2(\overline{KG} - T_m)}{g R_D} \cos\phi \tag{10.6}$$

and for an unknown radius with

$$\ell_{TC} = \frac{c_D \cdot V_{max}^2(\overline{KG} - 0.5T_m)}{g \cdot L_{DWL}} \cos\phi \tag{10.7}$$

where V_D is the mean speed in the turning circle, in ms^{-1}, but not less than $0.8V_{max}$, and L_{DWL}, the length of the design waterline, in m, R_D, the radius of the turning circle, and $c_D = 0.3$. The coefficient c_D can be used in the design stage, but for existing ships it should be determined from sea trials. The meaning of the coefficient c_D can be explained as follows. Usually, in the first design stages neither the speed in turning, V_{TC}, nor the radius of the turning circle, R_{TC}, are known. The speed in turning is smaller than the speed in straight-line sailing; therefore, let us write

$$V_{TC} = c_V V, \quad c_V < 1$$

The radius of the turning circle is usually a multiple of the ship length. Let us write

$$R_{TC} = c_R L_{DWL}, \quad c_R > 1$$

The factor V_{TC}^2/R_{TC} in the equation of the centrifugal force (see Section 6.4) can be written as

$$\frac{c_V^2 V^2}{c_R L_{DWL}} = c_D \frac{V^2}{L_{DWL}}$$

with $c_D = c_V^2/c_R$.

To check stability in turning proceed as follows:

1. plot the curve of the righting arm in still water, $\overline{GZ}_S$;
2. plot the curve of the free-surface effect, ℓ_{F1} or ℓ_{F2};
3. plot the curve of the sum of the free-surface arm plus that of the heeling arm of a 40-knots wind, ℓ_V;
4. plot the curve of the sum of the previous curve value plus that of the heeling arm of centrifugal force, ℓ_{TC}.

The stability in turning is considered satisfactory if the resulting heel angle does not exceed 15°.

10.2.9 Other Heeling Arms

Other heeling arms can act on the ship, for instance hanging loads or crowding of passengers on one side. The following data shall be considered in calculating the latter. The mass of a passenger shall be taken equal to 85 kg. The previous regulations also specified that the centre of gravity of a person shall be assumed as placed at 1 m above deck. Finally, a passenger density of 5 men per square metre shall be considered in general, and only 3 passengers per square metre for craft in Group E. The 1030-1 version states that 80 kg/person equipment shall be taken into account for ratings, and 110 kg/person for officers. As to provisions, the designer should assume.

Fresh provisions. 3.50 kg/(person × endurance in days + 5 days);

Beverages and canteen goods. 4.00 kg/(person × endurance in days + 5 days).

Replenishment at sea requires some connection between two vessels. A transverse pull develops; it can be translated into a heeling arm. A transverse pull also can appear during towing. The German regulations contain provisions for calculating these heeling arms.

The heel angle caused by replenishment at sea or by crowding of passengers shall not exceed 15°.

10.3 Summary

In the preceding chapter we have shown that longitudinal and quartering waves affect stability by changing the instantaneous moment of inertia that enters into the calculation of the metacentric radius. This effect is taken into account in the stability regulations of the German Navy and it has been proposed to consider it also for merchant ships (Helas, 1982). As shown in Chapter 9, German researchers were the first to investigate parametric resonance in ship stability. They also took into consideration this effect when they elaborated stability regulations for the German Navy. These regulations require that the righting arm be calculated both in still water and in waves. The latter righting arm is the mean of the cross-curves in wave trough and on wave crest.

In the German regulations, the criterion of stability under wind regard the difference between the righting arm and the wind heeling arm. This difference, $h_{RES} = \overline{GZ} - k_w$, is called *residual arm*. If the angle of static equilibrium is ϕ_{ST}, stability shall be checked at a *reference angle*, ϕ_{REF}, defined by

$$\phi_{REF} = \begin{cases} 35° & \text{if } \phi_{ST} \leq 15° \\ 5° + 2 \cdot \phi_{ST} & \text{otherwise} \end{cases}$$

At this reference angle, the residual arm shall be not smaller than the value given by

$$h_{RES} = \begin{cases} 0.1 & \text{if } \phi_{ST} \leq 15° \\ 0.01 \cdot \phi_{ST} - 0.05 & \text{otherwise} \end{cases}$$

Finally, let us return to the influence of ship forms. Traditionally ship forms have been chosen as a compromise between contradictory requirements of reduced hydrodynamic resistance, good seakeeping qualities, convenient space arrangements, and stability in still water. The study of the Mathieu effect has added another criterion: small variation of righting arms in waves. A formulation of this subject can be found in Burcher (1979). Pérez and Sanguinetti (1994) experimented with models of two small fishing vessels of similar size but different forms. They show that the model with round stern and round bilge displayed less metacentric height variation in wave than the model with transom stern. The influence of new ship forms, such as used in container, RoRo, and RoPax ships was discussed in the preceding chapter.

10.4 Examples

Example 10.1 (Computer function for BV 1030-1). In this example we describe a function, written in MATLAB that automatically checks the wind criterion of *BV 1033*. The function was written initially for that version of the regulations and contains the lever arms notations usual at that time, i.e. k instead of ℓ. Here we have changed only the statements that print on the curve of statical stability (see Figure 10.1).

The input consists of four arguments: `cond, w, sail, V`. The argument `cond` is an array whose elements are:

1. the displacement weight, Δ, in kN;
2. the height of the centre of gravity above BL, $\overline{KG}$, in m;

3. the mean draft, T, in m;
4. the height of the metacentre above BL, $\overline{KM}$, in m;
5. the free-surface arm in upright condition, $\ell_F(0)$, in m.

The argument w is a two dimensional array whose first column contains heel angles, in degrees, and the second column, w, the lever arms ℓ_k in metres. For instance, the following lines are taken from Example 10.2:

```
Maestral = [
0       0
5       0.582
...     ...
90      5.493 ];
```

The argument sail is an array with two elements: the sail area, in m^2, and the height of the sail area centroid above BL, in m. Finally, the argument V is the prescribed wind speed, in knots. Only wind speeds specified by *BV 1030-1* are valid arguments.

After calling the function with the desired arguments, the user is prompted to enter the name of the ship under examination. This name will be printed within the title of the stability diagram and in the heading of an output file containing the results of the calculation. In continuation a first plot of the statical-stability curve is presented, together with a cross-hair. The user has to bring the cross-hair on the intersection of the righting arm and heeling arm curves. Then, the diagram is presented again, this time with the angle of equilibrium and the angle of reference marked on it. The output file, bv1030.out, is a report of the calculations; among others it contains a comparison of the actual residual arm with the required one.

```
function       [ phiST, hRES ] = bv1030(cond, w, sail, V)
%BV1033 Stability calculations acc. to BV 1030-1.

    clc                        % clean window
    Delta  = cond(1);          % displacement, kN
    KG     = cond(2);          % CG above BL, m
    T      = cond(3);          % mean draft, m
    KM     = cond(4)           % metacentre above BL, m
    kf0    = cond(5);          % free-surface arm, m
    heel   = w(:, 1)*pi/180;   % heel angle, deg
    lever  = w(:, 2);          % arm of form stability, m
    A      = sail(1);          % sail area, sq m
    z      = sail(2);          % its centroid above BL, m
```

```
GZ        =  lever - KG*sin(heel);   % righting arm
%            choose wind pressure acc. to wind speed
switch    V
          case 90
                  p = 1.5;
          case 70
                  p = 1.0;
          case 50
                  p = 0.5;
          case 40
                  p = 0.3;
          case 20
                  p = 0.1;
          otherwise
                  error('Incorrect wind speed')
end

kf        =  kf0*sin(heel);      % free surface arm, m
%            calculate wind arm in upright condition
kw0       =  A*(z - 0.5*T)*p/Delta;
%            calculate wind arm at given heel angles
kw        =  kw0*(0.25 + 0.75*cos(heel).^3);
%%%%%%%%%%%%%%%% Initialize output file %%%%%%%%%%%%%%%%
sname     =  input('Enter ship name ', 's')
fid       =  fopen('BV1033.out', 'w');
fprintf(fid, 'Stability of ship %s acc. to BV 1033\n', sname);
fprintf(fid, 'Displacement ................ %9.3f kN\n', Delta);
fprintf(fid, 'KG .......................... %9.3f m\n', KG);
GM        =  KM - KG;    % metacentric height, m
fprintf(fid, 'Metacentric height, GM ...... %9.3f m\n', GM);
fprintf(fid, 'Mean draft, T ............... %9.3f m\n', T);
fprintf(fid, 'Free-surface arm ............ %9.3f m\n', kf0);
fprintf(fid, 'Sail area ................... %9.3f sq m\n', A);
fprintf(fid, 'Sail area centroid above BL .. %9.3f m\n', z);
fprintf(fid, 'Wind pressure ............... %9.3f MPa\n', p);
phi       =  w(:, 1);                      % heel angle, deg
fprintf(fid, '   Heel   Righting   Heeling  \n');
fprintf(fid, '  angle     arm        arm    \n');
fprintf(fid, '   deg       m          m     \n');
harm      =  kf + kw;                % heeling arm, m
report    =  [ phi'; GZ'; harm' ];       % matrix to be printed
```

```
fprintf(fid, '%6.1f %11.3f %11.3f \n', report);
plot(phi, GZ, phi, kf, phi, harm, [ 0 180/pi ], [ 0 GM ])
hold on
t1      = [sname ', \Delta = ' num2str(Delta) ' kN, KG = ', ];
t1      = [ t1 num2str(KG) 1 ' m, T = ' num2str(T) ' m' ];
title(t1)
xlabel('Heel angle, degrees')
ylabel('Lever arms, m')
text(phi(5), 1.1*lf(5), 'l_f')
text(phi(7), 1.1*(kf(7)+kw(7)), 'l_f  + l_w')
text(phi (6), 1.1*GZ(6), 'GZ')
t2      = [ 'GM = ' num2str(GM) ' m' ];
text(59, GM, t2)
[ phiST, GZ_ST ] = ginput(1);
plot([ phiST phiST ], [ 0 GZ_ST ], 'k-')
text(phiST, -0.1, '\phi_{ST}')
phiREF  = 5 +2*phiST;          % reference angle, deg
plot([ phiREF phiREF ], [ 0 max(GZ) ], 'k-')
text(phiREF, -0.1, '\phi_{REF}')
hRESm   = 0.01*phiST - 0.05;  % min required residual arm, m
resid   = GZ - (kf + kw);      % array of residual arms, m
%       find residual arm at reference angle
hRES    = spline(phi, resid, phiREF);
if hRES > hRESm
        t0 = ' greater than'
elseif hRES == hRESm
        t0 = ' equal to'
else
        t0 = ' less than'
end
fprintf(fid, '             \n')
fprintf(fid, 'The angle of static equilibrium is %5.1f degrees.\n',phiST);
fprintf(fid, 'The residual arm is %5.3f m      \n', hRES);
fprintf(fid, 'at reference angle %5.1f degrees, that is\n',phiREF);
fprintf(fid, '%s the required arm %5.3f m.     \n', t0, hRESm);
hold off
fclose(fid)
```

The following example illustrates an application of the function bv1033 to a realistic ship.

Example 10.2 (An application of the wind criterion).   This example is based on an undergraduate project carried on by I. Ganoni and D. Zigelman, then students at the TECHNION (Zigelman and Ganoni, 1985). The subject of the project was the reconstitution and analysis of the hydrostatic and hydrodynamic properties of a frigate similar to the Italian Navy Ship *Maestrale*. The lines and other particulars were based on the few details provided by Kehoe et al. (1980). To distinguish our example ship from the real one, we shall call it *Maestral*; its main dimensions are:

Table 10.2 Frigate *Maestral*, average of cross-curves in 10 wave phases

Heel Angle (°)	w (m)	Heel Angle (°)	w (m)
0	0	50	4.769
5	0.582	55	5.034
10	1.159	60	5.249
15	1.726	65	5.416
20	2.272	70	5.531
25	2.785	75	5.595
30	3.265	80	5.610
35	3.706	85	5.576
40	4.104	90	5.493
45	4.459		

L_{pp} 114.000 m
B 12.900 m
D 8.775 m

Table 10.2 contains the average of the cross-curves of stability in 10 wave phases, for a volume of displacement $\nabla = 2943$ m^3.

Example 10.1 illustrates a MATLAB function that automatically checks the wind criterion of *BV 1033*. To run this function, the cross-curves of stability of the example ship were written to a file, maestrale.m, in the format.

```
Maestral = [
0      0
...    ...
90     5.493 ];
```

The following lines show how to prepare the input and how to invoke the function.

```
maestrale              % load the cross-curves
cond = [ 1.03*9.81*2943 5.835 4.097 6.681 0.06 ];
sail = [ 1166.55 8.415 ];
bv1033(cond, Maestral, sail, 70)
```

The resulting diagram of stability is shown in Figure 10.1, the report, printed to file bv1033.out, appears below.

```
Stability of ship Maestral acc. to BV 1033
Displacement ................. 29736.955 kN
KG ........................... 5.835 m
```

Metacentric height, GM ....   0.846 m
Mean draft, T ....   4.097 m
Free-surface arm ....   0.060 m
Sail area ....   1166.550 sq m
Sail area centroid above BL ..   8.415 m
Wind pressure ....   1.000 MPa

Heel angle deg	Righting arm m	Heeling arm m
0.0	0.000	0.249
5.0	0.073	0.252
10.0	0.146	0.251
15.0	0.216	0.246
20.0	0.276	0.238
25.0	0.319	0.227
30.0	0.348	0.214
35.0	0.359	0.199
40.0	0.353	0.185
45.0	0.333	0.171
50.0	0.299	0.158
55.0	0.254	0.147
60.0	0.196	0.138
65.0	0.128	0.131
70.0	0.048	0.126
75.0	−0.041	0.123
80.0	−0.136	0.122
85.0	−0.237	0.122
90.0	−0.342	0.122

The angle of static equilibrium is 17.0 degrees.
The residual arm is 0.168 m
at reference angle 39.1 degrees, that is
greater than the required arm 0.120 m.

Table 10.3 *Lido 9*, cross-curves in seaway, 44.16 m³, trim −0.325 m

Heel Angle (°)	Wave Trough (m)	Still Water (m)	Wave Crest (m)
0	0.000	0.000	0.000
5	0.360	0.397	0.395
10	0.713	0.770	0.773
15	1.055	1.111	1.124
20	1.375	1.421	1.445
25	1.671	1.704	1.727
30	1.946	1.967	1.966
35	2.200	2.206	2.166
40	2.429	2.410	2.336
45	2.622	2.582	2.477
50	2.766	2.735	2.588
55	2.867	2.868	2.671
60	2.934	2.950	2.729
65	2.959	2.960	2.756
70	2.955	2.932	2.767
75	2.925	2.875	2.744
80	2.856	2.789	2.678
85	2.756	2.679	2.582
90	2.637	2.548	2.458

10.5 Exercises

Exercise 10.1 (*Lido 9*, cross-curves in seaway).   Table 10.3 contains the ℓ_k levers of the vessel *Lido9*, for a volume of displacement equal to 44.16 m³ and the full-load trim −0.325 m. The data are calculated in wave trough, in still water, and on wave crest. According to the *BV 1033* stability regulations of the German Federal Navy the wave length equals the length between perpendiculars, that is $\lambda = 15.5$ m, and the wave height is calculated from

$$H = \frac{\lambda}{10 + \lambda/20} = 1.439 \text{ m}$$

Assuming that the height of the centre of gravity is $\overline{KG} = 2.21$ m, calculate and plot the diagrams of statical stability ($\overline{GZ}$ curves) for the three conditions: wave trough, still water, wave crest.

Using the same data as in Example 6.1 and the wind-arm prescribed by the BV 1030-1 regulations, check the range of positive residual arms in wave trough and on wave crest. According to the older version, BV 1033, the range of positive residual arms should be at least 10°, and the maximum residual arm not less than 0.1 m.

10.6 Annex—Densities of Liquids

Table 10.4   Densities of liquids

Liquid	Density (t/m^3)
Fresh water	1.000
Fresh water	1.025
Bilge water, sewage	1.005
Grey/black water	1.050
Diesel fuel	0.830
Aviation fuel	0.810
Lubricants	0.900
Foam compound	1.150

CHAPTER 11

Flooding and Damage Condition

Chapter Outline

11.1 Introduction

In the preceding chapters we discussed the buoyancy and stability of **intact** ships. Ships, however, can suffer damages during their service. Hull damages that affect the buoyancy can be caused by collision, by grounding or by enemy action. Water can enter the damaged compartment and cause changes of draught, trim, and heel. Above certain limits, such changes can lead to ship loss. We expect a ship to **survive** a reasonable amount of damage, that is an amount compatible with the size and tasks of the vessel. More specifically, we require that a ship that suffered hull damage, to an extent not larger than defined by pertinent regulations, should continue to float and be stable under moderate environmental conditions. Then, passengers and crew can be saved. Possibly the ship herself can return or be towed to a safe harbour.

To achieve survivability as defined above, the ship hull is subdivided into a number of watertight compartments. Some regulations specify the number of adjacent compartments that should be assumed flooded. This number depends on the size and the mission of the ship. The reason for considering adjacent compartments is simple. Collision, grounding, or single enemy action usually damage adjacent compartments. Flooding of adjacent compartments also can be more dangerous than flooding of two non-adjacent compartments. Adjacent compartments situated at some distance from the midship section can cause large trim and submerge openings above the deck, leading thus to further flooding. Also, submerging part of the deck reduces the waterplane area and can cause a substantial decrease of the metacentric radius. Flooding of non-adjacent compartments, for example one in the forebody, the other in the afterbody, can produce negligible trim. Then, even with relatively large draught increases, the deck does not submerge, the waterplane area is not reduced, and the metacentric height may be sufficient. If the deck does not submerge, no openings are submerged.

The need for international regulations governing the **subdivision** of the hull into watertight compartments became clear after the *Titanic* disaster, in April 1912. A meeting was convened in London leading to the adoption on 20 January 1914 of an **International Convention of the Safety of Life at Sea**. The convention is better known under its acronym, **SOLAS**. The first convention should have been applied in July 1915, but the First World War stopped the process. In 1929 a new conference was held in London. The adopted text entered into force in 1933. Technical developments made necessary a new conference; it was held in 1948. The next edition was the 1960 SOLAS Convention, organized this time by IMO (about IMO see Section 8.2). Several amendments were adopted in the following years. The 1974 SOLAS Convention was again held in London. Since then many important amendments were issued, some of them influenced by major marine disasters, such as those of the roll-on/roll-off passenger ferries *Herald of the Free Enterprise*, near Zeebrugge, in March 1987, and *Estonia*, on 28 September 1994. The latest major amendment has been the harmonization of the provisions on subdivision and damage stability for passenger and cargo ships based on the *probabilistic* method of determining damage stability. The new regulation has taken into account the results of the *Harder* research project (Harmonisation of Rules and Design Rational), a project undertaken by a consortium of European industrial, research, and academic institutions to study the probabilistic approach for assessing the ship's damage stability and to develop new criteria and indexes of subdivision based on probability of survival. At the moment of this publication SOLAS 1974 together with all its amendments is the convention in force (see SOLAS, 2009, and de Juana and Garcia, 2009). The provisions are meant for merchant ships and not for warships or ships transporting troops. However, in the last years a number of navies have cooperated with classification societies also in this direction. This implies problems some of which are discussed by Riola and Pérez (2009).

SOLAS prescriptions cover many aspects of ship safety, among them fire protection, life boats and rafts, radars, radio equipment, and emergency lighting. What interests us in this book are the prescriptions referring to subdivision and damage stability. A detailed history of SOLAS activities can be found on a website organized by *Metal Safe Sign International Ltd*, http://www.mss-int.com, or in the IMO website, http:www.imo.org. A short history of damage regulations appears in Gilbert and Card (1990). A commented history of the SOLAS achievements can be read in Payne (1994). Because of the overwhelming importance of the SOLAS regulations we give here the translations of the official title in four other languages:

Fr Convention internationale pour la sauvegarde de la vie humaine en mer
G Internationales Übereinkommen zum Schutz des menschlichen Lebens auf See
I Convenzione internazionale per la salvaguardia della vita umana in mare
S Convenio internacional para la seguridad de la vida humana en el mar

As mentioned above, the SOLAS regulations apply to merchant ships. Damage regulations for warships are provided in the same regulations that deal with their intact stability.

An alternative term used in damage considerations is **bilging**. Derrett and Barrass (2000) define it as follows: 'let an empty compartment be holed ... below the waterline to such an extent that the water may ... flow freely into and out of the compartment. A vessel holed in this way is said to be bilged.'

Roll-on/roll-off ships, shortly Ro/Ro, are particularly sensitive to damage. To enable easy loading and unloading of vehicles these vessels are provided with a deck space uninterrupted by bulkheads. Damage can easily cause deck flooding with consequences like $\overline{KG}$ increase, large free-surface effect and added weight. Little and Hutchinson (1995) quote, 'Over the past 14 years, 44 RO/RO vessels have capsized.' Pawlowski (1999) appreciates, 'Roll-on/roll-off (RO/RO) ships are considered by the maritime profession ... as the most unsafe ships in operation.' Statistics on loss of life due to RO/RO disasters are simply frightening. For example, Ross et al. (1997) quote 193 casualties in the case on the *Herald of Free Enterprise*, 910 in the *Estonia* disaster. A few RoRo's sank in one and a half minutes after an accident. No wonder that many studies have been dedicated to this type of vessel. As some of them refer to constructive measures, we think that their treatment belongs to books on Ship Design, not here. We cite, however, in this chapter, the papers whose contents are close to the subject of this chapter. Based on SOLAS 90 deterministic damage stability methodology, a requirement for damage stability was agreed for RoRo ships in 1993 among North West European nations to account for the risk of accumulation of water on deck. This requirement, known as the Stockholm Agreement, was discussed in the SOLAS Conference 1995. An upgraded version of this regulation is currently in force in Europe (see EC directive 2003/25/EC).

In this chapter we give the definitions related to flooding and explain the principles on which flooding and damage calculations are based. To illustrate these principles we apply them to box shaped vessels. We also summarize a few pertinent regulations and codes of practice. When performing calculations for real-life projects, the reader is advised to refer to the full text of the most recent edition of the regulations to be applied.

Flooding and damage stability calculations for real ship forms are rather complex and tedious. Finding the floating condition requires iterative procedures. Today, such calculations are performed on computers; therefore, we do not describe them. We also give in this chapter the translations of the most important terms introduced in it.

11.2 A Few Definitions

In this section we introduce a few terms defined in the SOLAS conventions; they are also used by other regulations.

The hull is subdivided into compartments by means of **watertight bulkheads**. This term is translated into four other languages as

> Fr   cloisons étanches   I   paratie stagne
> G   Schotten   S   mamparos estancos

The deck up to which these bulkheads extend is called in English **bulkhead deck**, in four other languages

> Fr   pont de cloisonnement   I   ponte delle paratie
> G   Schottendeck   S   cubierta de cierre

After flooding of a prescribed number of compartments the ship shall not submerge beyond a line situated at least 76 mm (3 in.) below the bulkhead deck at side (for an exception to this requirement see the regulations of the German Navy, Section 11.6.5). The said line is called in English **margin line**, in four other languages

> Fr   ligne de surimmersion   I   linea limite
> G   Tauchgrenze   S   línea de margen

The **floodable length** at a given point of the ship length is the maximum length, with the centre at that point, that can be flooded without submerging the ship beyond the margin line. This subject is treated in more detail in Section 11.4.1. The term "floodable length" is translated as

> Fr   longueur envahissable   I   lunghezza allagabile
> G   flutbare Länge   S   eslora inundable

In Figure 11.1 we see the sketch of a ship subdivided by four bulkheads. The three waterlines WL_1, WL_2, and WL_3 are tangent to the margin line. They are examples of limit lines beyond

Figure 11.1 A few definitions

which no further submergence of the damaged ship is admissible. If the bulkhead deck is not continuous, a continuous margin line can be assumed such as having no point at a distance less than 76 mm below the deck at side.

Let us suppose that calculating the volume of a compartment starting from its dimensions we obtain the value v. There is almost no case in which this volume can be fully flooded because almost always there are some objects in the compartment. Even in an empty tank there are usually structural members – such as frames, floors, and deck beams – sounding instruments and stairs for entering the tank and inspecting it. If we deduct the volumes of such objects from the volume v we obtain the volume of the water that can flood the compartment; let it be v_F. The ratio

$$\mu = \frac{v_F}{v} \tag{11.1}$$

is called **permeability**; it is often noted by μ. More correctly, we should talk about **volume permeability**, to distinguish it from a related notion that is the **surface permeability**. Indeed, because of the objects stored or located in a compartment, the free-surface area is smaller than that calculated from the dimensions of the compartment. Also the moment of inertia of the free-surface area is calculated on the basis of the dimensions of the compartment. For example, if the calculations are carried on by a computer programme, they are based on an input that describes only the geometry of the tank and not its contents. The moment of inertia of the surface free to heel is smaller than the value found as above because the area considered is partially occupied by fixed objects that do not contribute to the free-surface effect. Then, it is necessary to multiply the calculated value by the surface permeability.

Typical values of volume permeability can be found in textbooks and in various regulations. Examples of the latter are given in this chapter. When the recommended values do not seem plausible, it is necessary to calculate in detail the volume of the objects found in the compartment. When there are no better data, the surface permeability can be assumed equal to the volume permeability of the same compartment.

The term "permeability" is translated into four other languages as follows

Fr	(coefficient de) perméabilité	I	(coefficiente di) permeabilitá
G	Flutbarkeit	S	(coeficiente de) permeabilidad

Usually, permeabilities are given in percent, for example, 8.5 for machinery spaces. In calculations, however, we must multiply by 0.85, and not by 85. Moreover, some computer programmes require as input the number 0.85 and not 85. Therefore, in the following sections permeabilities are mainly given in the format 0.95, 0.85, etc., rather than as percentages.

11.3 Two Methods for Finding the Ship Condition After Flooding

There are two ways of calculating the effect of flooding. One way is known as the **method of lost buoyancy**, the other as the **method of added weight**.

The method of lost buoyancy assumes that a flooded compartment does not supply buoyancy. This is what happens in reality. If we refer to Figures 2.4 and 2.5, we can imagine that if there is open communication between a compartment and the surrounding water, the water inside the compartment exercises pressures equal to and opposed to those of the external water. Then, the buoyancy force predicted by the Archimedes' principle is cancelled by the pressure of the flooding water.

In the method of lost buoyancy the volume of the flooded compartment does not belong anymore to the vessel, while the weight of its structures is still part of the displacement. The 'remaining' vessel must change position until force and moment equilibria are reestablished. During the process not only the displacement, but also the position of the centre of gravity remains constant. The method is also known as **method of constant displacement**. As the flooding water does not belong to the ship, it causes no free-surface effect.

In the method of added weight the water entering a damaged compartment is considered as belonging to the ship; its mass must be added to the ship displacement. Hence the term 'added weight.' Following modern practice we actually work with masses; however, we keep the traditional name of the method, that is we use the word 'weight.' Another reason may be the need to avoid confusion with the term **added mass** mentioned in Section 6.13 and detailed in Chapter 12. The latter term does not belong to the theory of flooding and damage stability.

In the method of added weight the displacement of the flooded vessel is calculated as the sum of the intact displacement and the mass of the flooding water. The position of the centre of gravity of the damaged vessel is obtained from the sums of the moments of the intact vessel and of the flooding water. Becoming part of the vessel, the flooding water produces a free-surface effect that must be calculated and considered in all equations.

For very small trim and negligible heel changes we can write

$$\Delta_F = \Delta_I + \rho v$$
$$LCG_F \cdot \Delta_F = LCG_I \cdot \Delta_I + lcg \cdot \rho \cdot v$$
$$TCG_F \cdot \Delta_F = tcg \cdot \rho \cdot v \tag{11.2}$$

where the subscript F distinguishes the properties of the flooded vessel, and the subscript I those of the intact ship. By lcg we mean the longitudinal centre of gravity of the flooding water volume, v, and by tcg its transverse centre of gravity. We assume $TCG_I = 0$. The vertical centre of gravity, $\overline{KG}$, must be recalculated. When the trim and the heel are not negligible, we must consider the vertical coordinates of the centres of gravity of the intact ship and of the flooding water volume. Example 11.1 shows how to do this for non-zero trim and zero heel.

To exemplify the above principles we follow an idea presented in *Handbuch der Werften* and later used by Watson (1998). While the latter solves algebraically the general problem, we prefer to solve it numerically and allow thus the reader to visualize the differences between methods and those between the intact and the damaged vessel. We choose the very simple example of the pontoon shown in Figure 11.2. Two transverse bulkheads subdivide the hull into three watertight compartments. In the following two subsections we assume that

Figure 11.2 A simple pontoon—intact condition

Compartment 2 is damaged and calculate the consequences of its flooding. We choose deliberately a compartment symmetric about the midship transverse plane of symmetry of the pontoon. Thus, the flooding of Compartment 2 produces no trim. Also, the compartment extends for the full ship breadth and its flooding produces no heel. The only change of position is **parallel sinking**. Thus, the complex calculations necessary for conventional ship forms, for large trim, or for unsymmetrical flooding, do not obscure the principles and it is possible to obtain immediately a good insight of the processes involved. For the same reasons we assume that the volume and surface permeabilities are equal to 1. We leave to an exercise the informal proof that taking permeability into account does not change the qualitative results. Although based on different physical models, calculations by the two methods yield the same final draught, as it should be expected. Moreover, the stability properties calculated by the two methods are identical, if we compare the initial righting moments. Here, the term "initial" has the meaning defined in Chapter 2 where we consider 'initial stability' as a property governing the behaviour of the floating body in a small heel range around the upright position. In that range the righting moment equals

$$M_R = \Delta \overline{GM} \sin \phi$$

As we are going to see, we obtain by the two methods the same M_R value. In the method of lost buoyancy the displacement remains equal to that of the intact vessel. In the method of added weight the displacement increases by the mass of the flooding water. To keep the product M_R constant, the other factor, $\overline{GM}$, must be smaller. At a first glance it may be surprising that the two methods yield different metacentric heights. The explanation given above shows that it should be so because the considered displacements are different. What should be kept in mind, after reading the examples, is that displacement and metacentric height have different significances in the two methods. Therefore, damage stability data should include the mention of the method by which they were obtained. Computer programmes use the method of lost buoyancy for final-stage calculations.

The length of the assumed pontoon is $L = 20$ m, the beam, $B = 5$ m, and the draught in intact condition, $T_I = 1.5$ m. Let the vertical centre of gravity be $\overline{KG_I} = 1.5$ m. The following calculations were carried on in MATLAB, using the full precision of the software. The results are rounded off to a reasonable number of decimal digits. We first find the data of the intact pontoon. The displacement volume is

$$\nabla_I = LBT_I = 20 \times 5 \times 1.5 = 150 \text{ m}^3$$

The mass displacement equals

$$\Delta_I = \rho \nabla_I = 1.025 \times 150 = 153.75 \text{ t}$$

The moment of inertia of the waterplane area about the centreline equals

$$I_I = \frac{B^3 L}{12} = \frac{5^3 \times 20}{12} = 208.3333 \text{ m}^4$$

and the resulting metacentric radius is

$$\overline{BM}_I = \frac{I_I}{\nabla_I} = \frac{208.3333}{150} = 1.389 \text{ m}$$

For such a simple form we could have found directly the metacentric radius as

$$\overline{BM}_I = \frac{B^3 L/12}{LBT_I} = \frac{B^2}{12T_I} = \frac{5^2}{12 \times 1.5} = 1.389 \text{ m}$$

The height of the centre of buoyancy is

$$\overline{KB}_I = \frac{T_I}{2} = 0.75 \text{ m}$$

and the metacentric height

$$\overline{GM}_I = \overline{KB}_I + \overline{BM}_I - \overline{KG}_I = 0.75 + 1.389 - 1.50 = 0.639 \text{ m}$$

For small heel angles the righting moment in intact condition is calculated as

$$M_{RI} = \Delta_I \overline{GM}_I \sin\phi = 153.75 \times 0.639 \times \sin\phi = 98.229 \sin\phi \text{ t m}$$

11.3.1 Lost Buoyancy

Four translations of the term 'method of lost buoyancy' are

> Fr La méthode des carènes perdues
> G   Methode des wegfallender Verdrängung
> I   Il metodo per perdita di galleggiabilità
> S   Método de la pérdida de empuje

In the method of lost buoyancy the flooded compartment does not supply buoyancy. As shown in Figure 11.3, the buoyant hull is composed only of Compartments 1 and 3. After loosing the central compartment, the waterplane area is equal to

$$A_L = (L - l)B = (20 - 4) \times 5 = 80 \text{ m}^2$$

To compensate for the loss of buoyancy of the central compartment the draught increases to

$$T_L = \frac{\nabla_I}{A_L} = \frac{150}{80} = 1.875 \text{ m}$$

The height of the centre of buoyancy increases to

$$\overline{KB}_L = \frac{T_L}{2} = \frac{1.875}{2} = 0.938 \text{ m}$$

Dimensions in m

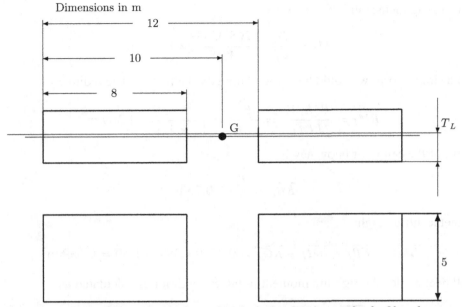

Figure 11.3   A simple pontoon—damage calculation by the method of lost buoyancy

We calculate the moment of inertia of the waterplane as

$$I_L = \frac{B^3(L-l)}{12} = \frac{5^3(20-4)}{12} = 166.6667 \text{ m}^4$$

and the metacentric radius as

$$\overline{BM}_L = \frac{I_L}{\nabla_I} = \frac{166.6667}{150} = 1.111 \text{ m}$$

Finally, the metacentric height is

$$\overline{GM}_L = \overline{KB}_L + \overline{BM}_L - \overline{KG}_I = 0.938 + 1.111 - 1.5 = 0.549 \text{ m}$$

and the righting moment for small heel angles, in the lost-buoyancy method

$$M_{RL} = \Delta_I \overline{GM}_L \sin\phi = 153.75 \times 0.549 \sin\phi = 84.349 \sin\phi \text{ t m}$$

11.3.2 Added Weight

The translations of the term "added-weight method" in four other languages are

    Fr   La méthode par addition de poids
    G   Methode des hinzukommenden Gewichts
    I   Il metodo del peso imbarcato
    S   Método del peso añadido

Dimensions in m

Figure 11.4 A simple pontoon—damage calculation by the method of added weight

For this subsection see Figure 11.4. Because of the added weight of the flooding water the draught of the pontoon must increase by a quantity δT. The volume of flooding water equals

$$v = lB(T_I + \delta T) \tag{11.3}$$

The additional buoyant volume of the vessel, due to parallel sinking, is

$$\delta\nabla = LB\delta T \tag{11.4}$$

To obtain the draught increment, δT, we equate the two volumes, that is we write $v = \delta\nabla$. Algebraic manipulation and numerical calculation yield

$$\delta T = \frac{lT_I}{L - l} = \frac{4 \times 1.5}{20 - 4} = 0.375 \text{ m}$$

The draught after flooding is

$$T_A = T_I + \delta T = 1.500 + 0.375 = 1.875 \text{ m}$$

The volume of flooding water is calculated as

$$v = lBT_A = 4 \times 5 \times 1.875 = 37.5 \text{ m}^3$$

and the height of its centre of gravity

$$\overline{kb} = \frac{T_A}{2} = \frac{1.875}{2} = 0.938 \text{ m}$$

Table 11.1 $\overline{KG}$ by the method of added weight

	Volume	Centre of Gravity	Moment
Initial	150.0	1.5	225.000
Added	37.5	0.938	35.156
Total	187.5	1.388	260.156

The displacement volume of the flooded pontoon is

$$\nabla_A = LBT_A = 20 \times 5 \times 1.875 = 187.5 \text{ m}^3$$

We consider the flooding water as an added weight; therefore, we must calculate a new centre of gravity. The calculations are shown in Table 11.1.

The moment of inertia of the damage waterplane is the same as in the initial condition, that is $I_A = 208.333 \text{ m}^4$. Then, the metacentric radius equals

$$\overline{BM}_A = \frac{I_A}{\nabla_A} = \frac{208.333}{187.5} = 1.111 \text{ m}$$

In this method the flooding water is considered as belonging to the displacement. Therefore, if there is a free surface its effect must be calculated. The moment of inertia of the free surface in the flooded compartment equals

$$i = \frac{B^3 l}{12} = \frac{5^3 \times 4}{12} = 41.667 \text{ m}^4$$

and the lever arm of the free surface effect is

$$\ell_F = \frac{\rho i}{\rho \nabla_A} = \frac{41.667}{187.5} = 0.222 \text{ m}$$

The height of the centre of buoyancy is yielded by

$$\overline{KB}_A = \frac{T_A}{2} = \frac{1.875}{2} = 0.938 \text{ m}$$

The corresponding metacentric height is calculated as

$$\overline{GM}_A = \overline{KB}_A + \overline{BM}_A - \overline{KG}_A - \ell_F = 0.938 + 1.111 - 1.388 - 0.222 = 0.439 \text{ m}$$

With the mass displacement

$$\Delta_A = \rho \nabla_A = 1.025 \times 187.5 = 192.188 \text{ t}$$

we obtain the righting moment for small angles of heel, in the added-weight method

$$M_{RA} = \Delta_A \overline{GM}_A \sin \phi = 192.188 \times 0.439 \sin \phi = 84.349 \sin \phi \text{ t m}$$

Table 11.2 Flooding calculations—a comparison of methods

	Intact Condition	Damaged, Lost Buoyancy	Damaged, by Added Weight
Draught, m	1.500	1.875	1.875
∇, m^3	150.000	150.000	187.500
Δ, t	153.750	153.750	192.188
$\overline{KB}$, m	0.750	0.938	0.938
$\overline{BM}$, m	1.389	1.111	1.111
$\overline{KG}$, m	1.500	1.500	1.388
$\overline{GM}$, m	0.639	0.549	0.439
$\Delta\overline{GM}$, t m	98.229	84.349	84.349

11.3.3 The Comparison

Table 11.2 summarizes the results of the preceding two subsections. As expected, both the method of lost buoyancy and that of added weight yield the same draught 1.875 m, and the same initial righting moment, 84.349 sin ϕ t m. The displacements and the metacentric heights are different, but their products, $\Delta\overline{GM}$, are the same. As happens in most cases, the righting moment in damage condition is less than in intact condition.

11.4 Damage Conditions Assessment

The survivability in damage condition can be assessed in different ways. In this section, we will review the three main approaches that have been considered in international regulations. In Section 11.6 we will outline their application.

11.4.1 Assessment of Floodable Lengths

The survivability in damage condition is directly related to the main watertight subdivision below the bulkhead deck. As previously defined, the floodable length at a given point of the ship length is the maximum length, with the centre at that point, that can be flooded without submerging the ship beyond the margin line. A first approach to the assessment of the damage condition is to find the floodable length at a given point of the ship length, multiplied by a number called *factor of subdivision*. A factor of subdivision equal to 1 means that the margin line should not submerge if one compartment is submerged, while a factor of subdivision equal to 0.5 means that the margin line should not submerge when two adjacent compartments are flooded. This assessment had been included in the SOLAS provisions, but was finally superseded by the last 2009 amendments. This approach is still used for design purposes in order to identify any issues regarding the location of bulkheads before carrying on further and more complex calculations. The assessment is based on two assumptions:

- The ship is subject to symmetrical flooding only as the hull is subdivided only by transverse bulkheads. This is a simplification as the ship subdivision is usually more complex and includes also longitudinal bulkheads that play a very important role in the flooding process.
- The method takes into consideration only floatability without checking the residual stability after damage. It may happen that for certain waterlines tangent to the margin line the residual stability is insufficient.

For the above considerations, SOLAS added an additional requirement regarding the deterministic assessment of damage stability, on top of the floodable-length requirement. As shown in Section 11.6.5, the German BV 1030-1 regulations include requirements for the stability in damage conditions, as their older versions also did.

11.4.2 Deterministic Assessment of Damage Stability

The actual ship compartmentation and all possible damage scenarios should be taken into account. The *standard damage dimensions* are defined by a damage length, a vertical extent and a penetration (the transverse extension from the shell inwards). These dimensions depend on the respective application. For example, the Offshore Supply Vessel Code specifies a penetration of 760 mm as the damage of this kind of vessel in an offshore scenario and low speeds is very limited. Passenger ships, however, are more likely to suffer damage with larger penetrations. Therefore, the old SOLAS 90 requirements specified a penetration equal to $B/5$, while the new SOLAS provisions changed the value to $B/10$. The US, UK, and German Navy regulations consider penetrations up to the centreline plane, that is $B/2$, as warships are likely to sustain major damages in their mission. Minor damage that may be more onerous should also be considered. The location of damage may be anywhere along the ship length, including positions between transverse bulkheads.

The applicable residual stability criteria consist of a set of requirements on the $\overline{GZ}$-curve, such as to heel angle, range of positive $\overline{GZ}$-values, or maximum $\overline{GZ}$, but may also consider other properties like margin line immersion. The criteria depend on the application. For example, a heel angle of 25° after damage may be acceptable for a cargo vessel with professional crew aboard that is able to evacuate the vessel if need arises. For passenger vessels, however, the criteria may limit the heel angle to 7°. On the other hand, for warships there may be limitations that allow the use of damage control equipment such as watertight doors or pumps. The deterministic approach to damage stability is meant to ensure minimum safety in all foreseen scenarios. The approach should also provide the ship Master with information on the damage scenarios the ship can survive. This approach is considered very reliable and is used in many regulations and design standards, as exemplified later in this chapter. On the other hand, if two ships comply with the deterministic approach, we would not be able to know which is the safest one. Therefore, the deterministic approach does not provide a measure of ship safety against capsize.

11.4.3 The Probabilistic Assessment of Flooding and Damage Stability

Wendel (1960a) introduces the notion of **probability of survival after damage**. A year later a summary in French appears in Anonymous (1961).This paper mentions a translation into French of Wendel's original paper (in Bulletin Technique du Bureau Veritas, February 1961) and calls the method "une nouvelle voie," that is "a new way." Much has been written since then on the probabilistic approach; we mention here only a few publications, such as Rao (1968), Wendel (1970, 1977), Abicht and Bakenhus (1970), and Abicht et al. (1977). Over the years Wendel used new and better statistics to improve the functions of probability density and probability introduced by him. The general idea is to consider the probability of occurrence of a damage of length y and transverse extent t, with the centre at a position x on the ship length. Statistics of marine accidents should allow the formulation of a function of probability density, $f(x, y, t)$. The probability itself is obtained by triple integration of the density function. The IMO regulation A265 introduces probabilistic regulations for passenger ships, and SOLAS 1974, Part B1, defines probabilistic rules for cargo ships. These regulations, together with the applicable SOLAS deterministic requirements for passenger ships, were harmonized in a revised SOLAS 2009 probabilistic damage stability approach that will be explained later in this chapter.

The probabilistic approach uses the concept of probability of survival after damage as a measure of ship safety in damage condition. This probability is referred to by SOLAS as the **attained subdivision index A**. The philosophy behind the probabilistic approach is that *two different ships with the same attained index are of equal safety*. Therefore, this attained index should be based on all possible damage scenarios. An **attained subdivision index** shall be calculated as

$$A = \Sigma p_i s_i$$

where p_i represents the probability that the ith compartment or group of compartments may be flooded, and s_i is the probability of survival after flooding the ith compartment or group of compartments. The attained subdivision index, A, should not be less than the required subdivision index, R. This attained index usually depends on the ship length and the number of passengers.

Early details of the standard for subdivision and damage stability of dry cargo ships are given by Gilbert and Card (1990). A critical discussion of the IMO 1992 probabilistic damage criteria for dry cargo ships appears in Sonnenschein and Yang (1993). The probabilistic SOLAS regulations are discussed in some detail by Watson (1998) who also exemplifies them numerically. Ravn et al. (2002) exemplify the application of the rules to Ro-Ro vessels.

Serious criticism of the SOLAS probabilistic approach to damage can be found in Björkman (1995). Quoting from the title of the paper, 'apparent anomalies in SOLAS and MARPOL requirements.' Watson (1998) writes, 'There would seem to be two main objections to the probabilistic rules. The first of these is the extremely large amount of calculations required, which although acceptable in the computer age is scarcely to be welcomed. The other objection is the lack of guidance that it gives to a designer, who may be even driven to continuing use of the deterministic method in initial design, changing to the probabilistic later—and hoping this does not entail major changes!' Another objection is that a ship may comply with the required index, but it could have suffered some minor damages that result in capsizing. For this reason, it may be necessary to include a deterministic 'minor damage' requirement on top of the probabilistic regulations. Thus it is possible to avoid the design of ships with what might be perceived as unacceptably vulnerable spots in some part of their length.

In May 2005 the MSC Committee adopted a revised Chapter 2.1 of the SOLAS convention with the intention of harmonizing the provisions on subdivision and damage stability for passenger and cargo ships. The revised provisions were meant for ships built after 1 January 2009. The regulations take into account wave effects, heeling moments, cargo shifting, transient effects, and arrangements against unsymmetrical flooding.

11.5 Details of the Flooding Process

The free surface in a compartment open to the sea behaves differently than that in an intact tank. In Figure 11.5a $W_I L_I$ is the waterline in upright position, $W_\phi L_\phi$ the waterline in a heeled position. We assume that the water level in the side tank is the same as the external water level. In the heeled position the water surface in the tank changes to FS, a line parallel to $W_\phi L_\phi$. The volume of water in the tank remains constant. In Figure 11.5b the side tank is damaged and in open communication with the sea. If the waterline in the heeled position is $W_\phi L_\phi$, this is also the water level in the damaged tank. The water volume is no longer constant, but varies with the heel angle. For the case shown in the figure, the volume increases by the slice comprised between the lines $W_\phi L_\phi$ and FS. This change of volume must be taken into account in the added-weight method.

Figure 11.5b shows a case of **unsymmetrical flooding**. This kind of flooding can easily submerge the deck. The consequences may be a drastic reduction of stability and the submergence of openings such as vents. Therefore, care must be exercised when placing longitudinal bulkheads. Sometimes, to compensate unsymmetrical flooding it is necessary to open a connection between the damaged tank and a tank situated symmetrically on the other

Figure 11.5 Free surface in intact and in damaged tank

side of the ship. This action is called **cross-flooding**. The UK-Navy document SSP 24 warns against the potential danger presented by longitudinal bulkheads.

Cross-flooding takes some time and can cause a slow change of the ship position. Söding (2002) lists other slow-flooding processes such as occurring 'through open or non-watertight doors, hatches with non-watertight or partly open hatch covers, through pipes, ventilation ducts ...' In his paper Söding describes the mathematics of such water flows. Air can be trapped above the flooding-water surface. If the top envelope of the compartment is airtight, flooding is stopped. If not, it is only slowed down. Between the position of intact condition and the final damage position (provided that an equilibrium position can be found) the vessel can pass through intermediate positions more dangerous than the final one. According to some regulations it is necessary to check if such positions exist and if the ship can survive them. For instance, SOLAS considers that there is no need to study intermediate stages of flooding when the final stage is reached in less than 60 s (instantaneous flooding). However, if the damaged area contains decks, inner bulkheads, or structural, non-watertight elements of sufficient tightness and strength to seriously restrict the flow of water, intermediate-stage flooding calculations should be performed. To apply the above approach in a uniform manner, IMO has developed a standard method for evaluating cross-flooding arrangements IMO (2007a). This methodology enables the calculation of the time required from commencement of cross-flooding to the final equilibrium position, and provides a guide for dimensioning pipes that do not delay cross-flooding.

11.6 Damage Stability Regulations

11.6.1 SOLAS Requirements for Dry-Cargo and Passenger Ships

Dry-cargo and passenger ships are required to comply with the revised SOLAS probabilistic stability requirements. The index is obtained as the sum of indices calculated, as previously explained, for the ship in three arbitrary conditions that cover all the operational draught range:

Deepest subdivision draught, d_s. This corresponds to the maximum draught.
Light service draught, d_l. Corresponds to the lightest anticipated loading condition of the
  ship in service and it may include ballast water as necessary for stability and/or immersion.
Partial subdivision draught, d_p. This is the light service draught plus 60% of the difference
  between the light service draught and the deepest subdivision draught.

Using the same subscripts for the corresponding subdivision indexes, the formula to be used is

$$A = 0.4A_s + 0.4A_p + 0.2A_l$$

The subdivision is considered sufficient if the attained index, A, is not less than the required subdivision index, R, calculated according to the regulations, and if, in addition, the partial indices A_s, A_p, and A_l are not less than $0.9R$ for passenger ships, and $0.5R$ for cargo ships. The p_i factor represents the probability that a compartment or group of compartments may be flooded, disregarding horizontal subdivision; it is based upon the examination of collision cases for which information on both damage penetration and damage longitudinal extent was available. The expression of the survivability factor, s_i, is a general formulation designed to represent the probability that the ship may survive the flooding of a compartment or group of compartments. This survivability factor depends on the heel angle, the maximum $\overline{GZ}$-value, and the positive range of the $\overline{GZ}$-curve. The formulation for cargo ships is different from that for passenger ships. The minimum required $\overline{GM}$ value (or, alternatively, the maximum

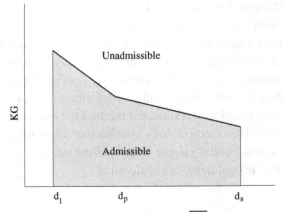

Figure 11.6 The limiting $\overline{KG}$ curve

Table 11.3 Damage scenarios

Number of Persons, N_p	Penetration	Damage Length	Location
$N_p \geq 400$	0.1B but not less than 0.75 m	0.03L but not less than 3 m	Anywhere along the ship length
$36 < N_p < 400$	Linear interpolation	Linear interpolation	Between transverse bulkheads but not less than 3 m
$N_p \leq 36$	0.05B but not less than 0.75 m	0.015L but not less than 3 m	Between transverse bulkheads but not less than 3 m

permissible $\overline{KG}$ value) for the three draughts d_s, d_l, and d_p are equal to the $\overline{GM}$ (or $\overline{KG}$) values corresponding to the loading cases considered for the calculation of the survivability factor, s_i. For intermediate draughts, the values to be used shall be obtained by linear interpolation of $\overline{GM}$ values only between the deepest and the partial subdivision draught, and between the partial load line and the light service draught. Then, for all loading conditions the operational point of the ship must lie in the admissible part of Figure 11.6, an adaptation of Figure 8.4.

A deterministic assessment has been added for passenger ships to ensure that they retain the capability to survive minor damage cases along their full length and still meet the currently accepted requirements for heel and minimum residual stability. This criterion is based on an s factor that is not less than 0.9. The damage scenario to be considered depends on the number of persons on board, as shown in Table 11.3. Obviously, the table refers to watertight, transverse bulkheads.

According to SOLAS, vessels should be designed with a *double bottom* extending from the aftermost to the foremost bulkhead. The double bottom is a watertight deck in the lower part of the hull and it serves as a redundant barrier in case of grounding or underwater damage. If this double bottom is not fitted, SOLAS provides a deterministic damage stability standard for demonstrating that the design has a sufficient level of safety against bottom damage. Following the sinking of the bulk carrier *Derbyshire* in September 1980, IMO adopted amendments to Chapter 12 of SOLAS; they include many design considerations referring to strength. The regulations state that any bulk carrier should have sufficient strength and stability in damage condition to withstand the flooding of any cargo hold.

11.6.2 MARPOL Requirements for Tankers

The tanker *Torrey Canyon* ran aground in the English Channel in March 1967 and spilled 120 000 t of crude oil into the sea, the biggest oil spill to that time. Consequently, an international conference was convened in 1973. The result was the first MARPOL Convention, the main international convention covering the prevention of pollution of the marine environment by ships. In our days, it is involved in many kinds of pollution (oil, chemicals, garbage, etc.) and it refers to all ship types. Oil tankers are required to meet damage stability requirements that ensure that they can survive side and bottom damage by collision or stranding. The scenarios to be considered depend on the ship length, as detailed below:

- For tankers longer than 225 m the damage can occur anywhere along the ship length.
- For tankers above 150 m, but not exceeding 225 m, the damage to be considered can occur anywhere along the ship length, but shall not involve the after and forward boundary bulkheads of the machinery space. This space shall be treated as a single floodable compartment.
- For tankers not exceeding 150 m the damage to be considered can occur between transverse bulkheads, anywhere along the ship length, but not in the machinery space. For tankers shorter than 100 m the Administration may allow relaxations from these requirements.

For oil tankers of 20 000 t deadweight and above the damage assumptions are supplemented by a bottom raking damage scenario which involves bottom damages with a longitudinal extent up to $0.6L$ for ships of 75 000 t deadweight and above, and up to $0.4L$ for ships of less than 75 000 t.

11.6.3 The US Navy

The regulations of the US Navy are contained in a document known as DDS-079-1. Part of the regulations are classified, part of those that are not classified can be found in Nickum (1988) or Watson (1998).

For a ship shorter than 30.5 m (100 ft) the flooding of any compartment should not submerge her beyond the margin line. Ships longer than 30.5 m and shorter than 91.5 m (300 ft) should meet the same submergence criterion with two flooded compartments. Ships longer than 91.5 m should meet the submergence criterion with a damage extent of $0.15L$ or 21 m, whichever is greater.

When checking stability under wind the righting arm, $\overline{GZ}$, should be reduced by $0.05 \cos \phi$ to account for unknown unsymmetrical flooding or transverse shift of loose material. As for intact condition (see Figure 8.5), the standard identifies two areas between the righting-arm and the wind-arm curves. The area A_1 is situated between the angle of static equilibrium and the angle of downflooding or 45°, whichever is smaller. The area A_2 is situated to the left, from the angle of static equilibrium to an angle of roll. The wind velocity and the angle of roll should be taken from DDS-079-1. As in the intact condition, the standard requires that $A_1/A_2 \geq 1.4$.

The US Navy uses the concept of **V lines** to define a zone in which the bulkheads must be completely watertight. We refer to Figure 11.7. Part (a) of the figure shows a longitudinal ship section near a bulkhead. Let us assume that after checking all required combinations of flooded compartments, the highest waterline on the considered bulkhead is WL; it intersects the bulkhead at O. In part (b) of the figure we show the transverse section AB that contains the bulkhead. The intersection of WL with the bulkhead passes though the point Q. The standard assumes that unsymmetrical flooding can heel the vessel by 15°. The waterline corresponding to this angle is W_1L_1. Rolling and transient motions can increase the heel angle by a value that depends on the ship size and should be taken from the standard. We obtain thus the waterline W_2L_2. Finally, to take into account the relative motion in waves (that is the difference between ship motion and wave-surface motion) we draw another waterline translated up by $h = 1.22$ m (4 ft); this is waterline W_3L_3. Obviously, unsymmetrical flooding followed by rolling can occur to the other side too so that we must consider the waterline W_4L_4 symmetrical of W_3L_3 about the centreline. The waterlines W_3L_3 and W_4L_4 intersect at the point P. We identify a V-shaped limit line, W_4PL_3, hence the term "V lines." The region below the V lines must be kept watertight; severe restrictions refer to it and they must be read in detail.

11.6.4 The UK Navy

The standard of damage stability of the UK Navy is defined in the same documents NES 109 and SSP 24 that contain the prescriptions for intact stability (see Section 8.4). We briefly discuss here only the rules referring to vessels with a military role. The degree of damage to be assumed depends on the ship size, as follows:

Waterline Length	Damage Extent
$L_{WL} < 30$ m	Any single compartment
$30 \leq L_{WL} \leq 92$	Any two adjacent main compartments, that is compartments of minimum 6-m length
> 92 m	Damage anywhere extending 15% of L_{WL} or 21 m, whichever is greater.

The permeabilities to be used are

Watertight, void compartment and tanks	0.97
Workshops, offices, operational and accommodation spaces	0.95
Vehicle decks	0.90
Machinery compartments	0.85
Store rooms, cargo holds	0.60

The wind speeds to be considered depend on the ship displacement, Δ.

Displacement Δ, Tonnes	Nominal Wind Speed, Knots
$\Delta \le 1000$	$V = 20 + 0.005\Delta$
$1000 < \Delta \le 5000$	$V = 5.06 \ln \Delta - 10$
$5000 < \Delta$	$V = 22.5 + 0.15\sqrt{\Delta}.$

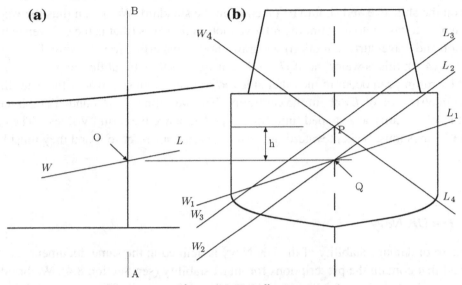

Figure 11.7 V lines

The following criteria of stability should me met (see also Figure 8.5):

1. angle of list or loll not larger than 30°;
2. righting arm $\overline{GZ}$ at first static angle not larger than 0.6 maximum righting arm;
3. area A_1 greater than A_{min} as given by

$$\Delta \le 5000\ \mathrm{t} \qquad A_{min} = 2.74 \times 10^{-2} - 1.97 \times 10^{-6}\Delta\ \mathrm{m\ rad}$$
$$5000 < \Delta < 50\,000\ \mathrm{t} \quad A_{min} = 0.164\Delta^{-0.265}$$
$$\Delta > 50\,000\ \mathrm{t} \qquad \text{Consult Sea Technology Group}$$

4. $A_1 > A_2$;
5. trim does not lead to downflooding;
6. $\overline{GM}_L > 0$

Like the US Navy, the UK Navy uses the concept of **V lines** to define a zone in which the bulkheads must be completely watertight; some values, however, may be more severe. We refer again to Figure 11.7. Part (a) of the figure shows a longitudinal ship section near a bulkhead. Let us assume that after checking all required combinations of flooded compartments, the highest waterline on the considered bulkhead is WL; it intersects the bulkhead at O. In part (b) of the figure we show the transverse section AB that contains the bulkhead. The intersection of WL with the bulkhead passes though the point Q. The standard assumes that unsymmetrical flooding can heel the vessel by 20°. The waterline corresponding to this angle is W_1L_1. Rolling and transient motions can increase the heel angle by 15°, leading to the waterline W_2L_2. Finally, to take into account the relative motion in waves (that is the difference between ship motion and wave-surface motion) we draw another waterline translated up by $h = 1.5$ m; this is waterline W_3L_3. Obviously, unsymmetrical flooding followed by rolling can occur to the other side too so that we must consider the waterline W_4L_4. The waterlines W_3L_3 and W_4L_4 intersect at the point P. Thus, we identify a V-shaped limit line, W_4PL_4, hence the term "V lines." The region below the V lines must be kept watertight; severe restrictions refer to it and they must be read in detail.

11.6.5 The German Navy

The BV 1003 regulations are rather laconic about flooding and damage stability. The main requirement refers to the extent of damage. For ships under 30-m length only one compartment should be assumed flooded. For larger ships a damage length equal to

$$0.18L_{WL} - 3.6 \text{ m}$$

but not exceeding 18 m, should be considered. Compartments shorter than 1.8 m should not be taken into account as such, but should be attached to the adjacent compartments. The leak may occur at any place along the ship, and all compartment combinations that can be flooded in the prescribed leak length should be considered. The damage extends transversely till the centreline, and vertically it is unlimited. The volume permeabilities to be taken into account are given in Table 11.4.

Damage stability is considered sufficient if

- the deck-at-side line does not submerge. This is a less severe requirement than that of IMO;
- without beam wind, and if symmetrically flooded, the ship floats in upright condition and the metacentric height is positive;

Table 11.4 Permeabilities

Liquid	Permeability
Control centre, accomodations, service spaces	0.95
Ammunition magazines	0.80
Provisions	0.50
Machinery	0.85
Bunkers, tanks	0.98
Void spaces	0.98

- in intermediate positions the list does not exceed 25° and the residual arm is larger than 0.05 m in the range up to 40°;
- under a wind of 40 knots openings of intact compartments do not submerge, the list does not exceed 25° and the residual lever arm is larger than 0.05 m in a range between the static and flooding angles.

If not all criteria can be met, the regulations allow for decisions based on a probabilistic factor of safety. For the practising Naval Architect it may be interesting to consult the prescribed permeabilities indicated in an appendix of this chapter.

11.6.6 A Code for Large Commercial Sailing or Motor Vessels

The code published by the UK Maritime and Coastguard Agency specifies that the free flooding of any one compartment should not submerge the vessel beyond the margin line. The damage should be assumed anywhere, but not at the place of a bulkhead. A damage of the latter kind would flood two adjacent compartments, a hypothesis not to be considered for vessels under 85 m. Vessels of 85 m and above should be checked for the flooding of two compartments.

In the damaged condition the angle of equilibrium should not exceed 7° and the range of positive righting arms should not be less than 15° up to the flooding angle. In addition, the maximum righting arm should not be less than 0.1 m and the area under the righting-arm curve not less than 0.015 m rad. The permeabilities to be used in calculations are

Stores	0.60
Stores, but not a substantial amount of them	0.95
Accommodation	0.95
Machinery	0.85
Liquids	0.95 or 0, whichever leads to worse predictions

The expression 'not a substantial amount of them' is not detailed.

11.6.7 A Code for Small Workboats and Pilot Boats

The code published by the UK Maritime and Coastguard Agency contains damage provisions for vessels up to 15 m in length and over, certified to carry 15 or more persons and to operate in an area up to 150 miles from a safe haven. The regulations are the same as those described for sailing vessels in Section 11.6.6, except that there is no mention of the two-compartment standard for lengths of 85 m and over.

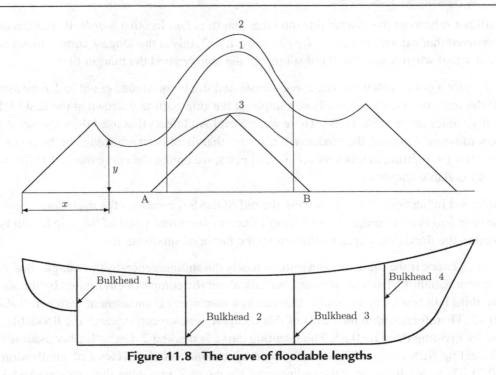

Figure 11.8 The curve of floodable lengths

11.7 The Calculation of the Curve of Floodable Lengths

Today computer programmes receive as input the descriptions of the hull surface and of the internal subdivision. In the simplest form, the input can consist of offsets, bulkhead positions and compartment permeabilities. Then, it is possible to check in a few seconds what happens when certain compartment combinations are flooded. If the results do not meet the criteria relevant to the project, we can change the positions of bulkheads and run flooding and damage stability calculations for the newly defined subdivision. Before the advent of digital computers the above procedure took a lot of time; therefore, it could not be repeated many times. Just to give an idea, manual flooding calculations for one compartment combination could take something like three hours. Usually, the calculations were not purely manual because most Naval Architects used slide rules, adding machines and planimeters. Still it was not possible to speed up much of the work. To improve efficiency, Naval Architects devised ingenious, very elegant methods; one of them produces the **curve of floodable lengths**. To explain it we refer to Figure 11.8. In the lower part of the figure we show a ship outline with four transverse bulkheads; above it we show a curve of floodable lengths and how to use it.

Let us consider a point situated a distance x from the aftermost point of the ship. Let us assume that we calculated the maximum length of the compartment having its centre at x and that will not submerge the margin line, and that length is L_F. In other words, if we consider a compartment that extends from $x - L_F/2$ to $x + L_F/2$, this is the longest compartment with centre at x that when flooded will not submerge the ship beyond the margin line.

Now, we plot a point with the given x-coordinate and the y-coordinate equal to L_F measured at half the scale used for x values. For example, if the ship outline is drawn at the scale 1:100, we plot y values at the scale 1:200. There were Naval Architects that used the same scale for both coordinates; however, the reader will discover that there is an advantage in the procedure preferred by us. Plotting in this way all $[x, L_F]$ pairs, we obtain the curve marked 1; this is the curve of floodable lengths.

As explained in Section 11.4.1, following the old SOLAS regulations the *maximum permissible length* of a compartment having its centre at a given point of the ship length is obtained as the floodable length multiplied by the factor of subdivision.

Now, let us check if the middle compartment meets the submergence-to-the-margin-line requirement. Counting from aft forward, we talk about the compartment limited by the second and the third bulkhead. Let us assume that this is a machinery compartment with permeability $\mu = 0.85$. Therefore, within the limits of this compartment we can increase the floodable lengths by dividing them by 0.85. The resulting curve is marked 2. Let us further assume that we are dealing with a ship subject to a "two-compartment" standard (factor of subdivision $F = 0.5$). Then, we divide by 2 the ordinates of the curve 2, obtaining the curve marked 3. This is the curve of permissible lengths. On the curve 3 we find the point corresponding to the centre of the machinery compartment and draw from it two lines at 45° degrees with the horizontal. The two lines intercept the baseline at A and B. Both A and B are outside the bulkheads that limit the machinery compartment. We conclude that the length of this compartment meets the submergence criterion. Indeed, as the y-coordinate of the curve of floodable lengths is equal to half the length L_F, we obtain on the horizontal axis a length $\overline{AB} = L_F/(\mu F)$, that is the permissible length. It is larger than the length of the compartment. To draw the lines at 45° we can use commercially available set squares (triangles). If we plot both x and y values at the same scale we must draw check lines at an angle equal to arctan 2; there are no set squares for this angle.

In Figure 11.8 we can identify the properties common to all curves of floodable lengths and give more insight into the flooding process.

1. At the extremities the curve turns into straight-line segments inclined 45° with respect to the horizontal. Let us choose any point of the curve in that region. Drawing from it lines at 45°, that is descending along the first or the last curve segment, we reach the extremities of the ship. These are indeed the limits of the floodable compartments at the ship extremities because there is no vessel beyond them.

2. The straight lines at the ship extremities rise up to local maxima. Then the curve descends until it reaches local minima. Usually the ship breadth decreases toward the ship extremities and frequently the keel line turns up. Thus, compartment volumes per unit length decrease toward the extremities. Therefore, floodable lengths in that region can be larger and this causes the local maxima.

3. As we go toward the midship the compartment volumes per unit length increase, while still being remote from the midship. Flooding of such compartments can submerge the margin line by trimming the vessel. Therefore, they must be kept short and this explains the local minima.

4. The curve has an absolute maximum close to the midship. Flooding in that region does not cause appreciable trim; therefore, floodable lengths can be larger.

The term "curve of floodable lengths" is translated as

<table>
<tr><td>Fr</td><td>Courbe des longueurs envahissable</td></tr>
<tr><td>G</td><td>Kurve der flutbaren Längen</td></tr>
<tr><td>I</td><td>curva delle lunghezze allagabile</td></tr>
<tr><td>S</td><td>curva de las esloras inundables</td></tr>
</table>

A very elegant method for calculating points on the curve of floodable lengths was devised by Shirokauer (1928). A detailed description of the method can be found in Nickum (1988), Section 4. A more concise description is given in 1928 by Shirokauer, Section 7.2. The procedure begins by drawing a set of waterlines tangent to the margin line. For each of these lines the Naval Architect calculates the volume and the centre of the volume of flooding water that would submerge the vessel to that waterline. The calculations are based on equations such as (11.3). The boundaries of the compartment are found by trial-and-error using the curve of sectional areas corresponding to the given waterlines lines.

11.8 Summary

Ships can be damaged by collision, grounding, or enemy action. A vessel can survive damage of some extent if the hull is subdivided into watertight compartments by means of watertight bulkheads. The subdivision should be designed to make sure that after the flooding of a given number of compartments the ship can float and be stable under moderate environmental conditions. The subdivision of merchant ships should meet criteria defined by the applicable International Conventions. Warships are subject to damage regulations defined by the respective navies.

The SOLAS convention defines as bulkhead deck the deck reached by the watertight bulkheads. The margin line is a line passing at least 76 mm (3 in.) below the side of the bulkhead deck. If the bulkhead deck is not continuous, the margin line should be defined as a continuous line that is everywhere at least 76 mm below the bulkhead deck. The term floodable length refers to a function of the position along the ship length. For a given position, say P, the floodable length is the maximum length of a compartment with the centre at P and whose flooding will not submerge the vessel beyond the margin line.

Let v be the volume of a compartment calculated from its geometrical dimensions. Almost always there are some objects in the compartment: therefore, the net volume that can be flooded, v_F, is less than v. We call the ratio $\mu = v_F/v$ volume permeability. The same objects that reduce the volume that can be flooded, reduce also the free-surface area that contributes to the free surface effect. We define a surface permeability as the ratio of the net free surface to the total free surface calculated from the geometric dimensions of the compartment. The moment of inertia of the free-surface calculated from the geometry of the compartment should be multiplied by the surface permeability.

There are two methods of calculating the properties of a flooded vessel: the method of lost buoyancy and the method of added weight. In the method of lost buoyancy we assume that a damaged compartment does not provide buoyancy. The displacement of the vessel and the centre of gravity do not change. The ship must change position until the undamaged compartments provide the buoyancy force and moments that balance the weight of the vessel. As the flooding water does not belong to the vessel, but to the surrounding environment, it does not cause a free-surface effect. This method corresponds to what happens in reality; it is the method used by computer programmes when calculating the final stage of flooding. In the method of added weight we consider the flooding water as a weight added to the displacement. The displacement and the centre of gravity change until the equilibrium of forces and moments is established and the level of flooding water is equal to that of the surrounding water. As the flooding water is now part of the vessel, it causes a free-surface effect. The two methods yield the same final equilibrium position and the same righting moment, $\Delta \overline{GM} \sin \phi$, in damage condition. As the displacements are different, the metacentric heights, $\overline{GM}$ also are different so as to yield the same product $\Delta \overline{GM}$.

Three different assessments of the damage condition have been explained in this chapter. The first approach is based on the notion of floodable length. For a given point on the ship length, the floodable length is the maximum length, with the centre at that point, which can be flooded without submerging the ship beyond the margin line. The assessment consists in checking that the length of a compartment with the centre at the given point is not longer than the floodable length multiplied by an appropriate subdivision number. A deterministic damage stability approach consists in assessing all possible damage scenarios, based on a given damage dimension, against the applicable residual stability criteria. Finally, a probabilistic approach defines the ship safety by an index based on all possible damage scenarios. This index is compared with a required safety index.

SOLAS and other codes of practice also prescribe damage stability criteria. For example, some criteria specify minimum value and range of positive residual arms and of areas under the righting-arm curve.

Flooding and damage stability can be studied on ship models, in test basins, or by computer simulation. A paper dealing with the former approach is that of Ross et al. (1997); it refers to Ro-Ro ferries. A few papers dealing with the latter approach are quoted in Chapter 13.

11.9 Examples

EXAMPLE 11.1 (Analysis of the flooding calculations of a simple barge). This example is taken from Schatz (1983). We consider the box-shaped barge shown in Figure 11.9, assuming as initial data $\nabla_I = 1824$ m^3, $\overline{KG} = 3.0$ m, and $LCG = 0$ m. These values were fed as input to the programme Archimedes, together with the information that Compartments 2.1 and 2.2 are flooded. The permeabilities of the two compartments are 1. Using various run options of the programme, we calculate the properties of the intact hull, of the flooded hull, and of the flooded volume. The results are shown in Table 11.5.

The programme ARCHIMEDES uses two systems of coordinates. A system xyz is attached to the ship. The ship offsets, the limits of compartments, the displacement, and the centre of gravity are input in this system. The programme is invoked specifying the numbers of the flooded compartments. The calculations are run in the lost-buoyancy method and the results are given in a system of coordinates, $\xi\eta\zeta$, fixed in space. In this example only the trim changed. A sketch of the coordinate systems involved is shown in Figure 11.10. The data of the damaged hull and of the flooded compartments, columns 3 and 4 in Table 11.5 are given in the $\xi\zeta$ system. To get more insight into the process let us check if the results fulfil the

Figure 11.9 A simple barge—damage calculation

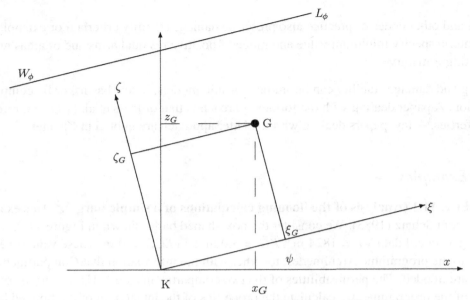

Figure 11.10   A simple barge—coordinate systems used in calculations

equations of equilibrium (11.2). To do this we must use data expressed in the same system of coordinates. For example, we transform the coordinates of the centre of gravity using an equation deduced from Figure 11.10:

$$\xi_G = x_G \cos \psi + z_G \sin \psi = LCG \cos \psi + \overline{KG} \sin \psi \qquad (11.5)$$

First, we calculate

$$\psi = \arctan \frac{\text{trim}}{L_{pp}} = \arctan \frac{-1.092}{76} = 0.823°$$

Table 11.5   Simple barge—compartments 2.1 and 2.2 flooded

	Intact Condition	Damaged Hull	Flooded Compartment
Draught, m	1.999	2.711	2.711
∇, m³	1824.000	2472.682	649.294
Δ, t	1869.600	2534.500	665.5628
$\overline{KG}$, m	3.000		1.285
LCG, m, from midship	0.000		−9.671
LCB, m, from midship	0.000	2.670	
Trim, m	0.000	−1.092	−1.092
$\overline{KB}$, m	0.750	1.337	0.915
$\overline{BM}$, m	1.389	4.427	1.139
$\overline{GM}$, m	4.001		0.454
FS moment of inertia			2736.276

The moment of the intact-displacement volume about the midship section, in the trimmed position, is

$$\nabla_I (LCG\cos\psi + \overline{KG}\sin\psi) = 1824(0\times\cos(-0.823°) + 3\sin(-0.823°) = -78.616\ \text{m}^4$$

The moment of the flooded compartment equals

$$v\cdot lcg = 649.294\times(-9.671) = -6279.322\ \text{m}^4$$

The moment of the flooded barge resulting directly from hydrostatic calculations is

$$\nabla_F\cdot LCF_F = 2472.682\times(-2.570) = -6354.793\ \text{m}^4$$

The deviation between the two moments is less than 0.05%; the equilibrium of moments is fulfilled. As to the equilibrium of forces, we can easily see that $1824 + 649.294$ is practically equal to 2472.682.

The programme ARCHIMEDES, like other computer programmes, carries on calculations by the lost-buoyancy method. Then the final displacement volume remains equal to the intact volume, 1824 m^3, while the calculated metacentric height, $\overline{GM}$, is 2.858 m The righting moment for small heel angles, in the lost-buoyancy method, is

$$M_{RL} = 1.025\times1824\times2.858\sin\phi = 5342.3\sin\phi\ \text{t m}$$

As an exercise let us compare this moment with that predicted by the added-weight method. Hydrostatic calculations for the damaged barge yield $\overline{KM} = 5.764$ m. Capacity calculations for the Compartments 2.1 and 2.2 give a total volume of flooding water equal to 649.294 m^3, with a height of the centre of gravity at 1.286 m. In Table 11.6 we calculate the damage displacement and the coordinates of its centre of gravity in the added-weight method. Capacity calculations for the flooded compartments yield a total moment of inertia of the free surfaces, $i = 2736.276$ m^4. The corresponding lever arm of the free surface is

$$l_F = \frac{i}{\nabla_A} = \frac{2736.276}{2473.294} = 1.106\ \text{m}$$

The resulting metacentric height is

$$\overline{GM}_A = \overline{KM} - \overline{KG}_A - l_F = 5.764 - 2.550 - 1.106 = 2.107\ \text{m}$$

and the righting moment for small heel angles, in the added-weight method

$$M_{RA} = \rho\nabla_A\overline{GM}_A\sin\phi = 1.025\times2473.294\times2.107\sin\phi = 5341.7\sin\phi\ \text{t m}$$

Due to errors of numerical calculations the values of M_{RL} and M_{RA} differ by 0.03%; in fact they are equal, as expected.

Table 11.6 Simple barge—added-weight calculations

	Volume (m^3)	kg (m)	Moment (m^4)	lcg (m)	Moment (m^4)
Intact hull	1824.000	3.000	5472.000	0.000	0.000
Flooding water	649.294	1.286	834.992	−9.671	−6279.322
Flooded hull	2473.294	2.550	6306.992	−2.539	−6279.322

Table 11.7 Flooding calculations—a comparison of methods considering permeability

	Intact Condition	Damaged, Lost Buoyancy	Damaged, by Added Weight
Draught, m	1.500	1.829	1.829
∇, m^3	150.000	150.000	182.927
Δ, t	153.750	153.750	187.500
$\overline{KB}$, m	0.750	0.915	0.915
$\overline{BM}$, m	1.389	1.139	1.139
$\overline{KG}$, m	1.500	1.500	1.395
$\overline{GM}$, m	0.639	0.554	0.454
$\Delta\overline{GM}$, m	98.229	85.104	85.104

11.10 Exercise

Exercise 11.1 (Comparison of methods while considering permeability). In Sections 11.3.1 and 11.3.2 we compared the lost-buoyancy method to the added-weight method, but, to simplify things, we did not consider permeabilities. This exercise is meant to show the reader that even if we consider permeabilities, the two methods yield the same draught and the same righting moment in damage condition. The reader is invited to redo the calculations in the mentioned sections, but under the assumption that the volume and surface permeabilities of the flooded compartment equal 0.9.

A hint for using the method of lost buoyancy is that the waterplane area, LB, is reduced by the floodable area of Compartment 2, μBl. The hint for the method of added weight is that the volume of flooding water equals μlBT_A, where T_A is the draught in damage condition.

The results should be those shown in Table 11.7.

CHAPTER 12

Linear Ship Response in Waves

Chapter Outline

12.1 Introduction

The title of the book is 'Ship hydrostatics and stability.' This chapter describes processes that are not hydrostatic, but can affect stability. Most modes of capsizing researched in the last years imply phenomena and theories that go beyond the classical field of ship hydrostatics. The research should lead to so called *second generation* criteria of stability. Much of the activity is carried on within the framework of IMO and we cannot ignore today these new developments. This chapter is meant as a bridge between the study of ship stability based on classical ship hydrostatics and the new theories.

We elaborate in this chapter on some reservations expressed in Section 6.13 and sketch the way toward more realistic models. First, we need a wave theory that can be used in the description of real seas. Therefore, we introduce the theory of **linear waves**. Next, we show how real seas can be described as a **superposition** of regular waves. This leads to the introduction of **sea spectra**. A floating body moves in six degrees of freedom. The oscillating body generates waves that absorb part of its energy. The integration of pressures over the hull surface yields the forces and moments acting on the body. We return here, without detailing, to the notions of **added mass** and **damping coefficients** introduced in Section 6.13. A full treatment would go far beyond the scope of the book; therefore, we limit ourselves to mentioning a few important results.

The problems of mooring and anchoring deserve special treatment and their importance has grown with the development of offshore structures. We cannot discuss here the behaviour of **compliant floating structures**, that is moored floating structures, but give an example of how the mooring can change the natural frequencies of a floating body.

We mention in this chapter a few methods of reducing ship motions, mainly the roll. This allows us to show that under very particular conditions, free water surfaces can help, a result that seems surprising in the light of the theory developed in Chapter 6.

With the exception of the simplest models of uncoupled motions, the other models introduced in this chapter are too complex to yield explicit mathematical expressions that can be directly applied in engineering practice. It is only possible to implement the models in computer programmes that yield numerical results. The input to such programmes is a statistical description of the sea considered as a **random process**. Correspondingly, the output, that is the ship **response**, is also a random process.

This chapter assumes the knowledge of more mathematics than the rest of the book. Mathematical developments are concise, leaving to the interested reader the task of completing them or to refer to specialized books. The reader who cannot follow the mathematical treatment can find in the summary a non-mathematical description of the main subjects.

12.2 Linear Wave Theory

In Section 10.2.3 we mentioned the theory of trochoidal waves. Trochoidal waves approximate well the shape of swells and were prescribed by certain codes of practice for stability and bending-moment calculations. Another wave theory is preferred for the description of real seas and for the calculation of ship motions; it is the theory of linear waves. The basic assumptions are:

1. the sea water is **incompressible**;
2. there is no viscosity, i.e., the sea water is **inviscid**;
3. there is no surface tension;
4. no fluid particle turns around itself, i.e., the motion is **irrotational**;
5. the wave amplitude is much smaller than the wavelength.

The first assumption, that of incompressibility, is certainly valid at the small depth and with the wave velocities experienced by surface vessels. This is a substantial difference from phenomena experienced in aerodynamics. Excepting roll damping, the second assumption, the

Figure 12.1 Two-dimensional waves: swell

lack of viscous phenomena, leads to results confirmed by experience. For roll, certain corrections are necessary; often they are done by empiric means. Surface tension plays a role only for very small waves, such as the ripple that can be seen on the surface of a swell. We shall see how the fourth hypothesis, that is irrotational flow, makes possible the development of an elegant **potential theory** that greatly simplifies the analysis. The fifth hypothesis, low-amplitude waves, is not very realistic; surprisingly, it leads to realistic results.

We consider two-dimensional waves, that is waves with parallel crests of infinite length, such as shown in Figure 12.1. The crests are parallel to the y-direction and we are only interested in what happens in the x- and z-directions. Let u be the horizontal and w the vertical velocity of a water particle. We note by ρ the water density. The theory of fluid dynamics shows that the rate of change of the mass of a unit volume of water is

$$\frac{\partial(\rho u)}{\partial x} + \frac{\partial(\rho w)}{\partial z}$$

The density of an incompressible fluid, ρ, is constant. Then, the condition that the mass of a unit volume of water does not change is expressed as

$$\frac{\partial u}{\partial x} + \frac{\partial w}{\partial z} = 0 \tag{12.1}$$

Equation (12.1) is known as the **equation of continuity**; it states that the **divergence** of the vector with components u, w is zero. The assumption of irrotational motion is expressed by the condition that the **curl** of the vector with components u, w is zero (for the interpretation of the curl see the Appendix of this chapter). In two dimensions this is

$$\frac{\partial w}{\partial x} - \frac{\partial u}{\partial z} = 0 \tag{12.2}$$

We define a **velocity potential**, Φ, such that

$$u = \frac{\partial \Phi}{\partial x}, \quad w = \frac{\partial \Phi}{\partial z} \tag{12.3}$$

These expressions verify, indeed, Eq. (12.2). Substituting Eq. (12.3) into Eq. (12.1) yields the **Laplace equation**

$$\frac{\partial^2 \Phi}{\partial x^2} + \frac{\partial^2 \Phi}{\partial z^2} = 0 \tag{12.4}$$

This equation must be solved together with a set of **boundary conditions**. Let $\zeta(x, z, t)$ be the **elevation** of the free surface and z the vertical coordinate measured from the mean water surface upwards. In simple terms, ζ represents the wave profile. The **kinematic condition**

$$\frac{\partial \zeta}{\partial t} = \frac{\partial \Phi}{\partial z}, \quad \text{at } z = 0 \tag{12.5}$$

states that the vertical velocity of the wave surface equals the vertical velocity of a water particle at the mean water level. This is an approximation acceptable for small wave amplitudes.

The **dynamic free-surface condition** states that the water pressure on the wave surface is equal to the atmospheric pressure

$$\frac{\partial \Phi}{\partial t} + g\zeta + \frac{1}{2}\left[\left(\frac{\partial \Phi}{\partial x}\right)^2 + \left(\frac{\partial \Phi}{\partial z}\right)^2\right] = 0 \quad \text{on } z = \zeta(x, y, z) \tag{12.6}$$

Assuming small wave amplitudes we can neglect the squares of particle velocities and thus we remain with the condition

$$g\zeta + \frac{\partial \Phi}{\partial t} = 0, \quad \text{at } z = 0 \tag{12.7}$$

From Eqs. (12.5) and (12.7) we obtain the linearized free-surface condition

$$\frac{\partial^2 \Phi}{\partial t^2} + g\frac{\partial \Phi}{\partial z} = 0 \tag{12.8}$$

Additional boundary conditions must be written for the sea bottom, for walls that limit the water domain, and for the surfaces of bodies floating in that domain. As the water does not pass through such boundaries, the velocity components normal to such boundaries should be zero.

Let the **wavelength** be λ, and the **wave number** $k = 2\pi/\lambda$. The vertical coordinate of a water particle is $z = 0$ at the mean sea level, and $z = -d$, at the depth d. We give the results of the theory for infinite-depth water as these are the most interesting for sea-going ships. We leave to an exercise the proof that these results fulfil the Laplace equation and the boundary conditions. The solution that interests us is the potential

$$\Phi = \frac{g\zeta_0}{\omega} \cdot e^{kz} \cos(\omega t - kx) \tag{12.9}$$

The equation of the sea surface is

$$\zeta = \zeta_0 \sin(\omega t - kx) \tag{12.10}$$

A MATLAB animation of this equation, as shown in Biran and Breiner (2002, p. 733), let us visualize the propagation of the wave form.

The potential defined by Eq. (12.9) fulfils the boundary condition of Eq. (12.7) if the following relationship exists between the wavelength, λ, and the **wave period**, T:

$$\lambda = \frac{g}{2\pi}T^2 \tag{12.11}$$

Figure 12.2 shows the propagation of the wave described by Eq. (12.10). The wave period is $T = 6.5$ s, and the wavelength given by Eq. (12.11) is $\lambda = 65.965$ m. The assumed wave height, $H = 2\zeta_0$, equals $\lambda/20$, a ratio often used in Naval Architecture.

The speed of propagation of the wave shape is called **celerity**, a term that comes from the Latin "celeritas," speed. Using Eq. (12.11) we find the celerity

$$c = \frac{\lambda}{T} = \sqrt{\frac{g\lambda}{2\pi}} \tag{12.12}$$

Figure 12.2 The propagation of a linear wave

We immediately see that long waves propagate faster than short waves. Therefore, we say that water waves are **dispersive**. Acoustic waves, for example, are not dispersive.

The components of the water-particle velocity are

$$u = \omega \zeta_0 e^{kz} \sin (\omega t - kx) \qquad (12.13)$$

$$w = \omega \zeta_0 e^{kz} \cos (\omega t - kx) \qquad (12.14)$$

We invite the reader to use the latter equations and prove that in infinite-depth water the particles move on circular orbits whose radii decrease with depth. At a depth equal to about one-half wavelength the orbital motion becomes negligible.

Figure 12.3 shows the orbit of a water particle at the surface of the wave represented in Figure 12.2. The orbital velocities, u and v, are shown at two time instants, namely $t = 1$ s and $t = 4$ s.

In preparation for the next section let us calculate the wave energy per unit sea-surface area; it is the sum of the kinetic and potential energies. For the former we consider an elementary volume of length dx, 1 m width (in the y-direction), and height dz. As for sailing ships we are mainly interested in infinite-depth seas, we first integrate between $z = -\infty$ and $z = 0$ neglecting the wave amplitude assumed to be relatively small. Next, we integrate with respect

$$T = 6.5 \text{ s}, \quad \lambda = 65.9653 \text{ m}$$

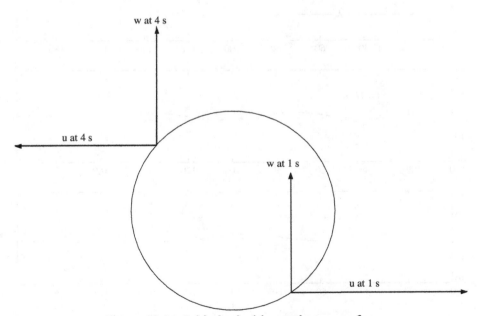

Figure 12.3   Orbital velocities at the sea surface

to x in an interval covering a wavelength. The kinetic energy contained in the given volume is

$$K = \int_0^\lambda \int_{-\infty}^0 \frac{\rho}{2}(u^2 + w^2)dz\,dx = \frac{\rho\omega^2\zeta_0^2}{2}\int_0^\lambda \int_{-\infty}^0 e^{2kz}\,dz\,dx \qquad (12.15)$$

Taking into account Eq. (12.11), and with $H = 2\zeta_0$ for the wave height, we obtain

$$K = \frac{\rho g H^2 \lambda}{16} \qquad (12.16)$$

The variation of the potential energy of a volume element defined as above is proportional to its distance from the mean surface elevation, z. Therefore, we write for the potential energy

$$dU = \rho g z\, dx\, dz \qquad (12.17)$$

The potential energy of the fluid mass up to the mean level is a constant. Integrating from the mean level to the surface elevation and for a wavelength we write

$$U = \rho g \int_o^\lambda \int_0^\zeta z\, dz\, dx = \frac{\rho g \zeta_0^2}{2}\int_0^\lambda \sin^2(\omega t - kx)dx \qquad (12.18)$$

with the result

$$U = \frac{\rho g H^2 \lambda}{16} \qquad (12.19)$$

Finally, the total energy per unit area is

$$E_W = \frac{K + U}{\lambda} = \frac{\rho g H^2}{8} \qquad (12.20)$$

The wave energy is directly proportional to the square of the wave height.

12.3 Modelling Real Seas

We can register the elevation of the sea at a given point and obtain a function of time $\zeta = f(t)$. Alternatively, we can consider the sea surface at a given time instant, t_0, and a given coordinate y_0. Then, we can register the elevations along the x-axis and obtain a function $\zeta = g(x)$. Both representations have an **irregular** aspect in the sense that there is no pattern that repeats itself. The linear wave theory allows us to represent the sea surface as the superposition of a large number of sine waves, that is

$$\zeta = \sum_{i=1}^N A_i \sin(\omega_i t - k_i x + \epsilon_i) \qquad (12.21)$$

where A_i is the wave amplitude, ω_i the angular frequency, k_i the wave number, and ϵ_i the phase of the ith wave. We assume that the numbers ϵ_i are random and uniformly distributed

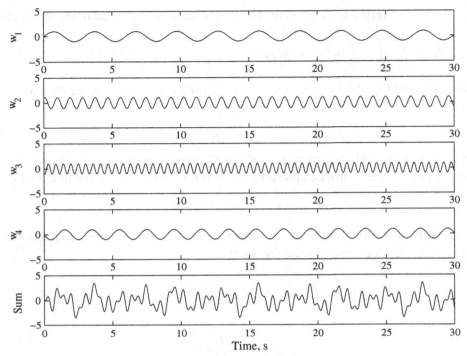

Figure 12.4 The superposition of four waves

between 0 and 2π. To explain how the superposition of sine waves can produce an irregular sea we refer to Figure 12.4. The lower curve represents the sum of the four sine waves plotted above it. A periodical pattern can still be detected; however, as the number of components increases, any periodicity disappears and there is no pattern that repeats itself.

As the wave phases, ϵ_i, are random, the sea surface is a random process. Let us consider a segment of a wave record, such as in Figure 12.5. We distinguish two types of trough-to-crest heights. When measuring the height H_1, the trough and the crest lie on two sides of the mean sea level, while H_2 is measured between two points on the same side of the mean sea level. Experience shows that heights of the first type, H_1, follow approximately the **Rayleigh distribution**

$$f(H) = \frac{H}{4m_0} \cdot e^{-\frac{H^2}{8m_0}} \tag{12.22}$$

The **mean height** is

$$H_m = \int_0^\infty Hf(H)dH = \sqrt{2\pi m_0} \tag{12.23}$$

Figure 12.5 Significant wave height

An important characteristic is the **significant wave height** defined as the **mean of the highest third of the wave heights**

$$H_{1/3} = \int_{H_0}^{\infty} H f(H) dh \qquad (12.24)$$

where H_0 is defined by

$$\int_{H_0}^{\infty} f(H) dh = 1/3 \qquad (12.25)$$

The significant height allows the calculation of other characteristics, for example the sea spectrum. A natural question arises: given the significant height, $H_{1/3}$, what is the maximum wave height, H_{max}, that can be expected? It appears that the larger the number of waves considered, the higher the maximum wave height that can be expected. Using data in Bonnefille (1992) we find that $H_{max}/H_{1/3}$ varies from 1.2 for a sample of 10 waves, to 1.92 for 1000 waves.

Let us return now to Eq. (12.21). It can be shown that the total energy of N wave components, per unit sea area, is proportional to

$$E_0 = \frac{1}{2} \sum_{i=1}^{N} A_i^2 \qquad (12.26)$$

The proof of Eq. (12.26) is related to Parseval's theorem (see, for example, Kreyszig, 2006). To define the **wave spectrum**, $S(\omega)$, we consider a band of frequencies extending from ω_j to $\omega_j + \Delta\omega$ and write

$$S(\omega_j)\Delta\omega = \frac{1}{2}A_j^2 \qquad (12.27)$$

where A_j is the amplitude of the wave component in the frequency band considered by us. For example, in Figure 12.6 we consider the band of breadth $\Delta\omega$ centred around 0.8 rad s^{-1}. In this case $A_j^2/2 = 0.08$ m^2 s. The area of this band, like the whole area under the spectrum curve, is measured in m^2.

The wave spectrum describes the distribution of wave energy versus wave angular frequency. At the end of Section 12.6 we shall find an important use of this concept.

Wave spectra can be obtained from measurements. A number of formulae have been proposed for calculating **standard spectra** on the basis of a few given or measured sea characteristics. We shall give only one example, the **Pierson-Moskovitz spectrum** as described by Fossen (1994)

$$S = A\omega^{-5}e^{-B\omega^{-4}}, \quad \text{m}^2 \text{ s} \qquad (12.28)$$

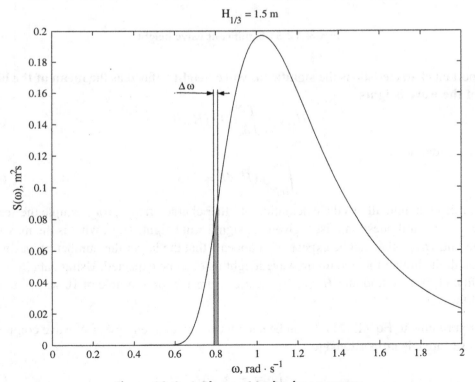

Figure 12.6 A Pierson-Moskovitz spectrum

where

$$A = 8.1 \times 10^{-3} g^2, \quad \text{m}^2 \text{ s}^{-4}$$

$$B = 0.0323 \left(\frac{g}{H_{1/3}} \right)^2, \quad \text{s}^{-4}$$

This spectrum corresponds to fully developed seas recorded in the North Atlantic; an example is shown in Figure 12.6. Other sea spectra are described, for example, in Molland (2008, Section 1.7).

The theory of linear waves exposed in this section is a first-order approximation in which the wave shape moves, but there is no mass transport. This approximation is sufficient for moving ships as their speed is usually larger than the 'drift' caused by waves. For stationary structures it may be necessary to consider higher-order approximations that predict a drift.

12.4 Wave Induced Forces and Motions

Like any other free body, a ship moves in six degrees of motion; we describe them with the aid of Figure 12.7. The six motions of a ship have traditional names that were adopted in the previous century also for airplanes and cars. We follow the notation of Faltinsen (1993). Three motions are linear; they are described below:

> **Surge**, along the x-axis; we note it by η_1.
> **Sway**, in the direction of the y-axis; we use the notation η_2.
> **Heave**, along the z-axis; we note it by η_3.

The other three degrees of freedom define angular motions, as detailed below:

> **Roll**, around the x-axis; we note it by η_4.
> **Pitch**, around the y-axis; we use the notation η_5.
> **Yaw**, around the z-axis; it is noted by η_6.

The motion of any point on a floating body is the resultant of all six motions

$$s = \eta_1 \mathbf{i} + \eta_2 \mathbf{j} + \eta_3 \mathbf{k} + \omega \times \mathbf{r} \tag{12.29}$$

where $\mathbf{i}$ is the unit vector on the x-axis, $\mathbf{j}$ the unit vector on the y-axis, $\mathbf{k}$ the unit vector on the z-axis, and $\times$ denotes the vector product. The rotation vector is

$$\omega = \eta_4 \mathbf{i} + \eta_5 \mathbf{j} + \eta_6 \mathbf{k}$$

and the position vector of a point with coordinates x, y, z is

$$\mathbf{r} = x\mathbf{i} + y\mathbf{j} + z\mathbf{k} \tag{12.30}$$

Figure 12.7   Ship motions—definitions

For example, the vertical motion of the point with coordinates x, y, z is the resultant of the heave, roll, and pitch motions

$$(\eta_3 + y\eta_4 - x\eta_5)\mathbf{k}$$

12.5 Uncoupled Motions

For particular purposes we can write an equation of motion in one degree of freedom, without considering the influence of the motions in the other degrees of freedom. We say that such equations describe **uncoupled** motions. Thus, in Section 6.7 we developed a non-linear equation of roll, the non-linear term being $\rho\Delta\overline{GZ}$. In Section 6.8 we linearized the equation for small roll angles. We neglected the damping term that for roll is non-linear. One way of deriving an uncoupled roll equation with linear damping and a forcing term due to a beam wave is based on Figure 12.8 and the following assumptions:

- the displacement mass, Δ, is constant during the roll;
- the wavelength is sufficiently large with respect to the ship beam so that the wave profile across the ship can be approximated by a plane;
- the damping is proportional to the angular speed, $\dot{\phi}$;
- the buoyancy and the ship weight act perpendicularly to the wave surface.

Figure 12.8   Rolling in beam waves

Devauchelle (1986) bases the last assumption on a proof that, with an error of less than 10%, the vector of apparent weight of a body floating in trochoidal waves can be considered normal to the wave surface. In Exercise 12.3 we invite the reader to prove for a simple body that when the water surface is an inclined plane, the hydrostatic force acting on that body is perpendicular to the inclined waterplane.

Let the heel angle measured from the true horizontal be ϕ, and the wave slope α. Then, the righting arm is $\overline{GM} \sin (\phi - \alpha)$. With the notation η_4 for the roll angle we can write the equation

$$J\frac{d^2\eta_4}{dt^2} + b\frac{d\eta_4}{dt} + g\Delta\overline{GM} \sin (\eta_4 - \alpha) = 0 \qquad (12.31)$$

For small angles and with the definition of the radius of gyration, i, given in Eq. (6.24), we rewrite the equation as

$$\Delta i^2\frac{d^2\eta_4}{dt^2} + b\frac{d\eta_4}{dt} + g\Delta\overline{GM}(\eta_4 - \alpha) = 0 \qquad (12.32)$$

Finally, dividing by Δi^2, noting $n = b/(2\Delta i^2)$, and transferring the term in α to the right-hand side, we obtain

$$\frac{d^2\eta_4}{dt^2} + 2n\frac{d\eta_4}{dt} + \omega_{n4}^2\eta_4 = \omega_{n4}^2\alpha \qquad (12.33)$$

where n is a linear damping coefficient and ω_{n4} is the ship natural angular frequency in roll (see Section 6.8). The angle α is the wave slope; to find an expression for it we refer to

Eq. (12.10) and obtain:

$$\frac{d\zeta}{dy} = -\zeta_0 k \cos(\omega_w t - ky) \tag{12.34}$$

where ω_w is the wave angular frequency. To employ the system of coordinates used throughout this book, we consider the wave crests parallel to the ship x-axis. Then, for $y = 0$, and with the definition of the wave number $k = 2\pi/\lambda$,

$$\alpha = \frac{2\pi \zeta_0}{\lambda} \cos \omega_w t = \frac{2\pi \zeta_0}{\lambda} \sin(\omega_w t - \pi/2) \tag{12.35}$$

so that the linear equation of rolling in beam seas becomes

$$\frac{d^2 \eta_4}{dt^2} + 2n\frac{d\eta_4}{dt} + \omega_{n4}^2 \eta_4 = \omega_{n4}^2 \frac{2\pi \zeta_0}{\lambda} \sin(\omega_w t - \pi/2) \tag{12.36}$$

The general solution of this equation is the sum of the solution of the equation without right-hand side (the homogeneous equation) and a particular solution of the complete equation. Devauchelle (1986) analyzes in detail both parts; we are going to treat them more succinctly and derive only the results that interest us. The characteristic equation of the homogeneous equation is

$$r^2 + 2nr + \omega_{n4}^2 = 0$$

with the discriminant

$$\sqrt{n^2 - \omega_{n4}^2}$$

For conventional ships the damping is very small so that the roots of the characteristic equation are always imaginary, i.e.,

$$r = -n \pm i\sqrt{\omega_{n4}^2 - n^2}$$

Noting $\omega_0^2 = \omega_{n4}^2 - n^2$, the solution of the homogeneous equation is

$$\eta_{4t} = e^{-nt}(C_1 \cos \omega_0 t + C_2 \sin \omega_0 t) \tag{12.37}$$

The constants C_1 and C_2 can be found immediately from initial conditions. What interests us is that this part of the ship response dies out rapidly, so that the *steady state* ship response is given by the particular solution of the complete Eq. (12.36). We assume for this a solution of the form

$$\eta_{4s} = \Phi_0 \cos(\omega_w t - \epsilon) \tag{12.38}$$

To find the amplitude, Φ_0, and the phase, ϵ, we can substitute the above solution into Eq. (12.36) and assume for the argument $(\omega_w t - \epsilon)$ the values 0 and $\pi/2$. In this book we considered only steady state solutions. In the investigation of some modes of capsizing, such as broaching to, one should consider the complete, *transient state* solution, for example

$$\eta_4 = \eta_{4t} + \eta_{4s}$$

The roll amplitude reaches a maximum for $\omega_w = \omega_0$. When the ship rolls close to this frequency we talk about **synchronous rolling**.

A full treatment of the above equations can be found also in Francescutto (2004). For a more realistic analysis of the roll motion, first of all one must take into account the added mass in roll, A_{44}, so that the factor that multiplies $d^2\eta_4/dt^2$ becomes $(J + A_{44})$. Next, the roll equation should consider the non-linear natures of damping and of the restoring force. The number of papers that adopt this approach is considerable. We cite as example Francescutto and Contento (1999) and Bulian (2005). Further, rolling is coupled to other motions (see, for example, Francescutto, 2002; IMO, 2004a). A comprehensive review of the formulations adopted by various researchers can be read in Kröger (1987). Non-linear models of roll are used nowadays also in the analysis of parametric resonance (see, for example, Holden et al., 2007).

Equations for uncoupled pitch motion can be developed in the same way as those of roll, substituting $\overline{GM_L}$ for $\overline{GM}$. For example, Schneekluth (1988) gives the following equation for undamped pitch:

$$\frac{d^2\eta_5}{dt^2} + \frac{g\overline{GM_L}}{i_{55}^2}\eta_5 - \frac{g\overline{GM_L}}{i_{55}^2}\gamma \sin\frac{2\pi t}{T_E} = 0 \qquad (12.39)$$

where i_{55} is the radius of inertia of the ship mass about the Oy-axis, γ is the maximum pitch amplitude, and T_E is the period of encounter. Obviously

$$\omega_{n5} = \frac{\sqrt{g\overline{GM_L}}}{y_{55}} \qquad (12.40)$$

is the ship natural, angular frequency in pitch, and

$$\omega_E = \frac{2\pi}{T_E}$$

is the angular frequency of encounter.

We use Figure 12.9 to develop an equation of the uncoupled heave motion

$$(m + A_{33})\ddot{\eta}_3 + b\dot{\eta}_3 + \rho g A_W \eta_3 = \rho g A_W \zeta_0 \cos\omega_E t \qquad (12.41)$$

Above, we assumed that the wavelength is large compared to the dimensions of the waterplane. In Figure 12.9b we see a mass-dashpot-spring analogy of the heaving body. This analogy holds only for the form of the governing equations. In Figure 12.9b the damping coefficient, b, is a constant. In Figure 12.9a the added mass in heave, A_{33}, and the damping coefficient, b, are functions of the frequency of oscillation. After the extinction of transients, that is in steady state, the frequency of oscillation is equal to the exciting frequency, that is the wave frequency, ω_w, for a body that does not move, and the frequency of encounter, ω_E, for a moving, floating body.

$$(m + A_{33})\ddot{\eta}_3 + b\dot{\eta}_3 + k\eta_3 = F$$

(a)
$$k = \rho g A_W$$
$$F = \rho g A_W \, \zeta_0 \cos \omega t$$

(b)
$$m\ddot{x} + b\dot{x} + kx = F_0 \cos \omega t$$

$\zeta = \zeta_0 \cos \omega t$

Wave surface

W_0

L_0

m

k

b

m

$A_W =$
Waterplane
area

x

$F = F_0 \cos \omega t$

Figure 12.9 A heaving, floating body as a second-order dynamic system. (a) A floating body, (b) A mass-spring-damper system

Let us return to the terms that are proportional to motion in the equations of roll, pitch, and heave

$$g\Delta \overline{GM}\eta_4, \quad g\Delta \overline{GM}_L\eta_5, \quad \rho g A_W \eta_3$$

These terms represent two hydrostatic moments and one hydrostatic force that oppose the motion and tend to return the floating body to its initial position. The collective name for these moments and force are **restoring forces**. Only the roll, pitch, and heave motion are opposed by hydrostatic restoring forces. There are no hydrostatic restoring forces that oppose surge, sway, or yaw.

12.6 Coupled Motions

The equations of uncoupled motions are simplified models that allow us to reach a few important conclusions. In reality, certain couplings exist between the various motions. Thus, we already know that during roll the centre of buoyancy moves along the ship causing pitch. As pointed out by Schneekluth (1988), the combination of roll and pitch motions causes an oscillation of the roll axis and induces yaw. Also, the combination of roll and pitch induces heave. Moreover, one motion can influence the added masses and the damping coefficients of other motions. The most complete model of **coupled motions** is

$$(\mathbf{M} + \mathbf{A})\ddot{\eta} + \mathbf{B}\dot{\eta} + \mathbf{C}\eta = \mathrm{Re}(\mathbf{F}e^{-i\omega_E t}) \tag{12.42}$$

Above, $\mathbf{M}$ is a 6-by-6 matrix whose elements are the ship mass and its moments of inertia about the three axes of coordinates, and $\mathbf{A}$ is a 6-by-6 matrix of added masses (general term describing all added masses and added moments of inertia). The vectors of motions, speeds, and accelerations are

$$\eta = \begin{bmatrix} \eta_1 \\ \eta_2 \\ \eta_3 \\ \eta_4 \\ \eta_5 \\ \eta_6 \end{bmatrix}, \quad \dot{\eta} = \begin{bmatrix} \dot{\eta}_1 \\ \dot{\eta}_2 \\ \dot{\eta}_3 \\ \dot{\eta}_4 \\ \dot{\eta}_5 \\ \dot{\eta}_6 \end{bmatrix}, \quad \ddot{\eta} = \begin{bmatrix} \ddot{\eta}_1 \\ \ddot{\eta}_2 \\ \ddot{\eta}_3 \\ \ddot{\eta}_4 \\ \ddot{\eta}_5 \\ \ddot{\eta}_6 \end{bmatrix}$$

The expression $Re(\mathbf{F}e^{-i\omega_E t})$ means the real part of the vector of sinusoidal exciting forces and moments.

For a ship displaying port-to-starboard symmetry a part of the elements of the matrix $\mathbf{M}$ are zero, and another part are symmetric. The system of six ordinary differential equations can be simplified in many practical situations. Thus, for a floating structure presenting symmetry about the xOz plane, and with the centre of gravity in the position $(0, 0, z_G)$, Faltinsen (1993) shows that the matrix of inertias becomes

$$\mathbf{M} = \begin{bmatrix} M & 0 & 0 & 0 & Mz_G & 0 \\ 0 & M & 0 & -Mz_G & 0 & 0 \\ 0 & 0 & M & 0 & 0 & 0 \\ 0 & -Mz_G & 0 & I_4 & 0 & -I_{46} \\ Mz_G & 0 & 0 & 0 & I_5 & 0 \\ 0 & 0 & 0 & -I_{46} & 0 & I_6 \end{bmatrix}$$

where M is the mass of the floating body, I_4 the moment of inertia about the x-axis, I_{46} the product of inertia about the x- and z-axes, and I_6 the moment of inertia about the z-axis. Certain symmetries also can appear in the matrices of added masses, $\mathbf{A}$, and damping coefficients, $\mathbf{B}$. Remember, added masses and damping coefficients are functions of the frequency of oscillation. For a structure symmetric about the xOz plane the motions of surge, heave, and pitch (vertical-plane motions) can be uncoupled from those of sway, roll, and yaw.

The equations shown above are linear. Then, if for a wave amplitude equal to 1 the resulting motion amplitude is η_a, for a wave amplitude equal to A the motion amplitude will be $A\eta_a$. Further, the principle of superposition applies to motions as it applies to waves. The response to the sum of several waves is the sum of the responses to the individual waves. Then, if we characterize the exciting waves by their spectrum, we can characterize the resulting motion by a motion spectrum.

In Section 6.9.5 we introduced the concept of transfer function for a simple case of roll motion. The transfer function obtained from a differential equation such as those shown in this chapter is a function of frequency. Let the transfer function of the ith motion be $Y_i(\omega)$. The spectrum of the respective motion, $S_{\eta_i}(\omega)$, is related to the wave spectrum, $S_\omega(\omega)$, by the relationship

$$S_{\eta_i}(\omega) = \left[Y_i(\omega)Y_i(-\omega)\right] S_\omega(\omega) \qquad (12.43)$$

The expression between square brackets is called **response amplitude operator**, shortly **RAO**. The response amplitude operators of the various motions can be obtained from the coupled equations of all motions. All motions occur at the frequency of the exciting force, but have different phases. Therefore, the various motions at one point of the ship should be added vectorially. The resultant is a vector that rotates with the angular speed ω_E.

12.7 Dangerous Situations and Modes of Capsizing

In the beginning of this chapter we mentioned that the study of some dangerous situations and modes of capsizing needs more tools than those provided by classical ship hydrostatics. In this section we give a summary of these cases using the terminology employed by IMO and related researchers (IMO, 2008a).

The weather criteria described in Chapter 8 correspond to the **Dead ship condition**, a condition in which all machines are out of operation and there is no possibility of restarting the propulsion machines. Once most ships had superstructures approximately symmetric about the midship section. In dead ship condition such vessels turn so that they roll in beam seas. The influence of waves and wind is then maximal and large-amplitude roll angles can develop. Many modern ships present no more the above symmetry. In the dead ship condition these vessels turn into a position that makes an angle with waves and wind. Therefore, they are subjected to motions in more degrees of freedom and the situation may become dangerous. Belenky et al. (2011) treat the dead ship condition in Chapter 5 of their work. A proposal for the assessment of stability in the dead sea condition was submitted by Japan in IMO (2006).

In Chapter 9 we explain the physics of *parametric resonance* and the mathematical basis for studying it. Various researchers show that the phenomenon can be enhanced by other ship motions, especially by non-linear couplings with them. Thus, advanced studies of parametric resonance belong to the general field of ship motions and to applied *chaotic* dynamics.

As defined by Belenky et al. (2011), **surf-riding** occurs when a wave, approaching from the stern, captures a ship on the front slope of the wave and accelerates her to the wave celerity. This can lead to *broaching-to* (or simply **broaching**), a term we introduced in Section 6.14. In this situation control of the ship is lost and the resulting yaw can be so violent that the centrifugal force causes a large heel angle and even capsizing. This phenomenon belongs not only to ship motions, but also to the field of *manoeuvering*. According to Belenkin et al. (2011), surf-riding is probable when the wavelength equals between 0.75–2L and the ship speed is around 0.75 the wave celerity. The cited authors study broaching and surf-riding in the phase plane. Referring to small vessels, Tuite and Renilson (1997) develop a six degrees of freedom, non-linear model and a method for designing a rudder efficient in following waves. Other studies are due to Maki et al. (2010) and Hashimoto et al. (2011). In Chapter 8 we explained that ship safety depends not only on design, but also on operation. To help in this direction, IMO issued several documents for **operational guidance**, a concept defined in IMO (2008a) as 'the recommendation, information or advice to an operator aimed at decreasing the likelihood of failures and/or their consequences.' Simple means for identifying the risk of surf-riding and broaching are described in IMO (2007), IMO (2009a), and MCA (2010).

12.8 A Note on Natural Periods

If a linear mass-dashpot-spring system, such as that shown in Figure 12.9b, is excited by a force whose period is close to that of the system, the response amplitude can be very large; we talk about **resonance**. Theoretically, at zero damping the response is unbounded. In practice any physical system is damped to a certain extent and this limits the response to bounded values. Large-amplitude oscillations reduce the performance of the crew and of the equipment and, therefore, they should be avoided.

A very efficient means of avoiding resonance is to ensure that the natural period of the floating body is remote from that of the waves prevailing in the region of operation. In general, it is not possible to change the natural periods of ships because their designs must meet other important requirements. It is possible to change the natural periods of moored platforms, such as those used in offshore technology. To show an example let us refer to Figure 12.9a. The natural period of the undamped and uncoupled heave motion of the shown body is

$$T_{n3} = 2\pi \sqrt{\frac{M + A_{33}}{\rho g A_W}}$$

Let us assume that the floating body is moored as shown in Figure 12.10. The mooring cable is tensioned; it pulls the floating body down increasing its draught beyond the value corresponding to its mass, M. Thus, if we note by V the submerged volume, and by T_c the tension in the cable, we can write

$$\rho g V = g M + T_c$$

If the floating body is an offshore platform, we call it **tension-leg platform**, shortly **TLP**. When the floating body oscillates vertically, the hydrostatic force that opposes the heave motion is that predicted in Figure 12.9. An additional force develops in the cable; its value, according to the theory of elasticity, is

$$\frac{AE}{\ell} \cdot \eta_3 \tag{12.44}$$

where A is the sectional area of the cable, E the Young modulus of the material of the cable, and ℓ the cable length. This second force is usually much larger than the hydrostatic force.

Figure 12.10   A tension-leg floating body

Then, a good approximation of the uncoupled, undamped natural period of heave is

$$T_{n3} = 2\pi \sqrt{\frac{M + A_{33}}{AE/\ell}} \qquad (12.45)$$

and it can differ much from that of the unmoored platform.

Lateral mooring lines act like non-linear springs and can change the periods of other motions. Natural periods can change temporarily when a ship enters confined waters. The added masses are influenced by close vertical walls and by a close bottom. Schneekluth (1988) cites the case of a barge with a B/T ratio equal to 2. When performing the roll test in a depth equal to $1.25T$, the added mass in roll was found to be 2.7 times larger than in deep water. The measured roll period appeared larger than in deep water, leaving the impression that the stability was worse than in reality. Schneekluth appreciates that the added mass in roll, A_{33}, is approximately 15% of the ship mass, M, and that bilge keels increase the added mass by approximately 6%.

12.9 Roll Stabilizers

There are many systems of reducing roll amplitude; their aim is to produce forces whose moment can be added to the righting moment. The simplest and cheapest system is represented by the **bilge keels**; they are steel profiles assembled on part of the ship length, close to the bilge. Bilge keels act in two ways. First, a hydrodynamic resistance force develops on them; it is opposed to the roll motion. Second, bilge keels cause vortexes that increase the viscous damping of the roll motion. As shown in the previous chapters, some codes of stability acknowledge the contribution of bilge keels and provide for corresponding corrections of some requirements. Bilge keels are **passive devices**.

Roll fins are wing-shaped bodies that extend transversely; usually they can be rotated by a control system that receives as input the roll angle, velocity, and acceleration. The forward ship velocity causes hydrodynamic forces on the wings, forces that oppose the roll motion. No helpful forces are produced at low ship speeds. Rudders can be used as active anti-roll devices. Their action is coupled with other motions and influences manoeuvering.

We do not expand on the devices mentioned above, but prefer to concentrate on another possibility because its relation to stability is evident and because it contradicts to some extent the theory that any liquid free surface endangers stability. We mean **anti-roll tanks**. To explain their action we use a simple mechanical analogy. We consider a classical oscillating system composed of a mass, a spring and a dashpot. If a smaller mass is attached to the main mass by a spring, and if the second mass and spring are properly dimensioned, their vibration damps the oscillations of the main mass. This is the principle of the **Frahm vibration absorber**. In a similar mode, if two tanks, one on starboard, the other on the port side, are connected by a pipe, and water flows between them in a certain phase to the roll motion, this cross-flow opposes the roll motion. The main mass-spring-damper system above is the analog of the ship, the small mass-spring system is the analog of the anti-roll tanks.

We consider in Figure 12.11a a system composed of the mass m_1, the linear spring k_1, and the viscous damper (dashpot) c, and an auxiliary system composed of the mass m_2 and the linear spring k_2. A sinusoidal force, $F_0 \sin \omega t$, acts on the main mass, m_1. The position of the mass m_1 is measured by the variable x_1, that of the mass m_2 by the variable x_2. If properly "tuned," the auxiliary system $[k_2, m_2]$, 'absorbs' the forced vibrations of the main system. To show this we first write the equations that govern the behaviour of the composed system. The first Eq. (12.46) describes the forces that act on the mass m_1, and the second equation refers to the forces acting on the mass m_2,

$$m_1 \frac{d^2 x_1}{dt^2} + c \frac{dx_1}{dt} + k_1 x_1 + k_2 (x_1 - x_2) = F_0 \sin \omega t$$

$$m_2 \frac{dx_2}{dt} + k_2 (x_2 - x_1) = 0 \qquad (12.46)$$

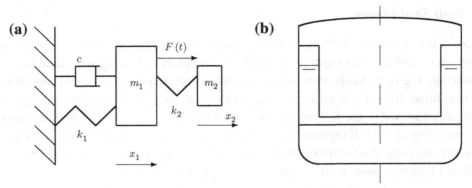

Figure 12.11   (a) A Frahm vibration absorber, (b) flume tanks

We assume that the initial conditions are all zero, that is $x_1 = 0$, $dx_1/dt = 0$, $x_2 = 0$, $dx_2/dt = 0$. Taking Laplace transforms and noting with s the Laplace-transform variable, with $X_1(s)$ the Laplace transform of $x_1(t)$, and with $X_2(s)$ that of x_2, we obtain

$$[m_1 s^2 + cs + k_1 + k_2]X_1(s) - k_2 X_2(s) = \frac{F_0 \omega}{s^2 + \omega^2}$$
$$k_2 X_1(s) - (m_2 s^2 + k_2)X_2(s) = 0 \qquad (12.47)$$

Eliminating $X_2(s)$ from Eq. (12.47) we arrive at

$$X_1(s) = \frac{F_0 \omega}{s^2 + \omega^2} \cdot \frac{m_2 s^2 + k_2}{(m_2 s + k_2)(m_1 s^2 + cs + k_1 + k_2) - k_2^2} \qquad (12.48)$$

Let us choose $k_2/m_2 = \omega^2$, that is we tune the auxiliary system to the exciting frequency ω. Then, the Laplace transform of the amplitude of oscillation of the main mass, m_1, becomes

$$X_1(s) = \frac{F_0 m_2 \omega}{(m_1 s^2 + k_2)(m_1 s^2 + cs + k_1 + k_2) - k_2^2} \qquad (12.49)$$

Churchill (1958) shows that the roots of the denominator (poles) have negative real parts so that the oscillation $x_1(t)$ is damped. A simulation of a system with a Frahm vibration absorber is shown in Example 1.

In Figure 12.11(b) we sketch a section through a ship equipped with **flume tanks**. A transverse pipe connects the two tanks. The flow of water between the two sides can be controlled by throttling the pipe or by acting on the outflow of air above the free surfaces. The water in the flume tanks causes a free-surface effect. Therefore, a trade-off is necessary between the benefits of roll stabilizing (i.e., the reduction of roll amplitude) and the disadvantage of reducing the effective metacentric height.

Figure 12.12 **Brown-NPL passive tank stabilizer: (a) Stern view of ship with passive tank rolled to starboard. The water is moving in the direction shown. (b) Ship rolling to port. The water in the tank on the starboard side provides a moment opposing the roll velocity. (c) Ship at the end of its roll to port. The water is providing no moment to the ship. (d) Ship rolling to starboard. The water in the tank on the port side provides a moment opposing the roll velocity (Reproduced from McGeorge (2002) by courtesy of Butterworth-Heinemann describes the action of a passive tank stabilizer)**

Shimon Lipiner, a friend of the first author, described years ago an experience carried on at the University of Glasgow. Tests on the model of a Ro/Ro ship were meant to show how disastrous the effect of water on the uninterrupted car deck can be. For the particular parameters involved in that experience the observed effect was a reduction instead of an increase of the roll amplitude. The water on deck acted then as a Frahm stabilizer.

Figure 12.12 reproduced from McGeorge (2002) by courtesy of Butterworth-Heinemann describes the action of a passive tank stabilizer.

12.10 Summary

To calculate the motion of a floating body in real waves we need an adequate description of a real sea. Therefore, we consider the real sea as the result of the superposition of a large number of linear waves. The theory of linear waves is based on the following assumptions:

1. the sea water is incompressible;
2. the sea water is inviscid (no viscous effects);
3. surface tension plays no role;
4. no water particle turns around itself (irrotational motion);
5. the wave amplitude is small compared to the wavelength.

The above assumptions allow the development of an elegant theory in which the velocities of water particles can be derived from a velocity potential.

The record of sea elevations in a fixed point is a function of time in which we cannot find any pattern that repeats itself. We can, however, characterize the sea by statistical quantities. One important example is the significant wave height defined as the mean of the highest third of trough-to-crest heights. The heights are measured between trough and crests situated on different sides of the sea level.

Another statistical characteristic of the sea is the wave spectrum, actually the distribution of wave energy as function of the wave frequency. Sea spectra can be measured or can be calculated on the basis of sea characteristics, such as the significant wave height. Formulae for standard spectra have been proposed for various ocean or sea regions.

Floating bodies move in six degrees of freedom. Three motions are linear: surge along the x-axis, sway along the y-axis, and heave along the z-axis, where the axes of coordinates are those defined in Chapter 1. The other three motions are angular: roll around the x-axis, pitch around the y-axis, and yaw around the z-axis.

We can write a differential equation for one particular motion without considering the influence of other motions. We say then that the motion is uncoupled. In reality certain couplings exist between motions. For example, we know from Chapter 2 that roll induces pitch. Moreover, one motion can influence the added masses and damping coefficients of other motions. The most general representation of motions in six degrees of freedom is by a system of six ordinary differential equations. The port-to-starboard symmetry of many floating structures simplifies the matrices of inertia, added masses and damping coefficients and allow the decoupling of equations. Then, for example, we can write a system of three equations for the vertical-plane motions, heave, surge, and pitch, and another system for sway, roll, and yaw.

Moorings can change the natural frequencies of motions. An example is that of tension-leg platforms. As the name says, the mooring 'tendons' are tensioned so that they pull down the platform and increase its draught beyond that corresponding to the platform mass. An elastic force develops in the tensioned tendons; it opposes heave and is much larger than the hydrostatic force developed by the added submerged volume in heave. The natural period in heave is changed so that it is remote from that of the waves prevailing in the region of operation. This principle was applied also to floating cages for offshore fish farming.

Natural periods of ships can change in confined waters because of the proximity of vertical walls and bottom. This effect must be avoided when performing roll tests.

The roll amplitude can be reduced by passive devices, such as bilge keels, or by active devices, such as roll fins. A frequently used roll stabilizer employs two tanks (flume tanks) connected by a transversal pipe. When properly tuned, the cross-flow between the two tanks opposes the roll motion. This is a case in which a free surface helps. However, a trade-off must be done between the good effect on roll and the reduction of effective metacentric height due to the free-surface effect of the water in the flume tanks.

The theory of ship motions is used today in the study of dangerous situations and modes of capsizing that require more tools than those provided by classical ship hydrostatics. Such phenomena include parametric rolling coupled to other motions, surf-riding and broaching. Many of the developed models are non-linear and may also involve the theory of chaotic dynamics. The related researches contribute to the development of second-generation stability criteria and to guidance for ship masters.

12.11 Examples

Example 12.1 (Simulating a Frahm vibration absorber). Let us simulate the behaviour of a system provided with a vibration absorber, such as described in Section 12.9. Dividing both sides of the first Eq. (12.46) by m_1 and both sides of the second equation by m_2, we obtain

$$\ddot{x}_1 + \frac{c}{m_1}\dot{x}_1 + \frac{k_1}{m_1}(x_1 - x_2) = \frac{F_0}{m_1}\sin\omega t$$

$$\ddot{x}_2 + \frac{k_2}{m_2}(x_2 - x_1) = 0 \tag{12.50}$$

We note by $\omega_o^2 = k_1/m_1$ the square of the natural angular frequency of the undamped main system. According to the theory developed in Section 12.9 we set $k_2/m_2 = \omega^2$, that is the square of the exciting frequency. We transform the factor k_2/m_1 as follows:

$$\frac{k_2}{m_1} = \frac{k_2}{m_2}\cdot\frac{m_2}{m_1} = \omega^2\frac{m_2}{m_1}$$

With the above notations we rewrite Eq. (12.50) as

$$\ddot{x}_1 + \frac{c}{m_1}\dot{x}_1 + \omega_0^2 x_1 + \omega^2\frac{m_2}{m_1}(x_1 - x_2) = \frac{F_0}{m_1}\sin\omega t$$

$$\ddot{x}_2 + \omega^2(x_2 - x_1) = 0 \tag{12.51}$$

For numerical integration we must convert the above system of two second-order differential equations into a system of four first-order differential equations. To do so we define the four variables

$y_1 = \dot{x}_1$ the speed of mass m_1
$y_2 = x_1$ the motion of mass m_1
$y_3 = \dot{x}_2$ the speed of mass m_2
$y_4 = x_2$ the motion of mass m_2

Using these notations the system of first-order differential equations becomes

$$\dot{y}_1 = -\frac{c}{m_1}y_1 - \omega_o^2 y_2 - \omega^2 \frac{m_2}{m_1}(y_2 - y_4) - \frac{F_o}{m_1}\sin \omega t$$

$$\dot{y}_2 = y_1$$

$$\dot{y}_3 = -\omega_o^2(y_4 - y_2)$$

$$\dot{y}_4 = y_3 \qquad\qquad (12.52)$$

As shown, for example, in Biran and Breiner (2002), Chapter 14, or Biran (2011), Chapter 9, we write the model as the following function Frahm:

```
%FRAHM Model of a Frahm vibration absorber.
   function yd = Frahm (t, y, rm)
% Input arguments: t time, y variable, rm m2-to-m1 ratio
% meaning of derivatives
% yd(1)      speed of main mass m1
% yd(2)      displacement of main mass m1
% yd(3)      displacement of absorbing mass m2
% yd(4)      displacement of absorbing mass m2
w0 = 2*pi/14.43;        % natural frequency of main system
w = 2*pi/7;             % wave frequency, rad/s
c_m = 0.1;              % damping coefficient, c-to-m1 ratio
F_m = 1;                % exciting amplitude, F-to-m1 ratio
yd = zeros(size(y));    % allocate space for y
% derivatives
yd(1) = -c_m*y(1)- w0^2*y(2) - w^2*rm*(y(2) - y(4)) -
F_m*sin(w*t);
yd(2) = y(1);
yd(3) = -w^2*(y(4) - y(2));
yd(4) = y(3);}
```

The ratio m_2/m_1 appears as an input argument, *rm*. Thus, it is possible to play with the *rm* value and visualize its influence. To call the function Frahm we write a script file, call_Frahm; its beginning may be

```
%CALL_FRAHM Calls ODE23 with Frahm derivatives.
%       Integrates the model of the Frahm damper.
t0 = 0.0; % initial time, s
```

```
tf = 100; % final integration time
y0 = [ 0; 0; 0; 0 ] % initial conditions
% call integration function for system without absorber
[ t, y ] = ode23(@Frahm, [ t0, tf ], y0, [], 0);
subplot(3, 1, 1), plot(t, y(:, 2))
  axis([ 0 100 −5 5 ])
  Ht = text(80, 3.5, 'r_m = 0');
  set(Ht, 'FontSize', 12)
  Ht = title('Displacement of main mass');
  set(Ht, 'FontSize', 14)
% call integration function with mass ratio 1/10
...
```

The results of the simulation are shown in Figure 12.13. The larger the *rm* ratio, the more effective the absorber is. On a ship, however, large flume tanks mean a serious reduction of the effective metacentric height and of the cargo. Hence the need for a trade-off between advantages and disadvantages.

12.12 Exercises

Exercise 12.1 (Potential wave theory). Prove that Eqs. (12.13) and (12.14) fulfil Eq. (12.3).

Exercise 12.2 (Vertical motion). Draw a sketch to prove that the vertical motion of a ship point with coordinates x, y, z is, indeed, given by the equation on page 302. In other words, show that the vector of the vertical motion is the resultant of three vectors produced by heave, roll, and pitch.

Exercise 12.3 (Apparent vertical in beam waves). In Figure 12.14 we show a parallelepipedic floating body in a beam wave of sufficient length so that we can consider that within the body breadth the wave is plane. The figure also shows the distribution of pressures on the submerged sides and bottom. Show that

1. the vertical component of the resultant force is equal to the submerged volume multiplied by the specific gravity of the surrounding liquid;
2. the resultant of the vertical and horizontal components of the hydrostatic force is perpendicular to the liquid surface;
3. considering the hydrostatic reaction of the liquid, the apparent weight of the body is also perpendicular to the liquid surface.

Exercise 12.4 (A Frahm vibration absorber – 1). Referring to Example 12.1, change the value of c_m in function Frahm and study the influence of the damping value.

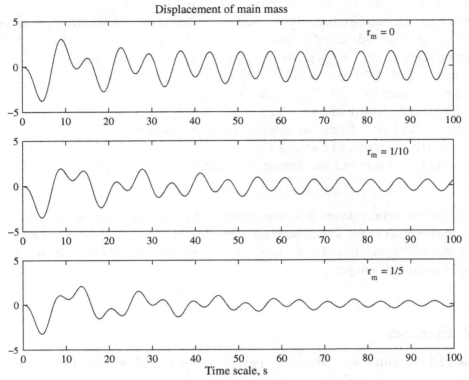

Figure 12.13 The simulation of a Frahm vibration absorber

Figure 12.14 Hydrostatic forces in beam waves

Exercise 12.5 (A Frahm vibration absorber–2). Referring to Example 12.1 modify the file `call_Frahm` so as to plot also the motion of the absorbing mass m_2.

A. Appendix—The Relationship Between Curl and Rotation

In Figure 12.15 we consider an infinitesimal square whose sides are dx and dz. The horizontal speed of the lower left corner is u, and the vertical speed w. Then, the horizontal velocity of the upper left corner is

$$u + \frac{\partial u}{\partial z}dz$$

and the vertical velocity of the lower right corner is

$$w + \frac{\partial w}{\partial x}dx$$

The difference of velocities between the lower left and the lower right corner of the square causes a counter-clockwise rotation around the y-axis with the angular speed

$$\frac{\partial w}{\partial x}$$

The difference of horizontal speeds between the lower left and the upper left corners causes a clockwise rotation with the angular speed around the y-axis

$$\frac{\partial u}{\partial z}$$

The resulting mean angular speed is

$$\frac{1}{2}\left(\frac{\partial w}{\partial x} - \frac{\partial u}{\partial z}\right)$$

Figure 12.15   The relationship between curl and rotational motion

In three-dimensional space the curl of the vector of velocities $[u, v, w]$ is calculated from the determinant

$$\text{curl}([u, v, w]) = \begin{bmatrix} \mathbf{i} & \mathbf{j} & \mathbf{k} \\ \dfrac{\partial}{\partial x} & \dfrac{\partial}{\partial y} & \dfrac{\partial}{\partial z} \\ u & v & w \end{bmatrix} \tag{12.53}$$

where $\mathbf{i}, \mathbf{j}, \mathbf{k}$ are the unit vectors in the x-, y-, and z-directions, respectively. One can see immediately that Eq. (12.2) says that there is no rotation around the y-axis.

The terms corresponding to "curl" in continental Europe are different, for example:

Fr   roteur
G   rotor
I   rotore
S   rotacional

CHAPTER 13

Computer Methods

Chapter Outline

13.1 Introduction

The large amount of multiplications, summations, and integrations required in hydrostatic calculations made necessary a systematic approach and the use of mechanical computing devices. Amsler invented in 1856 the **planimeter**, a mechanical instrument that yields the area enclosed by a given curve. The planimeter is an **analog computer**. Other examples of mechanical, analog computers once widely used in Naval Architecture are the **integraph** and the **integrator**. The integraph draws the integral curve, $\int_{x_0}^{x} f(\xi)\,d\xi$, of a given curve,

$y = f(x)$ (see Section 3.4). The integrator yields the area, the first and second moments of the area bounded by a closed curve. When digital computers appeared, they gradually replaced the mechanical instruments. To our knowledge, the first publication of a digital computer programme for Naval Architecture is due to Kantorowitz (see Kantorowitz, 1955). More programmes for hydrostatic calculations appeared in the following years. Today digital computers are used extensively in modern Naval Architecture and computer programme are commercially available.

With the arrival of computer graphics Naval Architects understood that they can apply the new techniques to solve the problems of hull definition. Today, some of the most sophisticated software packages are used for this purpose. Once the hull surface is defined, the programme uses this definition to perform hydrostatic and other calculations.

In this chapter we discuss concisely a few ways of using computers for the treatment of the subjects described in the book. A detailed treatment would require a dedicated book (for Naval-Architectural graphics see Nowacki et al., 1995). Besides this, computer software changes so rapidly that it would be necessary to update the book at short intervals.

One of the first subjects treated in the book is the definition of the hull surface. It is natural to begin this chapter by showing how computers are used for this definition. To do so we first introduce a few elementary concepts of differential geometry and of computer graphics, and afterwards we give a few simple examples of application to hull-surface definition.

The next subjects discussed in the book are hydrostatic and weight calculations. Correspondingly, we give in this book a few examples of computer implementations of these matters. We end this chapter by explaining what simulation is and give a simple example that uses SIMULINK^R, a powerful toolbox that extends the capabilities of MATLAB^R.

13.2 Geometric Introduction

13.2.1 Parametric Curves

The ellipse shown in Figure 13.1 can be described by the following equation

$$\frac{x^2}{a^2} + \frac{y^2}{b^2} = 1 \tag{13.1}$$

where $2a$ is called **major axis** and $2b$ **minor axis**. In the particular case shown in Figure 13.1, $a = 3$ and $b = 2$. We cannot use this **implicit equation** to draw the curve by means of a computer. We can, however, derive the **explicit equation**

$$y = \pm b\sqrt{1 - (x/a)^2} \tag{13.2}$$

Now, we can draw the ellipse in MATLAB using the following commands

Major axis, 2a = 6, minor axis, 2b = 4

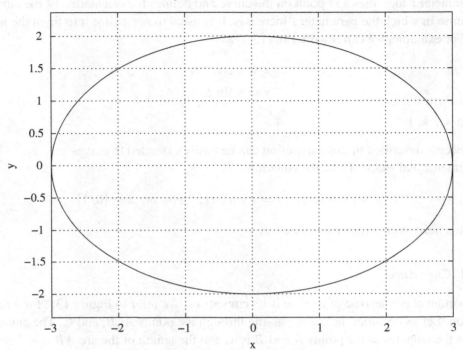

Figure 13.1 The plot of an ellipse

```
a  = 3;  b  = 2;  x = −3: 0.01: 3;
y1  = b*(1 − (x/a).^2).^(1/2);  y2 = -y1;
plot(x, y1, 'k-', x, y2, 'k-'), axis equal
```

There is another way of plotting the ellipse, namely by using a **parametric equation** of the curve. An easy-to-understand example is

$$x = a \cos t$$
$$y = b \sin t \tag{13.3}$$

where t is a **parameter** running from 0 to 2π. We invite the reader to show that Eq. (13.1) can be obtained from Eqs. (13.3). The MATLAB commands that implement Eqs. (13.3) are

```
a = 3; b = 2; t = 0: pi/60: 2*pi;
x = a*cos(t);
y = b*sin(t);
plot (x, y, 'k-'), axis equal
```

The parameter t identifies any point on the curve and defines the orientation of the curve, that is the sense in which the parameter t increases. It is usual to normalize it to lie in the interval [0, 1]. For example, we can rewrite Eq. (13.3) as

$$x = a \cos 2\pi t$$
$$y = b \sin 2\pi t \tag{13.4}$$

where $0 \leq t \leq 1$.

The concepts described in this subsection can be easily extended to curves in three-dimensional space. Thus, the equations

$$x = r \cos 2\pi t, \quad y = r \sin 2\pi t, \quad z = pt, \quad t = [0, 1]$$

describe a helix with radius r and pitch p.

13.2.2 Curvature

An important characteristic of a curve is its **curvature**. We refer to Figure 13.2 for a formal definition. Let us consider the curve passing through the points A, B, and C. The angle between the tangents at the points A and B is α, and the length of the arc AB is s. Then

$$k = \lim_{s \to 0} \frac{\alpha}{s} = \frac{d\alpha}{ds} \tag{13.5}$$

is the curvature at the point A. In words, the curvature is the rate of change of the curve slope.

For a curve defined in the explicit form $y = f(x)$ the curvature is given by

$$k = \frac{\frac{d^2 y}{dx^2}}{\left[1 + \left(\frac{dy}{dx}\right)^2\right]^{3/2}} \tag{13.6}$$

We see that the curvature is directly proportional to the second derivative of y with respect to x. The curvature of a circle with radius r is constant along the whole curve and equal to $1/r$. For other curves the curvature may vary along the curve. The **radius of curvature** is the inverse of curvature, that is $1/k$. A most important example is the metacentric radius, $\overline{BM}$, defined in Section 2.8.2; it is the radius of curvature of the curve of centres of buoyancy.

The curvature has a strong influence on the shape of the curve. Fairing the lines of a ship means in a large measure taking care of curvatures. For a three-dimensional curve we have to define a second quantity, **torsion**, which is a measure of how much it bends outside of a plane. More details can be found in books on differential geometry.

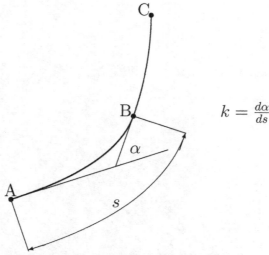

$$k = \frac{d\alpha}{ds}$$

Figure 13.2   The definition of curvature

13.2.3 Splines

In Naval Architecture the term **spline** designs a wood, metal, or plastic strip used to draw the curved lines of the ship. According to the *Webster's Ninth New Collegiate Dictionary*, the origin of the word is unknown and it first appeared in 1756. It can be shown that, when forced to pass through a set of given points, a spline bends so that its shape can be described by a cubic polynomial. According to Schumaker (1981) Schoenberg adopted in 1946 the term **spline functions** to describe a class of functions that approximate the behaviour of 'physical splines.'

Spline functions use polynomials to describe curves. It is easy to calculate, differentiate, or integrate polynomials. On the other hand, it may be difficult to fit a single polynomial to a large number of points. A set of n points defines a polynomial of degree $n - 1$. When $n = 3$ the fitted curve is a parabola that connects the three points without oscillating. For $n = 4$ the curve may show a point of inflection and as n increases the curve may oscillate wildly between the given points. Runge (German, 1856–1927) described the phenomenon of **polynomial inflexibility**; an example in MATLAB is shown in Biran and Breiner (2002, 428–429). The general idea of the spline functions is to solve the problem by subdividing the given set of points into several subsets, to fit a polynomial to each subset, and to ensure certain continuity conditions at the junction of two polynomials. For example, let us suppose that we have to fit a spline over the interval $[x_a, \ x_b]$, and we subdivide it into two at x_i, where, by definition, $x_a < x_i < x_b$. Let $y_1(x)$ be the polynomial fitted over the interval $[x_a, \ x_i]$ and $y_2(x)$ the polynomial fitted over the interval $[x_i, \ x_b]$, Obviously, we impose the condition

$$y_1(x_i) = y_2(x_i)$$

Figure 13.3   Points along a ship station

For slope continuity we also require that

$$\left[\frac{dy_1}{dx}\right]_{x=x_i} = \left[\frac{dy_2}{dx}\right]_{x=x_i}$$

A nicer curve is obtained when the curvature too is continuous, that is

$$\left[\frac{d^2y_1}{dx^2}\right]_{x=x_i} = \left[\frac{d^2y_2}{dx^2}\right]_{x=x_i}$$

Additional conditions can be imposed on the slopes of the curve at the beginning and the end of the interval $[x_a, x_b]$. The set of conditions makes possible the writing of a system of linear equations that yields all the coefficients of the two polynomials. The extension to more subintervals is straightforward.

Let us consider in Figure 13.3 a set of points arranged along a ship station. If the curve passes through all given points, as in Figure 13.4, we say that the curve is an **interpolating spline**. Figure 13.4 was drawn with the MATLAB `spline` function. A detailed explanation of how the MATLAB spline function works can be found in Biran (2011, 157–165). In ship design we may be less interested in passing the curve through all the given points, than in obtaining a fair curve. The fitted curve is then an **approximating spline**. An example obtained with the MATLAB `polyfit` and `polyval` functions is shown in Figure 13.5. In this case the curve is a single cubic polynomial fitted over seven points so that the sum of the squares of deviations is minimal, that is a **least-squares fit**. The two solutions described in this paragraph

Figure 13.4   An interpolating spline

Figure 13.5   An approximating spline

do not allow the user to change interactively the fit; other solutions enable this and they are introduced in the following sections.

13.2.4 Bézier Curves

Working at Citröen, Paul de Faget de Casteljau (French, born 1930, see Bieri and Prautzsch, 1999; De Casteljau, 1999) introduced a kind of curves that were further developed at Renault by Pierre Bézier (French, 1910–1999). These curves, called now **Bézier curves**, are defined by a set of **control points**, B_0, $B_1, \ldots, B_n$, so that the coordinates of any point, $P(t)$, on the curve, are weighted averages of the coordinates of the control points. On the other hand, the coordinates are functions of a parameter $t = [0, 1]$. The curve begins at $t = 0$ and ends at $t = 1$.

The simplest Bézier curve is a straight line that connects the two points

$$\mathbf{B_0} = \begin{bmatrix} x_0 \\ y_0 \end{bmatrix}, \quad \mathbf{B_1} = \begin{bmatrix} x_1 \\ y_1 \end{bmatrix} \tag{13.7}$$

The coordinates of a point on the segment $\overline{B_0 B_1}$ are given as functions of the parameter t

$$\mathbf{P(t)} = \begin{bmatrix} x \\ y \end{bmatrix} = (1 - t)\mathbf{B_0} + t\mathbf{B_1}, \quad t = [0, 1] \tag{13.8}$$

The above equation is in fact a formula for linear interpolation. A second-degree curve is defined by three points, $\mathbf{B_0}, \mathbf{B_1}, \mathbf{B_2}$, and its equation is

$$\mathbf{P}(t) = (1 - t)^2 \mathbf{B_0} + 2(1 - t)t\mathbf{B_1} + t^2\mathbf{B_2} \tag{13.9}$$

It can be shown that Eq. (13.9) describes a parabola.

A cubic Bézier curve is defined by four control points, $\mathbf{B_0}, \ldots, \mathbf{B_3}$, and its equation is

$$\mathbf{P}(t) = (1 - t)^3 \mathbf{B_0} + 3(1 - t)^2 t\mathbf{B_1} + 3(1 - t)t^2 \mathbf{B_2} + t^3 \mathbf{B_3} \tag{13.10}$$

An example is shown in Figure 13.6. We concentrate on cubic polynomials for the simple reason that cubics are the lowest-degree curves that display inflection points. Thus, cubic curves can reproduce the change of curvature sign present in some ship lines. Increasing the degree of polynomials above 3 can cause fluctuations (see above 'polynomial inflexibility') and make computation more complex. In Example 13.1 we give the listing of a MATLAB function that plots a cubic Bézier curve. An interactive example can be found in Biran and Breiner (2002, 255–258).

The following properties of Bézier curves are given here without proof.

Property 1. The curve passes through the first and the last control point only. In Figure 13.6 the curve passes, indeed, through the points $\mathbf{B_0}$ and $\mathbf{B_3}$ only.

Figure 13.6 A cubic Bézier spline

Property 2. The curve is tangent to the first and last segment of the control polygon. In Figure 13.6 the curve is tangent to the segments $\overline{B_0 B_1}$ and $\overline{B_2 B_3}$.

Property 3. The sum of the coefficients that multiply the coordinates of the control points equals 1. In spline theory the functions that produce these coefficients are called **blending** or **basis functions**.

Property 4. Moving one control point influences the shape of the whole curve. Thus, in Figure 13.7 the point B_3 was moved horizontally until it lies on the line $\overline{B_1 B_2}$. We see that the curve eventually becomes a straight line.

As the point B_3 is moved further to the right, a point of inflexion appears as in Figure 13.8.

The property of the tangents at the ends of a Bézier curve allows us to join two Bézier curves so that the continuity of the first derivative is achieved. For example, in Figure 13.9 two Bézier curves are joined at point B_3, while the point B_4 lies on the straight line defined by the points B_2 and B_3.

The general form of a Bézier curve of degree n is

$$\mathbf{P}(t) = \sum_{i=0}^{n} \mathbf{B_i} J_{n,i}(t), \quad 0 \leq t \leq 1 \tag{13.11}$$

Figure 13.7 Another cubic Bézier spline

where the blending function is

$$J_{n,i}(t) = \frac{n!}{i!(n-i)!} t^i (1-t)^{n-i} \qquad (13.12)$$

and $0^0 = 1,\ 0! = 1$. The blending functions of Bézier curves are also known as **Bernstein polynomials**.

More degrees of freedom can be obtained by using **rational Bézier curves** defined by

$$\mathbf{P}(t) = \frac{\sum_{i=0}^{n} \mathbf{B_i} W_i J_{n,i}(t)}{\sum_{i=0}^{n} W_i J_{n,i}(t)} \qquad (13.13)$$

The numbers W_i are called **weights**. We assume that all the weights are positive so that all denominators are positive. The numerator is a vector, while the denominator is a scalar. When all $W_i = 1$, the rational curves become the non-rational Bézier curves described in this subsection. Rational Bézier curves can describe accurately conic sections. The kind of curve depends on the chosen weights. An application of rational Bézier curves to hull-surface design is given by Kouh and Chau (1992). Examples of earlier uses of cubic or rational cubic splines to ship design can be found in Kouh (1987), Ganos (1988), and Söding (1990). Jorde (1997) poses a 'reverse' problem, how to define the ship lines to achieve given sectional area curves and coefficients of form.

Figure 13.8 A third cubic Bézier spline

13.2.5 B-splines

It is easy to calculate points along Bézier curves. On the other hand, moving a control point produces a **global** change of the curve. Another class of more sophisticated curves, the **B-spline curves**, do not have this disadvantage. Moving a point on the latter curves causes only a **local** change, that is a change that affects only the curve segment that neighbours the moved point. We give below the **recursive** definition of a B-spline.

Given $n + 1$ control points, $B_1, \ldots, B_{n+1}$, the position vector is

$$\mathbf{P}(t) = \sum_{i=1}^{n+1} \mathbf{B_i} N_{i,k}(t), \quad t_{min} \leq t \leq t_{max}, \qquad 2 \leq k \leq n+1 \qquad (13.14)$$

Here k is the **order** of the B-spline, and $k - 1$ the degree of the polynomials in t. The basis functions are

$$N_{i,t}(t) = \begin{cases} 1 & \text{if } t_i \leq t \leq t_{i+1} \\ 0 & \text{otherwise} \end{cases} \qquad (13.15)$$

and

$$N_{i,k}(t) = \frac{(t - t_i)N_{i,k-1}(t)}{t_{i+k-1} - t_i} + \frac{(t_{i+k-t} - t)N_{i+1,k-1}(y)}{t_{i+k} - t_{i+1}} \qquad (13.16)$$

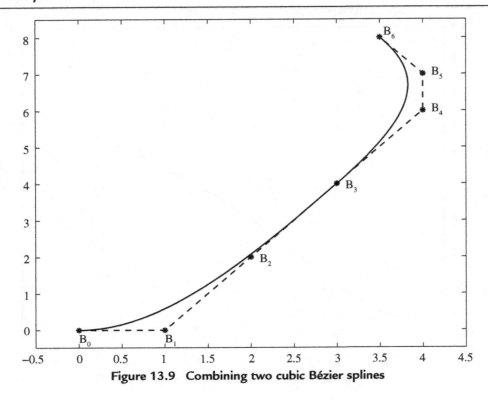

Figure 13.9   Combining two cubic Bézier splines

The set of t_i values is called **knot vector**. If the knot values are not equally spaced the B-spline is called **non-uniform**, otherwise it is called **uniform**. The sum of the **basis functions** is

$$\sum_{i=1}^{n+1} N_{i,k}(t) = 1 \qquad (13.17)$$

for all t.

The calculation of points along B-spline curves requires rather complex algorithms that are beyond the scope of this chapter.

The **NURBS**, or **non-uniform rational B-splines** are an extension of the B-splines; their definition is

$$\mathbf{P}(t) = \frac{\sum_{i=0}^{n+1} \mathbf{B_i} W_i M_{i,k}(t)}{\sum_{i=1}^{n=1} W_i N_{i,k}(t)}$$

As in Eq. (13.13), W_i are the weights. The basis functions, $N_{i,k}$, are defined by Eqs. (13.15) and (13.16). A book on splines that includes historical and biographical notes is that of Rogers (2001).

13.2.6 Parametric Surfaces

Surfaces can be defined by implicit equations such as

$$f(x, y, z) = 0$$

This form is not suitable for computer plots; a helpful form is an explicit equation like

$$z = f(x, y)$$

However, as for curves, the preferred form in computer graphics is a parametric representation of the form

$$x = x(u, w), \quad y = y(u, w), \quad z = z(u, w)$$

Two parameters are sufficient, indeed, to define any point on a given surface. As an example let us consider the upper half of an ellipsoid whose parametric equations are

$$x = a \cos \pi \frac{u}{2} \cos 2\pi w$$

$$y = b \cos \pi \frac{u}{2} \sin 2\pi w$$

$$z = c \sin \pi \frac{u}{2}, \quad u = [0, 1], \quad w = [0, 1] \tag{13.18}$$

When $a = b = c$ the ellipsoid becomes a sphere with centre in the origin of coordinates and radius 1. Then $\pi u/2$ is the analog of what is called in geography **latitude**, and πw is the analog of **longitude**.

Figure 13.10 shows a **wireframe view** of a surface obtained with Eq. (13.18). The curve that bounds the surface at its bottom corresponds to $u = 0$. A net composed of two **isoparametric** curve families is shown. The constant-u curves are marked $u = 0, 0.1, \ldots, 1$. The curve corresponding to $u = 1$ condenses to a single point, the Northern Pole in the case of a sphere.

Ellipsoid $x^2/a^2 + y^2/b^2 + z^2/c^2 = 1$, a = 5, b = 3 c = 2

Figure 13.10 The *u* and *w* nets on a parametric ellipsoidal surface

For the sake of visibility only part of the constant-w curves are marked: $w = 0.5, \ldots, 0.9$. As $\cos 0 = \cos 2\pi$, and $\sin 0 = \sin 2\pi$, the curves $w = 0$ and $w = 1$ coincide.

Figure 13.10 shows that a surface can be described by a net of *isoparametric* curves. One procedure for generating a surface can begin by defining a family of plane curves, for example ship stations, with the help of Bézier curves, non-rational or rational B-splines, or NURBS, with the parameter u. Taking then the points $u = 0$ on all curves we can fit them a spline of the same kind as that used for the first curves. Proceeding in the same manner for the points $u = 0.1, \ldots, u = 1$, we obtain a net of curves.

We explained that plane curves can be properly described by breaking them into spline segments and imposing continuity conditions at the junction points. Similarly, surfaces can be broken into **patches** with continuity conditions at their borders. The expressions that define the patches can be direct extensions of plane-curve equations such as those described in the preceding subsections. For example, a **tensor product Bézier patch** is defined by

$$\mathbf{P}(\mathbf{u}, \mathbf{w}) = \sum_{i=0}^{m} \sum_{j=0}^{n} \mathbf{B_{ij}} J_{i,m}(u) J_{j,n}(w), \quad u = [0, 1], \quad w = [0, 1]$$

where the control points, $\mathbf{B_{ij}}$ define a control polyhedron, and $J_{i,m}(u)$ and $J_{j,n}(w)$ are the basis functions we met in the subsection on Bézier curves. There are more possibilities and they are described in detail in the literature on geometric modelling. Summarized examples of using Maple or the dynamic-geometry programme Cinderella for calculating Bernstein polynomials and drawing Bézier curves and parametric surfaces can be found in Fernández-Jambrina and López-Pulido (2003).

13.2.7 Ruled Surfaces

A particular case is that in which corresponding points on two 3D-space curves are joined by straight-line segments. For example, in Figure 13.11 we consider three of the constant-w curves shown in Figure 13.10. Then, we draw a straight line from a $u = i$ point on the curve $w = 0.6$ to the $u = i$ point on the curve $w = 0.7$, for $i = 0, 0.1, \ldots, 1$. The surface patch bounded by the $w = 0.6$ and the $w = 0.7$ curves is a **ruled surface**. A second ruled-surface patch is shown between the curves $w = 0.7$ and $w = 0.8$. Ruled surfaces are characterized by the fact that it is possible to lay on them straight-line segments.

13.2.8 Surface Curvatures

In Figure 13.12 let $\mathbf{N}$ be the normal vector to the surface at the point P, and $\mathbf{V}$ one of the tangent vectors of the surface at the same point, P. The two vectors, $\mathbf{N}$ and $\mathbf{V}$, define a plane, π_1, normal to the surface. The intersection of the plane π_1 with the given surface is a planar curve, say C. The curvature of the surface at the point P is the **normal curvature of the**

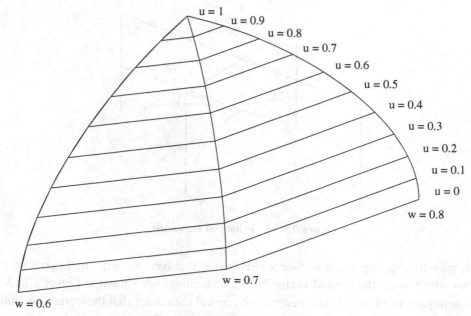

u = 1
u = 0.9
u = 0.8
u = 0.7
u = 0.6
u = 0.5
u = 0.4
u = 0.3
u = 0.2
u = 0.1
u = 0
w = 0.8
w = 0.7
w = 0.6

Figure 13.11 Two ruled surfaces

surface at the point P in the direction of V. We note it by k_n. A theorem due to Euler states that there is a direction, defined by the tangent vector $\mathbf{V_{min}}$, for which the normal curvature, k_{min}, is minimal, and another direction, defined by the tangent vector $\mathbf{V_{max}}$, for which the normal curvature, k_{max}, is maximal. Moreover, the directions $\mathbf{V_{min}}$ and $\mathbf{V_{max}}$ are perpendicular one to another. The curvatures k_{min} and k_{max} are called **principal curvatures**. For example, in Figure 13.12 the planes π_1 and π_2 are perpendicular one to the other and their

Figure 13.12 Normal curvatures

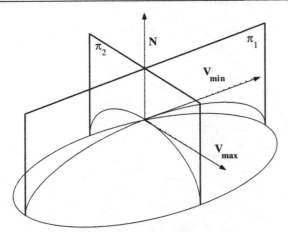

Figure 13.13   Principal curvatures

intersections with the ellipsoidal surface yields curves that have the principal curvatures at the point from which starts the normal vector **N**. The two curves are shown in Figure 13.13. As an illustration important in Naval Architecture we remind the reader that the transverse and the longitudinal metacentric radii are the inverses of the principal curvatures of the surface of the centres of buoyancy, B.

The product of the principal curvatures is known as **Gaussian curvature**:

$$K = k_{min} \cdot k_{max} \tag{13.19}$$

and the mean of the principal curvatures is known as **mean curvature**:

$$H = \frac{k_{min} + k_{max}}{2} \tag{13.20}$$

In Naval-Architecture curvatures are used for checking the fairness of surfaces and the possibilities of developing the hull surface. A surface with zero Gaussian curvature is **developable**. By this term we understand a surface that can be unrolled on a plane surface without stretching. In practical terms, if a patch of the hull surface is developable, that patch can be manufactured by rolling a plate without stretching it. Thus, a developable surface is produced by a simpler and cheaper process than a non-developable surface that requires pressing or forging. A necessary, but not sufficient condition for a surface to be developable is to be a ruled surface. Cylindrical surfaces are developable and so are cone surfaces. The sphere is not developable and this causes problems in mapping the earth surface. Readers interested in a rigorous theory of surface curvatures can refer to Davies and Samuels (1996) and Marsh (2000). The literature on splines and surface modelling is very rich. To the books already cited we would like to add Rogers and Adams (1990), Piegl (1991), Hoschek and Lasser (1993), Farin (1999), Mortenson (1997), and Piegl and Tiller (1997). An article on computer-aided design of developable surfaces is due to Konesky (2005).

13.3 Hull Modelling

13.3.1 Mathematical Ship Lines

De Heere and Bakker (1970) cite Chapman (Fredrik Henrik af Chapman, Swedish Vice-Admiral and Naval Architect, 1721–1808) as having described ship lines as early as 1760 by parabolae of the form

$$y = 1 - x^n$$

and sections by

$$y = 1 - z^n$$

In 1915 David Watson Taylor (American Rear Admiral, 1864–1940) published a work in which he used 5th degree polynomials to describe ship forms. Names of later pioneers are Weinblum, Benson, and Kerwin. More details on the history of mathematical ship lines can be found in De Heere and Bakker (1970), Saunders (1972, Chapter 49), and Nowacki et al. (1995). Kuo (1971) describes the state of the art at the beginning of the 1970s. Present-day Naval-Architectural computer programme use mainly B-splines and NURBS.

13.3.2 Fairing

In Section 1.4.3 we defined the problem of fairing. A major object of the developers of mathematical ship lines was to obtain fair curves. Digital computers enabled a practical approach. Some early methods are briefly described in Kuo (1971), Section 9.3. A programme used for many years by the Danish Ship Research Institute is due to Kantorowitz (1967a,b). Calkins et al. (1989) use one of the first techniques proposed for fairing, namely differences. Their idea is to plot the 1st and the 2nd differences of offsets. In addition, their software allows for the rotation of views and thus greatly facilitates the detection of unfair segments.

As mentioned in Sections 13.2.2 and 13.2.8, plots of the curvature of ship lines can help fairing. Surface-modelling programme, like FORAN, Maxsurf, MultiSurf, and SurfaceWorks (see next subsection), allow to do this in an interactive way. More about curvature and fairing can be read in Wagner et al. (1995), Tuohy et al. (1996), Pigounakis et al. (1996), and Farouki (1998). Rabien (1996) gives some features of the Euklid fairing programme.

13.3.3 Modelling with MultiSurf and SurfaceWorks

In this subsection we are going to describe a few steps of the hull-modelling process performed with the help of an older version of MultiSurf and SurfaceWorks, two products of AeroHydro. We have chosen this software because its relationship to geometric theory is immediately visible and the related documentation is excellent.

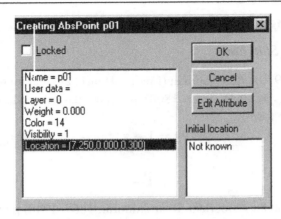

Figure 13.14   MultiSurf, the dialog box for defining an absolute 3D point

The programme described in this subsection are based on a concept developed by John Letcher; he called it **relational geometry** (see Letcher et al., 1995, Mortenson, 1997, Chapter 12, and Letcher, 2009). The idea is to establish a hierarchy of dependencies between the elements that are successively created when defining a surface or a hull surface composed of several surfaces. To model a surface one has to define a set of control, or **supporting** curves. To define a supporting curve, the user has to enter a number of supporting points; they are the control points of the various kinds of curves. Points can be entered giving their absolute coordinates, or the coordinate differences from given, **absolute** points.

Moreover, it is possible to define points constrained to stay on given curves or surfaces. When the position of a supporting point or curve is changed, any **dependent** points, curves or surfaces are automatically updated. Relational geometry considerably simplifies the problems of intersections between surfaces and the modification of lines.

Both MultiSurf and SurfaceWorks use a system of coordinates with the origin in the forward perpendicular, the x-axis positive toward aft, the y-axis positive toward starboard, and the z-axis positive upwards. When opening a new model file, a dialog box allows the user to define an axis or plane of symmetry, and the units. For a ship the plane of symmetry is $y = 0$.

We begin by "creating" a set of points that define a desired curve, for example, a station. Thus, in MultiSurf a first point, $p01$, is created with the help of the dialog box shown in Figure 13.14. The last line is highlighted; it contains the coordinates of the point, $x = 17.250$, $y = 0.000$, $z = 3.000$. There is a quick way of defining a set of points, such as shown in Figure 13.15. In this example all the points are situated along a station; they have in common the value $x = 17.250$ m.

To 'create' the curve defined by the points in Figure 13.15 the user has to select the points and specify the curve kind. A Bcurve (this is the MultiSurf terminology for B-splines) uses the **support** points as a control polygon (see Section 13.2.4), while a Ccurve (MultiSurf

Figure 13.15 MultiSurf, points that define a control curve, in this case a transverse section

Figure 13.16 MultiSurf, a curve that defines a transverse section

terminology for cubic splines) passes through all support points. Figure 13.16 shows the Bcurve defined by the points in Figure 13.15. The display also shows the point in which the curve parameter has the value 0, and the positive direction of this parameter.

Several curves, such as the one shown in Figure 13.16, can be used as support of a surface. To "create" a surface the user selects a set of curves and then, through pull-down menus, the user chooses the surface kind. An example of surface is shown in Figure 13.17. Any point on this surface is defined by the two parameters u and v. The display shows the origin of the parameters, the direction in which the parameter values increase, and a normal vector.

To exemplify a few additional features, we use this time screens of the SurfaceWorks package. In Figure 13.18 we see a set of four points along a station. The window in the lower, left corner of Figure 13.18 contains a list of these points. Figure 13.19 shows the B-spline that

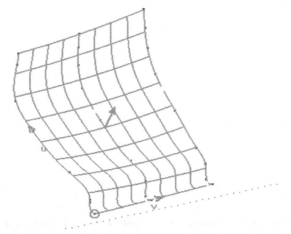

Figure 13.17   MultiSurf, a surface defined by control curves such as those in Figure 13.16

Figure 13.18   SurfaceWorks, points that define a control (supporting) curve for a surface)

uses the points in Figure 13.18 as control points. At full scale it is possible to see that the curve passes only through the first and the last point, but very close to the others. The display shows again the origin and the positive sense of the curve parameter.

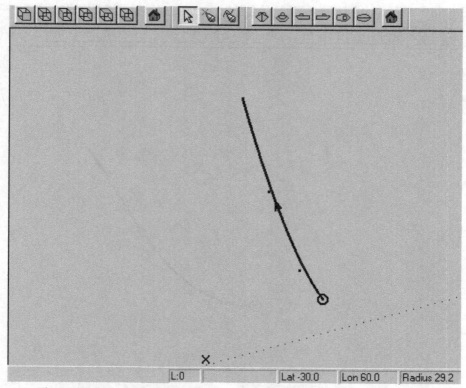

Figure 13.19 SurfaceWorks, a curve that defines a transverse section

Figure 13.19 is an axonometric view of the curve. Figure 13.20 is an orthographic view normal to the x-axis. In Figure 13.21 we see the same station and below it a plot of its curvature. In this case we have a simple third-degree B-spline; the plot of its curvature is smooth. In other cases the curve we are interested in can be a **polyline** composed of several curves. Then, the curvature plot can help in fairing the composed curve.

Usually it is not possible to define a single surface that fits the whole hull of a ship. Then, it is necessary to define several surfaces that can be joined together along common edges. A surface is defined by a set of supporting curves, for example, the bow profile, some transverse curves, etc.

Figure 13.22 shows a wireframe view of a powerboat. The hull surface is composed of the following surfaces: bow round, bulwark, bulwark round, hull, keel forward, keel aft, and transom.

The software enables the user to view the hull from any angle, for example, as in Figure 13.23. Other views can be used to check the appearance and the fairing of the hull. The **rendered**

Figure 13.20   The same curve projected in the direction of the x-axis

view may be very helpful; we do not show an example because it is not interesting in black and white.

Three plots of surface curvature are possible: normal, mean, or Gaussian. We have chosen the plot of normal curvature shown in Figure 13.24.

The `view—offsets` option displays the transverse stations to be used for hydrostatic calculations. The display for our powerboat is shown in Figure 13.25. A drawing of the ship lines produced by the programme is shown in Figure 13.26. A dialog box enables the user to choose the display. For example, Figure 13.26 is produced with the option `Body plan top right, same scale`.

13.4 Modelling with FORAN

As an example of a comprehensive, professional software system for ship design we describe a few highlights of the FORAN system. We have chosen this package as the Spanish

Figure 13.21   The curvature of the same curve

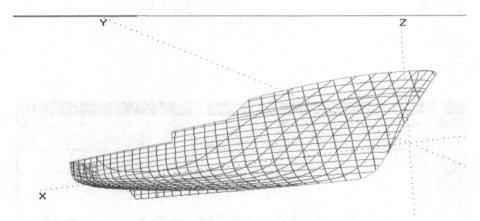

Figure 13.22   The wireframe view of a powerboat

co-authors of this book were involved in some of its developments and are familiar with it. In this example we can see well how present-day, specially developed computer systems cover almost all ship design and building phases. As written in a technical document of the distributing company, 'Everything began in the 1960s with Manuel Sendagorta, then Director of Sener (*Sener Ingenieria y Systemas*[R] all rights reserved), and his interest in the mathematical representation of hull forms of vessels. The research studies by Sendagorta and his collaborators showed that the use of a mathematical formulation to represent ships' hulls, combined with the use of computers, could serve not only to describe existing forms but also to generate new ones.'

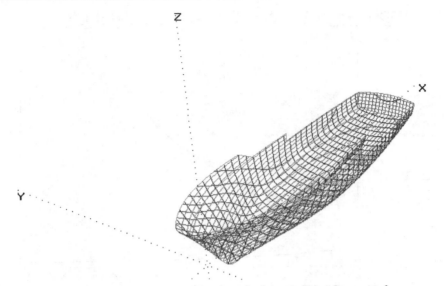

Figure 13.23   Rotating the wireframe view of a powerboat

Figure 13.24   A plot of normal curvatures

Figure 13.25   A plot of the offsets of a powerboat

Figure 13.26   The lines of a powerboat

The FORAN system consists of a large number of modules; those that cover matters treated in this book are briefly described below.

Forms definition, deck, and bulkheads. The module works with NURBS curves and patches. Figure 13.27 is a wireframe view of a forebody. A control polygon is shown in the lower part of one of the sections, and a plot of curvatures appears along the same section.

Figure 13.27   Fairing in FORAN. (© R, TM Sener Ingeniería y Sistemas. All rights reserved.)

For better readability in black-and-white we converted the original figure to its negative. A completed ship model is shown in Figure 13.28.

Hydrostatic calculations. This module calculates hydrostatic data, Bonjean curves, cross-curves of stability, floodable lengths, hydrostatic properties as functions of trim, and the curve of the areas of transverse sections. The latter is known as the curve of *sectional areas* and an example produced in FORAN is shown in Figure 13.29. The programme can perform calculations also in trochoidal or sinusoidal waves.

Space definition and management. The compartments of the ship can be defined as combinations of subspaces. The module allows also the definition of negative spaces, such as tunnels of bow thrusters.

Figure 13.28   A ship model in FORAN. (© R, TM Sener Ingeniería y Sistemas. All rights reserved.)

Space definition, capacities, tonnage. This module calculates the capacities and centres of gravity of loads in holds and tanks, sounding, and ullage tables, i.e., tables of the mentioned properties as functions of heeling height, grain heeling moments, and tonnage. Roughly the latter term means the volume of commercial spaces measured in *tonnes* of 100 cubic feet. FORAN shows the various loads, in a colour code, in a diagram such as that in Figure 13.30.

Loading conditions, stability, and longitudinal strength. This module calculates the results of inclining tests, loading conditions, the stability according to given criteria, and the strength in longitudinal bending.

Damage stability. The module calculates the flooded conditions either by the added weight or the lost buoyancy method, and damage stability by deterministic methods. Intermediate conditions are taken into account.

Subdivision and damage stability. The module calculates the attained subdivison index and the required index according to SOLAS (2009) or former regulations.

Launching and floating.

Powering, design of propeller, stern frame, and rudder.

Further modules are grouped in the following categories:

- Hull structure.
- Build strategy.
- Outfitting design and production.
- Electrical design.
- Draughting.
- Production links.
- Design review.

Figure 13.29 The curve of sectional areas. (© R, TM Sener Ingeniería y Sistemas. All rights reserved.)

- FORAN development tools.
- Database management.
- Change and access control.
- Product life management.

We invite the readers to consult the web addresses www.foran.es and www.sener.es.

Figure 13.30 The loaded spaces of a Ropax ship. (© R, TM Sener Ingeniería y Sistemas. All rights reserved.)

13.5 Recent Developments

The use of computers in ship design is nowadays routine practice. Ship designers make an extensive use of computers as they perform a myriad of calculations when they have to present several variants in the design stage. However, one has to note that for the time being computers are mostly used to work out designs that were once made without computers by carpenters and shipwrights. Computers in ship design are a question of 'how' rather than 'what.' Computer-aided Ship Design, shortly CASD, is today an emancipated, full discipline within the Computer-aided Design (CAD) domain. These matters and a history of their development are described in Nowacki (2010). Some general purpose CAD systems, such as SolidWorks, CATIA, Rhino or Grasshopper, have been used also in CASD. However, these fine examples of software are *solid modellers*, i.e., they work with three-dimensional objects. We have seen that surfaces, among them the hull surface, are two-dimensional objects as the definition of any point lying on them requires two parameters only. Therefore, Naval Architecture requires *surface modelling*. Examples of commercial packages that work on this principle are, in alphabetical order, Autoship, FORAN, Maxsurf, MultiSurf and SurfaceWorks, and NAPA. In the previous section we illustrated a few highlights of the

MultiSurf, SurfaceWorks, and FORAN systems. We invite the readers to consult the web for more Naval-Architectural software. Thus, for example, Maxsurf offers free demos and software for students, and Delft is a free to download programme.

Reviewing the existing software, Rodriguez and Fernández-Jambrina (2012) distinguish three approaches: geometric parametrization, parametrization by transformations, and global, or holistic parametrization. By the end of Chapter 4 we introduce the affine transformations; they conserve the coefficients of form and the relative positions of important points, such as B and F. Naval Architects use also more complex transformations with the aim of changing such properties and important contributions in this domain are due to Lackenby and Söding. Bole and Lee (2006) briefly describe various procedures of designing ship lines and Hefazi (2009) details computational methods for transforming multihull ship lines in order to achieve specified parameters, such as C_p and LCB.

13.6 Calculations Without and With the Computer

Before the era of computers the Naval Architect prepared a documentation that was later used for calculating the data of possible loading cases. The documentation included:

- hydrostatic curves;
- cross-curves of stability;
- capacity tables that contained the filled volumes and centres of gravity of holds and tanks, and the moments of inertia of the free surfaces of tanks.

For a given load case, the Naval Architect, or the ship Master, performed the weight calculations that yielded the displacement and the coordinates of the centre of gravity. The data for holds and tanks were based on the tables of capacity. The next step was to find the draught, the trim, and the height $\overline{KM}$ by interpolating over the hydrostatic curves. Finally, the curve of static stability was calculated and drawn after interpolating over the cross-curves of stability. It is in this way that **stability booklets** were prepared; they contained the calculations and the curves of stability for several pre-planned loadings. The same method was employed by the ship Master for checking if it is possible to transport some unusual cargo.

The above procedure is still followed in many cases, with the difference that the basic documentation is calculated and plotted with the help of digital computers, and the weight and $\overline{GZ}$ calculations are carried on with the aid of hand calculators, possibly with the help of an electronic spreadsheet. However, since the introduction of personal computers and the development of Naval-Architectural software for such computers, it is possible to proceed in a more efficient way. Thus, it is sufficient to store in the computer a description of the hull and of its subdivision into holds and tanks. The model can be completed with a description of the sail area necessary for calculating the wind arm. Then, the user can define a loading case by entering for each hold or tank a measure of its filling, for example, the filling height, and the specific gravity of the cargo. The computer programme calculates almost instantly the parameters of the floating conditions and the characteristics of stability, and it does so without rough approximations and interpolations. For example, in a manual, straightforward trim calculation one has to use the moment to change trim, MCT, read from the hydrostatic curves. Hydrostatic curves are usually calculated for the ship on even keel; therefore, using the MCT value read in them means to assume that this value remains constant within the trim range. Computer calculations, on the other hand, do not need this assumption. The floating condition is found by successive iterations that stop when the conditions of equilibrium are met with a given tolerance.

The ship data stored in the computer constitute a **ship model**; it can be organized as a **database**. In this sense, Biran and Kantorowitz (1986) and Biran et al. (1987) describe the use of relational databases. Johnson et al. (1990), Carnduff and Gray (1992), and Reich (1994) discuss more types of databases.

13.6.1 *Hydrostatic Calculations*

Some hydrostatic calculations are straightforward in the sense that we can perform them in a single iteration. For example, if we want to calculate hydrostatic curves we must perform integrations for a draught T_0, then for a draught T_1, and so on. Chapter 4 shows how to carry on such calculations. Other calculations can be carried on only by iterations. In a first example let us assume that we want to calculate the righting arm of a given ship, for a given displacement volume, ∇_0, and the heel angle ϕ_i. We do not know the draught, T_0, corresponding to the given parameters. We must start with an initial guess, T_{init}, draw the waterline, W_0L_0, corresponding to this draught and the heel angle ϕ_i, and calculate the actual displacement volume. If the guess T_{init} was not based on previous calculations, almost certainly we shall find a displacement volume $\nabla_1 \neq \nabla_0$. If the deviation is larger than an acceptable value, ϵ, we must try another waterplane, W_1L_1, parallel to the initial guess waterline, W_0L_0. This time we proceed in a more "educated" manner. Readers familiar with the Newton-Raphson procedure may readily understand why we use the derivative of the displacement volume with respect to the draught, that is the waterplane area, A_W. We calculate a draught correction

$$\delta T = \frac{\nabla_0 - \nabla_1}{A_W}$$

and we start again with a corrected draught

$$T_1 = T_{init} + \delta T$$

We continue so until the **stopping condition**

$$|\nabla_0 - \nabla_N| \leq \epsilon$$

is met.

A much more difficult, but frequent problem is that of finding the floating condition of a ship for a given loading. The input is composed of the displacement volume and the coordinates of the centre of gravity. The output is the triple of parameters that define the floating condition, that is the draught, the heel, and the trim. To solve the problem we can think of a Newton-like procedure in three variables. Such a procedure implies the calculation of a Jacobian whose elements are nine partial derivatives. Not less difficult is the problem of finding the floating condition of a damaged ship provided the ship can still float. The Naval Architect has to find the draught, trim, and heel for which the conditions described in Section 11.3 are met. In physical terms, the Naval Architect must find the ship position in which the water level in the flooded compartments is the same as that of the surrounding water and the centres of buoyancy and gravity lie on a common vertical. Some details of the above problems can be found in Söding (1978). The calculations of hydrostatic data from surface patches are discussed by Rabien (1985).

Many ingenious methods for solving the above problems have been devised; by elegant procedures they ensured satisfactory precision in reasonable calculating times. The methods based on mechanical computers are particularly interesting. Details can be found in older books. For example, an original publication of a method for calculating lever arms at large heel angles is due to Leparmentier (1899). Other methods for calculating cross-curves of stability are described by Rondeleux (1911), Dankwardt (1957), Attwood and Pengelly (1960), Krappinger (1960), Semionov-Tyan-Shansky (no year given), De Heere and Bakker (1970), Hervieu (1985), Rawson and Tupper (1996). Methods of flooding calculations are explained, for example, in Semionov-Tian-Shansky (no year given) and De Heere and Bakker (1970).

As mentioned, the first publication about a computer programme for Naval-Architectural calculations is that of Kantorovitz in 1955 (see Kantorovitz, 1958); it contains also an analysis of calculation errors. The first computer programmes worked in the **batch mode**; an input had to be submitted to the computer, the computer produced an output. For many years the input was contained in a set of punched cards, later it could be written on a file. An example of such a programme is ARCHIMEDES written at the University of Hannover (see Poulsen, 1980). The input consists of several sequences of numbers. One sequence defines the calculations to be performed, a second sequence describes the hull surface, a third sequence defines the subdivision into compartments and tanks, a fourth the longitudinal distribution of masses, a fifth defines run parameters such as the draught, trim, the wave characteristics, and the identifiers of the compartments to be considered flooded.

The programme ARCHIMEDES could be run for hydrostatic calculations, capacity calculations (compartment and tank volumes, centres of gravity, and free surfaces), cross-curves of stability, damage stability, and longitudinal bending. Many examples in this book were obtained with the ARCHIMEDES programme. A newer version of the software, ARCHIMEDES II, is described by Söding and Tonguc (1989).

Figure 13.31 The MultiSurf dialog box for entering the input for hydrostatic calculations

Recent programmes have a graphic interface that enables the user to build and change interactively the ship model, to define run parameters and run calculations. The output consists of tables and graphs.

Hydrostatic calculations can be performed in MultiSurf or SurfaceWorks after obtaining the offsets (see Figure 13.25). Figure 13.31 shows the dialog box in which the user has to input the height of the centre of gravity, under $Z.c.g$, the draught, under Sink, and the trim and the heel. A rich output is produced; Figure 13.32 shows only a fragment.

Computers have been used in Naval Architecture for performing more calculations, such as preliminary design (defining the main dimensions of the ship), resistance and powering, propeller design, structural design and strength calculations, and seakeeping.

13.7 Onboard Stability Calculators

The Intact Stability Code 2008 of IMO allows to have and use onboard calculators for checking the compliance with the approved stability booklet. It notes, however, that such an instrument, approved by the Administration, is a supplement and not a substitute for the booklet and should be used only to facilitate stability calculations. The document MSC.1/Circ.1229 (see IMO, 2007b) uses the term *stability instrument* for the computer plus

```
 34 stations,   6036 points
       Inputs
Sink            4.00        Spec. Wt.        64.02
Trim, deg.      0.00        Z c.g.           -3.00
Heel, deg.      0.00
      Dimensions
W.L. Length    18.50        W.L. Beam         5.11
W.L. Fwd. X    -1.80        Draft             6.00
W.L. Aft  X    16.70
      Displacement
Volume        681.8         Ctr.Buoy. X      14.68
Displ't.    43653.1         Ctr.Buoy. Y      -0.00
LCB (% w.l.)   89.1         Ctr.Buoy. Z      -3.14
      Waterplane
W.P. Area      12.52        Ctr.Flotn. X      0.33
LCF (% w.l.)   11.5
      Wetted Surface
Wetd.Area     613.29        Ctr. W.S. X      15.30
Ctr. W.S. Z    -3.07
      Lateral Plane
L.P. Area     132.81        Ctr. L.P. X      13.23
Ctr. L.P. Z    -3.43
      Initial Stability
Trans. GM       3.89        Trans.RMPD     2963.2
```

Figure 13.32   A fragment of the output of hydrostatic calculations carried on in Multisurf

the respective software, and specifies tolerances for printouts, with reference to data in the approved stability book or to results produced by an independent computer. In doing so the document distinguishes between various data. Pre-programmed input data include hydrostatic, stability, and compartment data and the tolerances for their printout are equal to zero. The tolerances for the printout of output calculated on the basis of pre-programmed input should be close to zero, but small differences such as those resulting from round-off can be accepted. When the output results from calculations based on a hull-surface model, and not on pre-programmed intermediate data, the tolerances should comply with those listed in a given table. For instance, the tolerance for the displacement is 2%, that for the vertical centre of buoyancy, $\overline{KB}$, and the transverse metacentric height, $\overline{GM}$, is 1% but not larger than 5 cm.

A circular issued by the American classification society, ABS, mentions that according to a SOLAS amendment passenger ships longer than 120 m and built after 1 January 2014 shall 'have onboard stability computers or access to shore-based support for the purpose of providing operational information to the Master for facilitating the safe return to port after a flooding casualty.' As an example of detailed specifications issued by a classification society we cite Det Norske Veritas (see DNV, 2012) that emitted a document for the classification of 'Loading computer systems (LCS) for stability and longitudinal strength.' It is clearly written that 'The loading computer system is regarded as supplementary to the Loading Manual and the Stability Booklet and if relevant the Grain Loading Manual which are always to be provided on board.' The document reproduces the tolerances specified by IMO and details the procedure for the certification of the system.

13.8 Simulations

The term **simulation** is frequently used in modern technical literature. The word derives from the Latin 'simulare,' which means to imitate, pretend, counterfeit. In our context, by simulation we understand computer runs that yield an approximation of the behaviour of a real-life system we are interested in. The steps involved in this activity are described below.

1. The building of a physical model that describes the most important features of the real-life system.
2. The translation of the physical model into a mathematical model. Many mathematical models are composed of ordinary differential equations that describe the evolution of physical quantities as functions of time.
3. The translation of the mathematical model into a computer programme.
4. The running of the computer programme and the output of results.

For several good reasons the physical model cannot describe all features of the real-life system. First, we may not be aware of some details of the phenomenon under study. Next, to use manageable mathematics we must accept simplifying assumptions. Last but not least, we must keep the computation time within reasonable limits and to achieve this we may be forced to accept more simplifying assumptions.

It follows that computer simulations do not exactly reproduce the behaviour of real-life systems; they only 'simulate' part of that behaviour. Better results can be certainly obtained by experiments, especially at full scale. It is easy to imagine that full-scale experiments on ships may be very expensive so that they cannot be carried out frequently. Dangerous experiments that can lead to ship loss are not possible. Such tests can be performed only on reduced-scale models. Still, basin tests too are expensive and their extent is usually limited by the available budget. Simulations may replace dangerous experiments, basin tests can be completed by simulations. Then, part of the possible cases can be simulated, part tested on basin models. The basin tests can be used to correct or validate the computer model.

It is possible to measure the motions of a ship model in a test basin equipped with a wave maker. Then, the motions are recorded as functions of time. It is also possible to simulate ship motions as functions of time, that is to simulate in the **time domain**. However, such measurements or simulations in the time domain have limitations. As explained in the previous chapter, the sea surface is a random process; therefore, ship motions are also random processes. To simulate a given spectrum in the basin or in a computer programme, it is necessary to draw a number of random phases. The resulting motions do not describe all possible situations, but are only an example of such possibilities. We say that we obtain a **realization** of the random process. Moreover, for practical reasons, the duration of a basin test is limited. Then, the time span may not be sufficient for the worst event to happen. Although we may afford simulation times longer than basin tests, they still may be insufficient for obtaining the worst events.

More results can be obtained by calculating motions as functions of frequency, that is calculating in the **frequency domain**. Programmes that perform such calculations are available both through universities and on the market. The software calculates the added masses and damping coefficients, for a series of frequencies, by using potential theory and certain simplifying assumptions. Next, the software calculates the response amplitude operators, RAOs, of various motions or **events**. For a wave frequency component, and given ship heading and speed, the programme calculates the frequency of encounter and transforms the spectra from functions of wave frequency to functions of the frequency of encounter. Response spectra are obtained as products of the spectra of encounter and RAOs. Statistics can be extracted from the spectra, for instance **root mean square**, shortly **RMS** values of the motions.

Taking into consideration the motion of the sea surface, the heave, and the pitch, the programme yields the motion of a deck point relative to the sea surface and calculates the probability of having waves on deck. Other events whose probability can be calculated are slamming and propeller racing, while the motions, velocities and accelerations of given ship points are obtained as combinations of motions in the various degrees of freedom.

An example of ship motions simulated in the time domain can be found in Elsimillawy and Miller (1986). Examples of studies of capsizing in the time domain are in Gawthrop et al. (1988) and de Kat and Paulling (1989). An example of simulation in the frequency domain is given by Kim et al. (1980).

13.8.1 A Simple Example of Roll Simulation

Section 9.4.2 shows how to implement in MATLAB a Mathieu equation and simulate the roll motion produced by parametric excitation. More complicated models can be simulated in a similar manner by writing the governing equations as systems of first-order differential equations and calling an integration routine. The more complex the system becomes, the more difficult it is to proceed in this way. The programmer must write more lines and arrange them in the order in which information must be passed from one programme line to another. Software packages have been written to make simulation easier. The common feature of the various packages is that the programmer does not have to care about the order in which information must be passed. Also, routines and functions frequently used in simulations are available in libraries from which the user can readily call them. The programmer has only to describe the various relationships, the software will detail the equations and arrange them in the required order. In this section we give one very simple example of the capabilities of modern simulation software. As we give in the book examples in MATLAB, it is natural to use here the related simulation toolbox, SIMULINK. Let us consider the following roll equation

$$\Delta i^2 \ddot{\phi} + g\Delta \overline{GZ} = M_H \qquad (13.21)$$

where Δ is the displacement mass, i the mass radius of inertia, $\overline{GZ}$ the righting arm, and M_H a heeling moment. We rewrite Eq. (13.21) as

$$\ddot{\phi} + \frac{g}{i^2}\overline{GZ} = \frac{M_H}{\Delta i^2} \qquad (13.22)$$

In this example we neglect added mass and damping, but use a non-linear function for $\overline{GZ}$ and can accept a variety of heeling moments. To represent this equation in SIMULINK we draw the **block diagram** shown in Figure 13.33 by dragging in blocks taken from the graphical-interface libraries of the software and connecting them by lines that define the relationships between blocks. At the beginning we put two blocks representing heeling moments, M_H. For the wind moment we use a **step function**. Initially the moment is zero, at a given moment it jumps to a prescribed value that remains constant in continuation. For the wave moment we use a sine function, but it is not difficult to input a sum of sines.

The next block to the right is a **switch**; it is used to select one of the heeling moments, M_H. The block called `Heeling arm` performs the division of the heeling moment by the displacement value supplied by the block called `displacement`. Follows a summation point. At this point the value $g\overline{GZ}$ is subtracted from the heeling arm. The output of the summation block is

$$\frac{M_H}{\Delta} - g\overline{GZ}$$

Continuing to the right we find a block that multiplies by $1/i^2$ the output of the summation block; the result is

$$\left(\frac{M_H}{\Delta} - g\overline{GZ}\right)\frac{1}{i^2}$$

We immediately see from Eq. (13.22) that the output of the block called `Product1` is the roll acceleration, $\ddot{\phi}$. This acceleration is the input to an **integrator**. The symbol

$$\frac{1}{s}$$

that marks the integrator block reminds the integration of Laplace transforms. The output of the integrator is the roll velocity, $\dot{\phi}$, in radians per second. The roll velocity is supplied as input to two output blocks. One block, above at right, is an oscilloscope, shortly scope, marked `Phase plane`. The other block, an integrator marked `Integrator 1`, outputs the roll angle, ϕ.

Figure 13.33   Simulating roll in SIMULINK

Following a path to the left, the roll angle becomes the input of a block called `Righting` `arm`. This block contains $\overline{GZ}$ values as functions of ϕ. In a **gain** block the $\overline{GZ}$ value is multiplied by the acceleration of gravity, g, and, at the summation point, the product is subtracted from the heeling arm. Following rightward paths, the roll angle is supplied directly to the scope `Phase plane`, while converted to degrees is input to the scope `Heel angle`. The scope `phase plane` displays the roll velocity versus the roll angle. The scope `angle` displays the roll angle versus time.

13.9 *Summary*

Ship projects require the drawing of lines that cannot be described by simple mathematical expressions, and extensive calculations, mainly iterated integrations. Interesting attempts have been made to use mathematical ship lines, but until the second half of the last century the procedures for drawing and fairing ship lines remained manual. As to calculations, many elegant methods were devised, not a few of them based on mechanical, analog computers, such as planimeters, integrators, and integraphs. As in other engineering fields, in the domain of Naval Architecture the advent of digital computers greatly improved the techniques and made possible important advances. Naval Architects were among the first engineers to use massive computer programmes.

The development of Computer Graphics has made possible the use of computers in the design of hull surfaces. In computer graphics curves are defined parametrically

$$x = x(t), \quad y = y(t), \quad z = z(t)$$

where the parameter, t, is frequently normalized so as to vary from 0 to 1.

The central idea in computer graphics is to define curves by piecewise polynomials. In simple words, the interval over which the whole curve should be defined is subdivided into subintervals, a polynomial is fitted over each subinterval and conditions of continuity are ensured at the junction of any two intervals. The conditions of continuity include the equality of coordinates at the junction point and the equality of the first, possibly also the second derivative at that point. The latter conditions mean continuity of tangent and curvature.

The simplest examples of curves used in Computer Graphics are the Bézier curves. The coordinates of a point on a Bézier curve are weighted means of the coordinates of n control points that form a control polygon. The degree of the polynomial representing the Bézier curve is $n - 1$. An extension of the Bézier curves are the rational Bézier curves; they can describe more curve kinds than the non-rational Bézier curves.

Moving a control point of a Bézier curve produces a general change of the whole curve. B-splines avoid this disadvantage by using a more complicated scheme in which the polynomials change between control points. Moving a control point of a B-spline produces only a local change of the curve. A powerful extension of the B-splines are the non-uniform rational B-splines, shortly NURBS. Computer programmes for ship graphics use mainly B-splines and NURBS.

Naval Architectural calculations involve many integrations. The calculations for hydrostatic curves can be performed straightforward. Other calculations can be carried on only by iterations, for example for finding the cross-curves of stability or the floating condition of a ship for a given loading, possibly also a given damage. Systematic and elegant methods were devised for performing the calculations with acceptable precision, in a reasonable time. Many methods used mechanical, analog computers. When digital computers became available it was possible to write computer programmes that performed the calculations in a faster and more versatile way. The first programmes worked in the batch mode. The input was first introduced on punched cards, later on files. The programme was run and the output printed on paper. Present-day programmes are interactive and graphic user interfaces facilitate the input and yield a better and pleasant output. The interface enables the user to build and change interactively the ship model. This model includes the definitions of the hull surface, of the subdivision into compartments, holds and tanks, the materials in holds and tanks, and the sail area required for the calculation of wind arms.

Another use of computer programmes is in the simulation of the behaviour of ships and other floating structures in waves or after damage. Thus, it is possible to study situations that would be too dangerous to experiment on real ships. Simulations can be carried on in the time domain or in the frequency domain. In the latter approach one input is a sea spectrum, the output consists of spectra of motions and probabilities of events such as deck wetness, slamming or propeller racing. Simulations are used also for studying the stability of ships in the presence of parametric excitation.

When the model used in simulation consists of ordinary differential equations the work can be greatly facilitated by using special simulation software. Then, the user employs a graphical interface to build the model with blocks dragged from libraries. The software produces the governing equations and arranges them in the order required for a correct information flow.

13.10 Examples

Example 13.1 (Cubic Bézier curve).

```
%BEZIER Produces the position vector of a cubic Bezier spline
    function P = Bezier (B0, B1, B2, B3)
% Input arguments are the four control points B0, B1, B2, B3 whose
% coordinates are given in the % format; x; y ]. Output is the
% position vector P with coordinates given in the same format.
```

```
% calculate array of coefficients, in fact
% Bernstein polynomials
t = [ 0: 0.02: 1 ]';                        % parameter
C0 = (1 - t).^3;
C1 = 3*t.*(1 - t).^2;
C2 = 3*t.^2.*(1 - t);
C3 = t.^3;
C = [ C0 C1 C2 C3 ];
% form control polygon and separate coordinates
B = [ B0 B1 B2 B3 ];
xB = B(1, :); yB  = B(2, :)
% calculate points of position vector
xP = C*xB'; yP = C*yB';
P = [ xP'; yP' ]
```

13.11 Exercises

Exercise 13.1 (Parametric ellipse). Write the MATLAB commands that plot an ellipse by means of Eqs. (13.4).

Exercise 13.2 (Bézier curves). Show that the sum of the coefficients in Eq. (13.9) equals 1 for all t values.

Answers

Exercise 2.7. Let the coordinate of the leftmost point be $g = B/2$, and the radius of the circular poop, $R = B/2$. The length of the rectangular part is $\ell = L - g - R - e$. We divide the schematic waterline into three simple geometrical forms:

1. a half of a circle;
2. a rectangle;
3. a triangle.

An Excel sheet that solves the problem is shown in Figure 2.30.

Exercise 2.8. Writing that the displacement mass is equal to the cylinder mass, and with the angle θ defined as in Figure 2.31, we obtain the equation

$$\rho_{water} \left(\pi r^2 - \frac{r^2}{2}(\theta - \sin\theta) \right) L = \rho_{wood}\pi r^2 L \tag{2.86}$$

Equation (2.86) reduces to

$$\theta - \sin\theta = 2\pi \left(1 - \frac{\rho_{wood}}{\rho_{water}} \right) \tag{2.87}$$

We see that the angle θ depends only on the density ratio, ρ_{wood}/ρ_{water}, and not on the dimensions of the cylinder. To solve Eq. (2.87) we use the algorithm described in Biran and Breiner (2002), Section 7.4. From Eq. (2.87) we deduce the *iteration function*

$$F(\theta) = \sin\theta + 2\pi \left(1 - \frac{\rho_{wood}}{\rho_{water}} \right) \tag{2.88}$$

The downloadable MATLAB function `Exer2.m` solves the problem. For the given data the output is

```
Exercise 2.8
INPUT
-----
Water density ........................... 1.025 t/m^3
Wood density ........................... 0.670 t/m^3
Cylinder diameter, B, ................. 0.550 m
Cylinder length ....................... 1.700 m
```

Item	Area	Centroid from		First moments		Own second moments		Steiner term		Total second moments	
		x-axis	y-axis	About x-axis	About y-axis	About x-axis	About y-axis	about x-axis	about y-axis	about x-axis	about y-axis
	m^2	m	m	m^3	m^3	m^4	m4	m4	m4	m4	m4
1	2	3	4	5	6	7	8	9	10	11	12
				2x3	2x4			5x3	6x4	7+9	8+10
1	157.08	0.00	15.76	0.00	2474.93	3926.99	1097.57	0.00	38994.61	3926.99	40092.18
2	1600.00	0.00	60.00	0.00	96000.00	53333.33	853333.33	0.00	5760000.00	53333.33	6613333.33
3	300.00	0.00	110.00	0.00	33000.00	5000.00	15000.00	0.00	3630000.00	5000.00	3645000.00
Total	2057.08	0.00	63.91	0.00	131474.93	62260.32	869430.90	0.00	9428994.61	62260.32	10298425.51

Given		L =	120		B =	20		e =	30
derived		l =	80		g =	10		R =	10

Reduce to centroidal axes						
	Ixx =	62260.32	m^4	Iyy =	1895417.73	m^4

Circumscribed rectangle								
	AreaR =	2400.00	m^2	IxR =	80000	m^4	IyR =	2880000	m^4
	Cwl	0.86						

Figure 2.30 An excel sheet for calculating area properties

Figure 2.31 Properties of a circular segment

```
OUTPUT
------

Subtended angle ....................... 151.780 deg
Waterline breadth ..................... 0.533 m
Draught, T ............................ 0.342 m
Vertical centre of buoyancy, KB ....... 0.194 m
Transverse metacentric radius ......... 0.081 m
Metacentric height .................... −0.000 m
```

Longitudinal metacentric radius ....   0.827 m
Longitudinal metacentric height, GM1 ,..   0.746 m
Block coefficient...... ....   0.851 m
Midship coefficient...... ....   0.851 m
Prismatic coefficient...... ....   1.000 m
Vertical prismatic coefficient..........   0.851 m
CHECKS

Relative error of sectional area ....   −0.002 %

Neutral Equilibrium—Analytic Proof

To simplify the calculations we take as origin, K, the centre of the circle. Then,

$$\overline{KB} = \frac{0 \cdot \pi r^2 - \frac{4r \sin^3 \theta/2}{3(\theta - \sin \theta)} \cdot \frac{r^2(\theta - \sin \theta)}{2}}{\pi r^2 - \frac{r^2}{2}(\theta - \sin \theta)} \tag{2.89}$$

which can be reduced to

$$\overline{KB} = \frac{-\frac{2}{3}\sin^3 \theta/2}{\pi - \frac{\theta - \sin \theta}{2}} r = -\frac{4}{3} \cdot \frac{\sin^3 \theta/2}{2\pi - \theta + \sin \theta} r \tag{2.90}$$

$$\overline{BM} = \frac{\frac{(2r \sin \theta/2)^3 L}{12}}{\pi r^2 - \frac{r^2}{2}(\theta - \sin \theta)L} r = \frac{4}{3} \cdot \frac{\sin^3 \theta/2}{2\pi - \theta + \sin \theta} r \tag{2.91}$$

$$\overline{GM} = \overline{KB} + \overline{BM} - \overline{GM} \tag{2.92}$$

$$= -\frac{4}{3} \cdot \frac{\sin^3 \theta/2}{2\pi - \theta + \sin \theta} r + \frac{4}{3} \cdot \frac{\sin^3 \theta/2}{2\pi - \theta + \sin \theta} r - 0 = 0 \tag{2.93}$$

We conclude that the cylinder floats in a condition of *neutral stability*. It is obvious that this conclusion is independent of the cylinder dimensions. In other words, the condition of neutral stability is a general property of homogeneous cylinders floating in the assumed position. We invite the reader to check what happens to a cylinder floating with its axis in a vertical position.

Neutral Equilibrium—A Geometric Proof

For all heeling angles the distance of the centre of buoyancy, B, from the centre of the circle is constant and equal to

$$y = \frac{4r \sin^3 \frac{\theta}{2}}{3(\theta - \sin \theta)} \tag{2.94}$$

Figure 6.26 Small cargo ship with displaced hanging load

We conclude that the B-curve is a circle with centre coinciding with that of the circular section of the cylinder. Its centre of curvature for all heel angles is the centre of the B-curve circle that is also the centre of gravity. It follows that the metacentre, M, coincides with the centre of gravity, G. This means $\overline{GM} = 0$. The resulting plot is shown in Figure 6.26.

Exercise 2.10. $\alpha = 0.72$, $T = 0.72$ m, $\Delta = 0.40$ t, $\overline{GM} = 0.34$ m.

$$\left(\frac{D}{H}\right)^2 = 4 > 4 \cdot \frac{1-\alpha}{\alpha} = 1.52$$

Downloadable file `ConeDown.m`.

Exercise 2.11.

(a) $\beta = 0.857$, $T = 0.143$ m, $\Delta = 0.064$ t, $\overline{GM} = -0.008$ m.

$$\left(\frac{D}{H}\right)^2 = 0.64 < 4 \cdot \frac{1-\beta}{\beta} = 0.668$$

(b) $\beta = 0.857$, $T = 0.143$ m, $\Delta = 0.081$ t, $\overline{GM} = 0.039$ m.

$$\left(\frac{D}{H}\right)^2 = 0.81 > 4 \cdot \frac{1-\beta}{\beta} = 0.668$$

Downloadable file `ConeUp.m`.

Exercise 6.5.   Downloadable file `Condition6.m`.

Bibliography

Abicht, W. (1971). Die Sicherheit der Schiffe im nachlaufenden unregelmäßigen Seegang. *Schiff und Hafen*, **23**, 12, 988–990.

Abicht, W. (1973). Unterteilung und Lecksicherheit. *Hansa*, **110**, No. 15/16, 1380–1382.

Abicht, W., and Bakenhus, J. (1970). Berechnung der Wahrscheinlichkeit des Überstehens von Verletzungen quer- und längsunterteilter Schiffe, in *Handbuch der Werften*, Vol. X. pp. 38–49

Abicht, W., Kastner, S., and Wendel, K. (1976). *Stability of ships, safety from capsizing, and remarks on subdivision and freeboard*, Rept. 19, Hannover: Technical University Hannover.

Abicht, W., Kastner, S., and Wendel, K. (1977). Stability of ships, safety from capsizing, and remarks on subdivision and freeboard, in *Proceedings of the Second West European Conference on Marine Technology*, 23–27 May, Paper No. 9, pp. 95–118.

ABS (2008). *Assessment of Parametric Roll Resonance in the Design of Container Carriers*. Updating of the 2004 version. Houston, Texas: American Bureau of Shipping.

Aláez Zazurca, J.A. (2004). *Teoria del buque*, Vol. 1. Madrid: Escuela Técnica Superior de Ingenieros Navales.

Anonymous (1872). The stability of the 'Captain', 'Monarch', and some other iron-clades, in *Naval Science* (Reed, E.J., ed). Vol. 1, 26 and following.

Anonymous (1961). Sécurité et compartimentage. *Bulletin du Bureau Veritas*, **45**, No. 5, May, 91.

Anonymous (2005). *Report of investigation into the sinking of M.V. Hui Long on 20 May 2005*, The Hong Kong Special Administration Region Marine Department.

Anonymous (2005a). *SLF48/4/4 – Revision of the intact stability code – Review of the MSC/Circ. 707 to include parametric rolling in head seas*, submitted by Australia and Spain.

Appel, P. (1921). *Traité de Mécanique Rationelle*, 3rd ed, Vol. 3. Paris: Gauthier-Villars et Cie.

Archer, C. *et al.* (2009). Dynamical models of extreme rolling of vessels in head seas, in *Proceedings of the 67th European Study Group Mathematics with Industry*, Wageningen University.

Arndt, B. (1960A). Ermittlung von Mindestwerten für die Stabilität. *Schiffstechnik*, **7**, No. 35, 35–46.

Arndt, B. (1960B). Die stabilitätserprobung des Segelschulschiffs "Gorch Fock". *Schiffstechnik*, **7**, No. 39, 177–190.

Arndt, B. (1962). Einige Berechnungen der Seegangsstabilität. *Schiffstechnik*, **9**, No. 48, 157–160.

Arndt, B. (1964). Systematische Berechnungen der Seegangsstabilität für ein Frachtschiff mit einer Völligkeit von 0,63. *Hansa*, No. 24, 2479–2491.

Arndt, B. (1965). Ausarbeitung einer Stabilitätsvorschrift für die Bundesmarine. *Jahrbuch der Schiffbautechnischen Gesellschaft*, **59**, 594–608. English translation 'Elaboration of a stability regulation for the German Federal Navy'. BSRA translation No. 5052.

Arndt, B. (1968). Schüttgut und Kentersicherheit. *Hansa*, **105**, Nov., 2013–2026.

Arndt, B., and Roden, S. (1958). Stabilität bei vor- und achterlichem Seegang. *Schiffstechnik*, **29**, No. 5, 192–199.

Arndt, B., Kastner, S., and Roden, S. (1960). Die Stabilitätserprobung des Segelschulschiffs "Gorch Fock". *Schiffstechnik*, **7**, No. 39, 177–190.

Arndt, B., Brandl, H., and Vogt, K. (1982). 20 years of experience – Stability regulations of the West-German Navy, in *Second International Conference on Stability of Ships and Ocean Vehicles*, Tokyo, pp. 111–121.

Arscott, F.M. (1964). *Periodic Differential Equations – An Introduction to Mathieu, Lamé and Allied Functions*. Oxford: Pergamon Press.

ASTM (2001). *Guide F1321-92 Standard Guide for Conducting a Stability Test (Lightweight Survey and Inclining Experiment) to Determine the light Ship Displacement and Centers of Gravity of a Vessel*. http://www.astm.org/DATABASE.CART/PAGES/F1321.htm.

Attwood, E.L., and Pengelly, H.S. (1960). *Theoretical Naval Architecture*, new edition expanded by Sims, A.J. London: Longmans.

Banchoff, T., and Lovett, S. (2010). *Differential Geometry of Curves and Surfaces*. Natick, Massachusetts: A.K. Peters.

Belenky, V., and Bassler, C.C. (2010). Procedures for early-stage naval ship design – Evaluation of dynamic stability: Influence of the wave crest. *Naval Engineers Journal*, **122**, No. 2, 93–106.

Belenky, V., Bassler, C.C., and Spyrou, K.J. (2011). Development of second generation intact stability criteria. *NSWCCD-50-TR-2011/06 Hydrodynamics Department Report*, Bethesda: Naval Surface Warfare Center.

Bieri, H.P., and Prautzsch, H. (1999). Preface. *Computer Aided Geometric Design*, **16**, 579–581.

Biran, A.B. (2011). *What Every Engineer Should Know about MATLAB and Simulink*. Boca Raton: CRC Press.

Biran, A., and Breiner, M. (2002). *MATLAB 6 for Engineers*. Harlow, England: Prentice Hall. Previous edition translated into German as: *MATLAB 5 für Ingenieure – Systematische und praktische Einführung* (1999). Bonn: Addison-Wesley. Greek translation 1999, Ekdoseis Tziola. French translation *MATLAB pour l'Ingénieur,* 2e éd. (2009), Pearson Education.

Biran, A., and Kantorowitz, E. (1986). Ship design system integrated around a relational data base, in *CADMO 86* (Keramidas, G.A., and Murthy, T.K.S., eds). Berlin: Springer-Verlag, pp. 85–94.

Biran, A., Kantorowitz, E., and Yanai, J. (1987). A relational data base for naval architecture, in *International Symposium on Advanced Research for Ships and Shipping in the Nineties, CETENA's 25th Anniversary*, S. Margherita Ligure, Italy, 1–3 October.

Birbănescu-Biran, A. (1979). A system's theory approach in Naval Architecture. *International Shipbuilding Progress*, **26**, No. 295, March, 55–60.

Birbănescu-Biran, A., translated and extended by (1982). *User's Guide for the Program System "Arhimedes 76" for hydrostatic calculations*. Technion. See also original, Poulsen (1980).

Birbănescu-Biran, A. (1985). *User's guide for the program STABIL for intact stability of naval vessels*, release 2. Haifa: Technion.

Birbănescu-Biran, A. (1988). Classification systems for ship items: A formal approach and its application. *Marine Technology*, **15**, No. 1, 67–73.

Björkman, A. (1995). On probabilistic damage stability. *The Naval Architect*, Oct., E484-5.

Bole, M., and Lee, B.S. (2006). Integrating parametric hull generation into early stage design. *Schiffstechnik*, **53**, 115–137.

Bonnefille, R. (1992). *Cours d'Hydraulique Maritime*, 3d ed. Paris: Masson.

Borisenko, A.I., and Tarapov, I.E. (1979). *Vector and Tensor Analysis with Applications*, translated and edited by Silverman, R.A. New York: Dover Publications.

Bouteloup, J. (1979). *Vagues, Marées, Courants Marins*. Paris: Presses Universitaires de France.

Bovet, D.M., Johnson, R.E., and Jones, E.L. (1974). Recent Coast Guard research into vessel stability. *Marine Technology*, **11**, No. 4, 329–339.

Brandl, H. (1981). Seegangsstabilität nach der Hebelarmbilanz auf Schiffen der Bundeswehr-Marine. *Hansa*, **118**, No. 20, 1497–1503.

Bulian, G. (2005). Nonlinear parametric rolling in regular waves – A general procedure for the analytical approximation of the GZ curve and its use in time domain simulations. *Ocean Engineering*, **32**, 309–330.

Bulian, G. (2010). Checking vulnerability to pure loss of stability in long crested following waves: A probabilistic approach. *Ocean Engineering*, **37**, 1007–1026.

Burcher, R.K. (1979). The influence of hull shape on transverse stability, in *RINA Spring Meetings*, paper No. 9.

Calkins, D.E., Theodoracatos, V.E., Aguilar, G.D., and Bryant, D.M. (1989). Small craft hull form surface definition in a high-level computer graphics design environment. *SNAME Transactions*, **97**, 85–113.

Cardo, A., Ceschia, M, Francescutto, A., and Nabergoj, R. (1978). Stabilità della nave e movimento di rollio: caso di momento sbandante non variabile. *Tecnica Italiana*, No. 1, 1–9.

Carnduff, T.W., and Gray, W.A. (1992). Object oriented computing techniques in ship design, in *Computer Applications in the Automation of Shipyard Operation and Ship Design IV, 1991*. Amsterdam: Elsevier Science, pp. 301–314.

Cartmell, M. (1990). *Introduction to Linear, Parametric and Nonlinear Vibrations*. London: Chapman and Hall.

Cesari, L. (1971). *Asymptotic Behaviour and Stability Problems in Ordinary Differential Equations*. Berlin: Springer-Verlag.

Chantrel, J.M. (1984). *Instabilités Paramétriques dans le Mouvement des Corps Flottants – Application au cas des bouées de chargement*, Thèse presentée à "Ecole Nationale Supérieure de Mécanique pour l'obtention du diplôme de Docteur-Ingénieur". Paris: Éditions Technip.

Churchill, R.V. (1958). *Operational Mathematics*, 2nd ed. New York: McGraw-Hill Book Company.

Cleary, C., Daidola, J.C., and Reyling, C.J. (1966). Sailing ship intact stability criteria. *Marine Technology*, **33**, No. 3, 218–232.

Comstock, J.P. ed. (1967), *Principles of Naval Architecture*. N.Y.: SNAME.

Costaguta, U.F. (1981). *Fondamenti di Idronautica*. Milano: Ulrico Hoepli.

Cunningham, W.J. (1958). *Introduction to Nonlinear Analysis*. New-York: McGraw-Hill Book Company.

Dahle, A., and Kjærland, O. (1980). The capsizing of M/S Helland-Hansen. The investigation and recommendations for preventing similar accidents. *Norwegian Maritime Research*, No. 3, 2–13.

Dahle, A., and Kjærland, O. (1980). The capsizing of M/S Helland-Hansen. The investigation and recommendations for preventing similar accidents. *The Naval Architect*, No. 3, March, 51–70.

Dahle, A.E., and Myrhaug, D. (1996). Capsize risk of fishing vessels. *Schiffstechnik/Ship Technology Research*, **43**, 164–171.

Dankwardt, E. (1957). Schiffstheorie, in *Schiffbautechnisches Handbuch*, Vol. I, Part 1 (Henschke, W., ed). Berlin: VEB Verlag Technink.

Davies, A., and Samuels, P. (1996). *An Introduction to Computational Geometry for Curves and Surfaces*. Oxford: Clarendon Press.

Deakin, B. (1991). The development of stability standards for U.K. sailing vessels. *The Naval Architect*, Jan., 1–19.

De Casteljau, P. de F. (1999). De Casteljau's autobiography: My time at Citroën. *Computer Aided Geometric Design*, **16**, 583–586.

De Heere, R.F.S., and Bakker, A.R. (1970). *Buoyanct and Stability of Ships*. London: George G. Harrap & Co.

de Juana, J. (2004). *Estudio del balance paramétrico en olas longitudinales*, Arias, C. thesis supervisor, Universidad Politécnica de Madrid, Escuela Técnica Superior de Ingenieros Navales.

de Juana, J., and Garcia, J.L. (2009). *Nuevo marco legislativo internacional de estabilidad de averias – SOLAS 2009*, Madrid, Ministerio de Fomento.

de Juana, J., Arias, C., and Pérez Rojas, L. (2005). On the parametric rolling of fishing vessels, in *International Conference on Marine Research and Transportation*, Naples, Sept. 2005.

de Kat, J.O. (1990). The numerical modeling of ship motions and capsizing in severe seas. *Journal of Ship Research*, **34**, No. 4, Dec., 289–301.

de Kat, J.O., and Paulling, R. (1989). The simulation of ship motions and capsizing in severe seas. *SNAME Transactions*, **97**, 1989, 139–168.

Den Hartog, J.P. (1956). *Mechanical Vibrations*, 4th ed. New York: McGraw-Hill Book Company.

Derrett, D.R., revised by Barrass, C.B. (2000). *Ship Stability for Masters and Mates*, 5th ed. Oxford: Butterworth-Heinemann. The 7th edition published in 2013 with authors Barrass, B., and Derrett, D.R.

Devauchelle, P. (1986). *Dynamique du Navire*. Paris: Masson.

DIN 81209-1 (1999). *Geometrie und Stabilität von Schiffen – Formelzeichen, Benennungen, Definitionen. Teil 1: Allgemeines, Überwasser-Einrumpfschiffe* (Geometry and stability of ships – Symbols for formulae, nomenclature, deffinitions. Part 1: General, surface monohull ships).

DNV (2012). Loading computer systems (LCS) for stability and longitudinal strength, *Rules for classification of ships*, Part 6, Chapter 9, Det Norske Veritas AS.

Dorf, R.C., and Bishop, R.H. (2011). *Modern Control Systems,* 12th ed. Boston: Pearson.

Douglas, J.F., Gasiorek, J.M., and Swaffield, J.A. (1979). *Fluid Mechanics*. London: Pitman Publishing Limited.

Doyère, Ch. (1927). *Théorie du Navire*, Baillière.

Dunwoody, A.B. (1989). *Roll of a ship in astern seas – Metacentric height spectra. Journal of Ship Research*, **33**, No. 3, Sept., 221–228.

EC (2003). Directive 2003/25/EC on specific stability requirements for Ro-Ro passenger ships. *Official Journal of the European Union*.

Elsimillawy, N., and Miller, N.S. (1986). Time simulation of ship motions: A guide to the factors degrading dynamic stability. *SNAME Transactions*, **94**, 215–240.

Erkin, Y. (2006). *Parametric wave excitation of traditional Turkish boats*, MSc thesis, supervisor, Ünsalan/. Izmir: Dukoz Eylül University, Graduate School of Natural and Applied Science.

Faltinsen, O.M. (1993). *Sea Loads on Ships and Offshore Structures*. Cambridge: Cambridge University Press.

FAO (2005). *Development of safety standards for decked fishing vessels of less than 12 m in length and undecked fishing vessels of any size*, IMO document SLF48/16, London: International Maritime Organization.

Farin, G. (1999). *NURBS from Projective Geometry to Practical Use*. Natick, Ma: A.K. Peters.

Farouki, R.T. (1998). On integrating lines of curvature. *Computer Aided Geometric Design*, **15**, 187–192.

Fernández-Jambrina, L., and López-Pulido, R. (2003). E-learning computer-aided geometric design, in *XIII ADM-XV INGEGRAF International Conference on Tools and Methods Evolution in Engineering Design, Cassino, Napoli, Salerno*.

Fernández-Jambrina, L., and López-Pulido, R. (2004). Web course on Computer-Aided Geometric Design, in Lucian and Neamţu, eds., *8th SIAM* Conference on Geometric Design and Computing, Seattle 2003, Brentwood, TN: Nashboro Press, pp. 177–188.

Ferreiro, L.D. (2010). *Ships and Science – The Birth of Naval Architecture in the Scientific Revolution, 1600–1800*. Cambridge, Massachusetts: The MIT Press.

Fog, N.G. (1984). Creative definition of ship hulls using a B-spline surface. *Computer-Aided Design*, **16**, No. 4, July, 225–229.

Fossen, T. (1994). *Guidance and Control of Ocean Vehicles*. Chichester: John Wiley and Sons.

France, W.N. *et al.* (2001). An investigation of head-sea parametric rolling and its influence on container lashing systems, in *SNAME Annual Meeting 2001*.

Francescutto, A. (2002). Roll-heave-sway coupling in beam waves, in *Proceedings of Twelfth International Offshore and Polar Engineering Conference*, Kitakyushu, Japan, pp. 281–287.

Francescutto, A. (2004). *Lezioni di statica della nave – Parte I*. Trieste: Universitá degli Studi di Trieste.

Francescutto, A. (2007). The intact stability code: Present status and future developments, in *ICMRT'07*, Session A, Naples, June 28th and June 30th, pp. 199–208. http://www.icmrt07.unina.it/Proceedings/Papers/A/5tento, G.

Francescutto, A., and Contento, A. (1999). Bifurcations in ship rolling: Experimental results and parameter identification technique. *Ocean Engineering*, **26**, 1095–1123.

Francescutto, A., and Umeda, N. (2010). *Current Status of New Generation Intact Stability Criteria Development*, Paper 1, ISSW 11th International Ship Stability Workshop, Wageningen.

Francescutto, A., and Papanikolaou, A.D. (2011). Buoyancy, stability and subdivision: From Archimedes to SOLAS 2009 and the way ahead. *International Journal of Engineering for the Maritime Environment*.

Furttenbach, Joseph (1968). *Architectura Navalis*, 2nd ed. Hamburg: Germanischer Lloyd.

Ganos, G.G. (1988). *Methodical series of traditional Greek fishing boats*, Dissertation for the degree of Doctor of Engineering, Loukakis, Th. supervisor. National Technical University of Athens, June.

Gawthrop, P.J., Kountzeris, A., and Roberts, J.B. (1988). Parametric excitation of nonlinear ship roll motion from forced roll data. *Journal of Ship Research*, **32**, No. 2, 101–111.

Gilbert, R.R., and Card, J.C. (1990). The new international standard for subdivision and damage stability of dry cargo ships. *Marine Technology*, **27**, No. 2, 117–127.

Godino, G.C. (1956). *Teoria del buque y sus aplicaciones – Estática del buque*, Barcelona: Ed. Gustavo Gili.

Gray, Alfred (1993). *Modern Differential Geometry of Curves and Surfaces*. Boca Raton, Florida: CRC Press.

Grimshaw, R. (1990). *Nonlinear Ordinary Differential Equations*. Oxford: Blackwell Scientific Publications.

Grochowalski, S. (1989). Investigation into the physics of ship capsizing by combined captive and free-running model tests. *SNAME Transactions*, **97**, 169–212.

Hahn, H.G. (1992). *Technische Mechanik*. München-Wien: Carl Hanser Verlag.

Hansen, E.O. (1985). An analytical treatment of the accuracy of the results of the inclining experiment. *Naval Engineers Journal*, **97**, No. 4, May, 97–115, discussion in No. 5, July, 82–84.

Hashimoto, H., Umeda, N., and Matsuda, A. (2011). Broaching prediction of a wave-piercing tumblehome vessel with screws and twin rudders. *Journal of Maritime Science and Technology*, **16**, 441–461.

Hass, C. (2002). *Darstellung des Stabilitätsverhaltens im Seegang*. Technische Universität Hamburg-Harburg.

Hefazi, H. (2009). *Geometric Modeling – Development and Validation of Multidisciplinary Design and Optimization Tools for Multihull Ships*. Long Beach, CA: California State University.

Helas, G. (1982). Stabilität von Schiffen in nachlaufenden Seegang. *Seewirtschaft*, **14**, No. 9, 440–441.

Hendrickson, R. (1997). *Encyclopedia of Word and Phrase Origins*, expanded edition. New York: Facts on File.

Henschke, W., ed (1957). *Schiffbautechnisches Handbuch*, 2nd ed, Vol. 1. Berlin: VEB Verlag Technik.

Hervieu, R. (1985). *Statique du Navire*. Paris: Masson.

Holden, C. (2011). *Modeling and Control of Parametric Resonance,* PhD thesis, Trondheim: Norwegian University of Science and Technology.

Holden, C., Galeazzi, R., Rodrigues, C., *et al.* (2007). Nonlinear container ship model for the study of parametric roll resonance. *Modeling, Identification and Control*, **28**, No. 4, 87–103.

Hoppe, H. (2005). Goal-based standards – A new approach to the international regulation of ship construction. *WMU Journal of Maritime Affairs*, **4**, No. 2, 169–180.

Hoschek, J., and Lasser, D. (1993). *Fundamentals of Computer Aided Geometric Design*, translated by Schumaker, L.L., Wellesley, MA: A.K. Peters.

Hsu, C.S. (1977). On nonlinear parametric excitation problems, in *Advances in Applied Mechanics* (Yih, Chia-Shun, ed), Vol. 17, pp. 245–301.

Hua, J. (1996). A theoretical study of the capsize of the ferry "Herald of Free Enterprise". *International Shipbuilding Progress*, **43**, No. 435, 209–235.

Hua, J., and Wang, W.-H. (2001). Roll motion of a RoRo ship in irregular following waves. *Journal of Marine Science and Technology*, **9**, 38–44.

Ilie, D. (1974). *Teoria Generală a Plutitorilor*. Bucharest: Editura Academiei Republicii Socialiste România.

ILO (2003). *Safety in Numbers. Pointers for Global Safety Culture of Work*. Geneva: International Labour Organization of the UN.

IMO (1991). *Adoption of the international code for the safe carriage of grain in bulk*, MSC.23(59), London: International Maritime Organization.

IMO (1995). *Code on Intact Stability for All Types of Ships Covered by IMO Instruments – Resolution A749(18)*. London: International Maritime Organization.

IMO (1997). *Interim guidelines for the application of formal assessment (FSA) to the IMO rule-making process – MSC/Circ.829, MEPC/Circ.335*, London: International Maritime Organization.

IMO (2004). *Goal-based new ship construction standards*, MSC 78/6/2, London: International Maritime Organization.

IMO (2004a). *Review of the intact stability code – Recordings of head sea parametric rolling on*, SLF 47/INF 5, London: International Maritime Organization.

IMO (2007). *Revised guidance to the master for avoiding dangerous situations in adverse weather and sea conditions*, MSC.1/Circ. 1228, London: International Maritime Organization.

IMO (2007a). *Recommendation on a standard method for evaluating cross-flooding arrangements*, MSC 25(83), London: International Maritime Organization.

IMO (2007b). *Guidelines for the approval of stability instruments*, MSC.1/Circ.1229, London: International Maritime Organization.

IMO (2008). *Explanatory notes to the international code on intact stability, 2008*, MSC.1/Circ.1281, London: International Maritime Organization.

IMO (2008a) *Revision of the intact stability code – Report of the working group*, SLF 51/WP.2, London: International Maritime Organization.

IMO (2009). *International Code on Intact Stability*. London: International Maritime Organization.

IMO (2009a) *Development of new generation intact stability criteria – Comments on surf-riding and broaching criteria*, SLF 52/3/4, London: International Maritime Organization.

IMO (2012). *Casualty statistics and investigation: Loss of life from 2006 to date*, FSI20/INF. 17, London: International Maritime Organization.

INSEAN (1962). *Carene di Pescherecci*, Quaderno n. 1. Roma: INSEAN (Vasca Navale).

INSEAN (1963). *Carene di Petroliere*, Quaderno n. 2. Roma: INSEAN (Vasca Navale).

ISO 7460 (1983). International standard *Shipbuilding – Shiplines – Identification of Geometric Data*.

ISO 7462 (1985). International standard *Shipbuilding – Principal Dimensions – Terminology and Definitions for Computer Applications*, 5th ed, English and French.

ISO 7463 (1990). International standard *Shipbuilding and Marine Structures – Symbols for Computer Applications*.

ITTC (2005). *Predicting the occurrence and magnitude of parametric rolling*, 7.5-0.2 07-04.3, International Towing Tank Conference.

Jakić, K. (1980). A new theory of minimum stability, a comparison with an earlier theory and with existing practice. *International Shipbuilding Progress*, **27**, No. 309, May, 127–132.

Johnson, B., Glinos, N., Anderson, N., *et al.* (1990). Database systems for hull form design. *SNAME Transactions*, **98**, 537–564.

Jons, O.P. (1987). Stability-related guidance for the commercial fisherman. *SNAME Transactions*, **95**, 215–237.

Jordan, D.W., and Smith, P. (1977). *Nonlinear Ordinary Differential Equations*. Oxford: Clarendon Press.

Jorde, J.H. (1997). Mathematics of a body plan. *The Naval Architect*, Jan., 38–41.

Kantorowitz, E. (1958). Calculation of hydrostatic data for ships by means of digital computers. *Ingeniøren International Edition*, No. 2, 21–25.

Kantorowitz, E. (1966). Fairing and mathematical definition of ship surface. *Shipbuilding and Shipping Record*, No. 108, 348–351.

Kantorowitz, E. (1967a). Experience with mathematical fairing of ship surfaces. *Shipping World and Shipbuilder*, **160**, No. 5, 717–720.

Kantorowitz, E. (1967b). *Mathematical definition of ship surfaces*, Report No. DSF-14, Danish Ship Research Institute.

Kastner, S. (1969). Das Kentern von Schiffen in unregelmäßiger längslaufender See. *Schiffstechnik*, **16**, No. 84, 121–132.

Kastner, S. (1970). Hebelkurven in unregelmäßigem Seegang. *Schiffstechnik*, **17**, No. 88, 65–76.

Kastner, S. (1973). Stabilität eines Schiffes im Seegang. *Hansa*, **110**, No. 15/16, 1369–1380.

Kastner, S. (1989). On the accuracy of ship inclining experiments. *Ship Technology Research – Schiffstechnik*, **36**, No. 2, 57–65.

Kauderer, H. (1958). *Nichtlineare Mechanik*. Berlin: Springer-Verlag.

Kehoe, J.W., Brower, K.S., and Meier, H.A. (1979). The Maestrale. *Naval Engineers' Journal*, Oct., 60–62.

Kerwin, J.E. (1955). Notes on rolling in longitudinal waves. *International Shipbuilding Progress*, **2**, No. 16, 597–614.

Kim, C.H., Chou, F.S., and Tien, D. (1980). Motions and hydrodynamic loads of a ship advancing in oblique waves. *SNAME Transactions*, **88**, 225–256.

Kiss, R.K. (1980). Mission analysis and basic design, in *Ship Design and Construction* (Taggart, R., ed). New York: SNAME.

Kleiman, A., and Gottlieb, O. (2011). Local and global bifurcation analysis based design of coupled roll-pitched-heave ship response to head seas, in *ENOC 2011*, Rome.

Kobyliński, L. (2007). System and risk approach to ship safety, with special emphasis of stability. *Archives of Civil and Mechanical Engineering*, **VII**, No. 4, 97–106.

Kobyliński, L.K. (2008). Stability and safety of ships: Holistic and risk approach. *Reliability & Risk Analysis: Theory and Applications*, **1**, 95–105.

Kobyliński, L. (2009). Stability of ships: Risk assessment due hazards created by forces of the sea. *Archives of Civil and Mechanical Engineering*, **VIII**, No. 1, 37–45.

Kobyliński, L.K., and Kastner, S. (2003). Stability and safety of ships: Regulations and operation, in *Elsevier Engineering Book Series*, Vol. 9. Amsterdam: Elsevier.

Konesky, B. (2005). Newer theory and more robust algorithms for computer-aided design of developable surfaces. *Marine Technology*, **42**, No. 2, April, 71–79.

Kouh, J.S. (1987). Darstellung von Schiffoberflächen mit rationalen kubischen splines. *Schiffstechnik*, **34**, 55–75.

Kouh, J.-S., and Chau, S.-W. (1992). Generation of hull surfaces using rational cubic Bézier curves. *Schiffstechnik – Ship Technology Reasearch*, **39**, 134–144.

Krappinger, O. (1960). Schiffstabilität und Trim, in *Handbuch der Werften*. Hamburg: Schiffahrts-Verlag "Hansa" C. Schroedter & Co, pp. 13–82.

Kreyszig, E. (2006). *Advanced Engineering Mathematics*, 9th ed. John Wiley & Sons.

Kröger, P. (1987). *Simulation der Rollbewegung von Schiffen im Seegangg*. Hamburg-Harburg: TUHH.

Krueger, S. (2002). *Performance based stability*, PhD course Stability of Ships, Lyngby: DCAMM – Danish Center for Applied Mathematics and Mechanics.

Krüger, S., Hinrichs, R., and Ihms, M. (2004). Bewertung der Kentersicherlichkeit von Schiffen durch numerissche Simulation. *Jahrbuch der Technischen Gesellschaft*, **98**, 27–33.

Kuo, Ch. (1971). *Computer Methods for Ship Surface Design*. London: Longman.

Kupras, L.K. (1976). Optimisation method and parametric study in precontracted ship design. *International Shipbuilding Progress*, May, 138–155.

Leiceaga, X.A. *et al.* (2007). Surface fairing for ship hull design, *GRAPHICA*, Curitiba, Paraná.

Leparmentier, M. (1899). Nouvelle méthode pour le calcul des carènes inclinées. *Bulletin de l''Association Technique Maritime*, **10**, 45 and following.

Letcher, J.S. Jr. (2009). The geometry of ships, *The Principles of Naval Architecture Series*. Jersey City, NJ: SNAME.

Letcher, J.S., Shook, D.M., and Shepherd, S.G. (1995). Relational geometric synthesis: Part 1 – Framework. *Computer-Aided Design*, **27**, No. 11, 821–832.

Lewis, E.V., ed (1988). *Principles of Naval Architecture – Second Revision*, Vol. I – Stability and Strength. Jersey City, NJ: The Society of Naval Architects and Marine Engineers.

Lindemann, K., and Skomedal, N. (1983). Modern hullforms and parametric excitation of the roll motion. *Norwegian Maritime Research*, No. 2, 2–20.

Little, P.E., and Hutchinson, B.L. (1995). Ro/Ro safety after the Estonia – A report on the activities of the ad hoc panel on Ro/Ro safety. *Marine Technology*, **32**, No. 3, July, 159–163.

Magnus, K. (1965). *Vibrations*. London: Blackie & Son Limited.

Maki, A. *et al.* (2010). Analytical formulae for predicting the surf-riding threshold for a ship in following seas. *Journal of Marine Science and Technology*, **15**, No. 3, 218–229.

Manning, G.C. (1956). *The Theory and Technique of Ship Design*. New York: The Technology Press of M.I.T. and John Wiley & Sons.

Mantari Laureano, J.L. (2010). *Stability of fishing vessels in waves and wind*, Dissertation for MSc. Lisbon: Instituto Superior Técnico.

Maritime and Coastguard Agency (1998). *The Code of Practice for Safety of Small Workboats & Pilot Boats*. London: The Stationery Office.

Maritime and Coastguard Agency (2001). *The Code of Practice for Safety of Large Commercial Sailing & Motor Vessels*. 4th impression. London: The Stationery Office.

Marsh, D. (2000). *Applied Geometry for Computer Graphics and CAD*. London: Springer.

MCA (2010). *Navigation safety: Guidance to the master for avoiding dangerous situations in adverse weather and sea conditions*, MIN 357 (M), London: Dept. of Transport, Maritime and Coastguard Agency.

McGeorge, H.D. (2002). *Marine Auxiliary Systems*. Oxford: Butterworth-Heinemann.

McLachlan, N.W. (1947). *Theory and Application of Mathieu Functions*. Oxford: Clarendon Press.

Meriam, J.R., and Kraige, L.G. (2008). *Engineering Mechanics – Statics*, 6th ed. John Wiley & Sons.

Merriam-Webster (1990). *Webster's Ninth New Collegiate Dictionary*. Springfield, MA: Merriam Webster.

Merriam-Webster (1991). *The Meriam-Webster New Book of Word Histories*. Springfield, MA: Merriam-Webster.

Mirokhin, B.V, Zhinkin, V.B., and Zil'man, G.I. (1989). *Teorya korablya*. Leningrad: Sudostroenie.

MoD (1999a). *Naval Engineering Standard NES 109 – Stability standard for surface ships – Part 1, Conventional ships*, Issue 4.

MoD (1999b). *SSP 24 – Stability of surface ships – Part 1 – Conventional ships*. Issue 2. Abbey Wood, Bristol: Defence Procurement Agency. Unauthorized version circulated for comments.

Molland, A., ed (2008). *The Maritime Engineering Reference Book*. Elsevier, Butterworth-Heinemann.

Moore, C.S. (1967). *Principles of Naval Architecture*, N.Y.: SNAME.

Morrall, A. (1980). The GAUL disaster: An investigation into the loss of a large stern trawler. *Transactions RINA*, 391–440.

Mortenson, M.E. (1997). *Geometric Modeling*. New York: John Wiley and Sons.

Myrhaug, D., and Dahle, E.Aa. (1994). Ship capsize in breaking waves, in *Fluid Structure Interaction in Ocean Engineering* (Chakrabarti, S.K., ed). Southampton: Computational Mechanics Publications, pp. 43–84.

Nayfeh, A.H., and Mook, D.T. (1995). *Nonlinear Oscillations*. New York: John Wiley and Sons.

Nicholson, K. (1975). Some parametric model experiments to investigate broaching-to, in *The Dynamics of Marine Vehicles and Structures in Waves International Symposium* (Bishop, R.E., and Price, W.G., eds). London: The Institution of Mechanical Engineers, Paper 17, pp. 160–166.

Nickum, G. (1988). Subdivision and damage stability, in *Principles of Naval Architecture*, 2nd revision (Lewis, E.V., ed). Vol. 1. Jersey: SNAME, pp. 143–204.

Norby, R. (1962). The stability of coastal vessels. *Transactions on RINA*, **104**, 517–544.

Nowacki, H. (2010). Five decades of computer-aided ship design. *Computer-Aided Design*, **42**, 956–969.

Nowacki, H., Bloor, M.I.G., Oleksiewicz, B., *et al.* (1995). *Computational Geometry for Ships*. Singapore: World Scientific.

Pauling, J.R. (1961). The transverse stability of a ship in a longitudinal seaway. *Journal of Ship Research*, **5**, No. 1, March, 37–49.

Pawlowski, M. (1999). Subdivision of Ro/Ro ships for enhanced safety in the damaged condition. *Marine Technology*, **36**, No. 4, Winter, 194–202.

Payne, S. (1994). Tightening the grip on passenger ship safety: The evolution of SOLAS. *The Naval Architect*, Oct., E482-7.

Pérez, N., and Sanguinetti, C. (1995). Experimental results of parametric resonance phenomenon of roll motion in longitudinal waves for small fishing vessels. *International Shipbuilding Progress*, **42**, No. 431, 221–234.

Pérez, N.A., and Sanguinetti, C.F.O. (2006). Scale model tests of a fishing vessel in roll motion – Parametric resonance, in Abramowski, P. *Sintesis Technólogica*, **3**, No. 1, 33–37.

Pérez-Arribas, F., Suárez-Suárez, J.A., and Fernández-Jambrina, L. (2006). Automatic surface modelling of a ship hull. *Computer-Aided Design*, **38**, 584–594.

Pérez-Arribas, F.L. *et al.* (2008). Parametric generation, modeling and fairing of simple hull lines with the use of nonuniform rational B-Spline surfaces. *Journal of Ship Research*, **52**, No. 1, March, 1–15.

Pérez Rojas, L., and Belenky, V.L. (2005). A review of the 8th International Conference on the Stability of Ships and Ocean Vehicles (STAB 20030). *Marine Technology*, **42**, No. 1, Jan., 21–30.

Piegl, L. (1991). On NURBS: A survey. *IEEE Computer Graphics & Applications*, Jan., 55–71.

Piegl, L.A., and Tiller, W. (1997). *The NURBS Book*, 2nd ed. Berlin: Springer.

Pierrottet, E., (1935). Standards of stability for ships, in *Trans. Institution of Naval Architects,* Vol. 77, pp. 208–222.

Pigounakis, K.G., Sapidis, N.S., and Kaklis, P.D. (1996). Fairing spatial B-splines curves. *Journal of Ship Research*, **40**, No. 4, Dec., 351–367.

Poulsen, I. (1980). *User's manual for the program system ARCHIMEDES 76*, ESS Report No. 36, Hannover: Technische Universität Hannover.

Price, R.I. (1980). Design for transport of liquid and hazardous cargos, in *Ship Design and Construction* (Taggart, R., ed). New York: SNAME, pp. 475–516.

Rabien, U. (1985). Integrating patch models for hydrostatics. *Computer-Aided Geometric Design*, 207–212.

Rabien, U. (1996). Ship geometry modelling. *Schiffstechnik – Ship Technology Research*, **43**, 115–123.

Rahola, J. (1939). The judging of the stability of ships and the determination of the minimum amount of stability especially considering the vessels navigating Finish waters. Ph.D. Thesis, Helsinki: Technical University of Finland.

Rao, K.A.V. (1968). Einfluß der Lecklänge auf den Sicherheitsgrad von Schiffen. *Schiffbautechnik*, **18**, No. 1, 29–31.

Ravn, E.S., Jensen, J.J., Baatrup, J., *et al.* (2002). Robustness of the probabilistic damage stability concept to the degree of details in the subdivision. Lecture notes for the Graduate Course *Stability of Ships* given at the Dept. of Mech. Engineering, Maritime Engineering, of the Technical University of Denmark, Lyngby, 10–18 June.

Rawson, K.J., and Tupper, E.C. (1996). *Basic Ship Theory*, 4th ed, Vol. 1. Harlow, Essex: Longman Scientific & Technical. New edition 2002.

Reed, E.J. ed. (1872), The stability of the 'Captain', 'Monarch', and some other iron-clads, in *Naval Science,* Vol. 1, Apr–Oct, pp. 26 and *ff*.

Reich, Y. (1994). Information Management for Marine Engineering Projects, in *Proceedings of the 25th Israel Conference on Mechanical Engineering*, Technion City, Haifa, May 25–26, pp. 408–410.

RINA (1978). *ITTC Dictionary of Ship Hydrodynamics*. London: The Royal Institution of Naval Architects.

Riola, M., and Pérez, R. (2009). Warship damage stability criteria case study. *Journal of Maritime Research*, **VI**, No. 3, 75–100.

Rodriguez, A., and Fernández-Jambrina, L. (2012). Programmed design of ship forms. *Computer-Aided Design*, **44**, 687–696.

Rogers, D.F. (2001). *An Introduction to NURBS with Historical Perspective*. San Francisco: Morgan Kaufmann Publishers.

Rogers, D.F., and Adams, J.A. (1990). *Mathematical Elements for Computer Graphics*, 2nd ed. New York: McGraw-Hill Publishing Company.

Rondeleux, M. (1911). *Stabilité du Navire en Eau Calme et par Mer Agitée*. Paris: Augustin Challamel.

Rose, G. (1952). *Stabilität und Trim von Seeschiffen*. Leipzig: Fachbuchverlag GMBH.

Ross, C.T.F., Roberts, H.V., and Tighe, R. (1997). Tests on conventional and novel model Ro-Ro ferries. *Marine Technology*, **34**, No. 4, Oct., 233–240.

Rumpl, G., and Sondershausen, H.D. (1994). Strength of materials, in *Handbook of Mechanical Engineering* (Beitz, W., and Küttner, K., eds), English edition. London: Springer Verlag.

Rusås, S. (2002). Stability of ships: Probability of survival. Lecture notes for the Graduate Course *Stability of Ships* given at the Dept. of Mech. Engineering, Maritime Engineering, of the Technical University of Denmark, Lyngby, 10–18 June.

Russel, Ph. (2011). *Theoretical investigation on the container loss of a small Panamax container vessel in heavy sea*. Schriftenreihe Schiffbau. Technische Univeristät Hamburg-Harburg.

Salin, D., and Martin, J. (1997). *La mécanique des fluides*. Paris: Éditions Nathan.

Santos Neves, M.A., Pérez, N., and Lorca, O. (2002). Analysis of roll motion and stability of a fishing vessel in head seas. *Ocean Engineering*, **30**, 921–935.

Sartori, G.A., and Podenzana-Bonvino, C. (1981). *Calcolo delle pressioni fluttuanti agenti sulla carena di una nave in mare ondoso*, Report No. 1150, Cetena.

Saunders, H.E. (1972). *Hydrodynamics in Ship Design*, Vol. 2, 2nd printing of the 1957 edition. New York: SNAME.

Schatz, E. (1983). *User's guide for the program DAMAGE*. Haifa: Techion – Department of Computer Sciences amd Faculty of Mechanical Engineering.

Schneekluth, H. (1980). *Entwerfen von Schiffen*, 2nd ed. Herford: Koehler.

Schneekluth, H. (1988). *Hydromechanik zum Schiffsentwurf*. Herford: Kohler.

Schneekluth, H., and Bertram, V. (1998). *Ship Design for Efficiency & Economy*, 2nd ed. Oxford: Butterworth-Heinemann.

Schumaker, L.L. (1981). *Spline Functions: Basic Theory*. New York: John Wiley and Sons.

Semyonov-Tyan-Shanski, V. (no year indicated). *Statics and Dynamics of the Ships*, translated from the Russian by Konyaeva, M. Moscow: Peace Publishers.

Sheikh, I.A. (2008). *Parametric roll instability of ships*, thesis for the MSc degree, University of Oslo, Faculty of Mathematics and Natural Sciences.

Shin, Y.S. *et al.* (2004). *Criteria for Parametric Roll of Large Containerships in Longitudinal Seas*. SNAME. http://www.eagle.org.news/press/parametricroll/SNAME2004-Paper-26.pdf.

Sjöholm, U., and Kjellberg, A. (1985). RoRo ship hull form: Stability and seakeeping properties. *The Naval Architect*, Jan., E12–E14.

Söding, H. (1978). *Naval Architectural Calculations*, in *WEGEMT 1978* (Buxton, I.L., ed). E2, pp. 29–50.

Söding, H. (1990). Computer handling of ship hull shapes and other surfaces. *Schiffstechnik*, **37**, 85–91.

Söding, H. (2002). Water ingress, down- and cross-flooding. Lecture notes for the Graduate Course *Stability of Ships* given at the Dept. of Mech. Engineering, Maritime Engineering, of the Technical University of Denmark, Lyngby, 10–18 June.

Söding, H., and Tonguc, E. (1989). Archimedes II – A program for evaluating hydrostatics and space utilization in ships and offshore structures. *Schiffstechnik*, **36**, 97–104.

SOLAS (2009). *SOLAS Consolidated Edition 2009 – Consolidated Text of the International Convention for the Safety of Life at Sea, 1974*. London: International Maritime Organization.

Soldà, G. (1964). Une propriété géométrique peu connue des flotteurs cylindriques originairement instables, in *Bulletin de l'ATMA*, Vol. 4. pp. 317–321.

Sonnenschein, R.J., and Yang, Ch. (1993). One-compartment damage survivability versus 1992 IMO probabilistic damage criteria for dry cargo ships. *Marine Technology*, **30**, No. 1, Jan., 3–27.

Spencer, C., and Tilsley, D. (2011). *Bulk cargo liquefaction (iron ore fines and nickel ore)*, in Standard Ore, Feb. http://www.standard-club.com/docs/StandardCargoLiquefactionFeb2011.pdf..

Spyrou, K. (1995). Surf-riding, yaw instability and large heeling of ships in following/quartering waves. *Schiffstechnik/Ship Technology Research*, **42**, 103–112.

Spyrou, K.J. (1996a). Dynamic instability in quartering seas: The behavior of a ship during broaching. *Journal of Ship Research*, **40**, No. 1, March, 46–59.

Spyrou, K.J. (1996b). Dynamic instability in quartering seas – Part II: Analysis of ship roll capsize for broaching. *Journal of Ship Research*, **40**, No. 4, Dec., 326–336.

Stoker, J.J. (1950). *Nonlinear Vibrations*. New York: Interscience Publishers.

Stoker, J.J. (1969). *Differential Geometry*. New York: Wiley Interscience.

Stoot, W.F. (1959). Some aspects of naval architecture in the eighteenth century. *Transactions of the Institution of Naval Architects*, **101**, 31–46.

Storch, R.L. (1978). Alaskan king crab boats. *Marine Technology*, **15**, No. 1, Jan., 75–83.

Struik, D.J. (1961). *Lectures on Classical Differential Geometry*. Reading, MA: Addison-Wesley Publishing Company.

Susbielles, G., and Bratu, Ch. (1981). *Vagues et Ouvrages Pétroliers en Mer*. Paris: Éditions Technip.

Svensen, T.E., and Vassalos, D. (1998). Safety of passenger/Ro-Ro vessels: Lessons learned from the North-West European R&D Project. *Marine Technology*, **35**, No. 4, Oct., 191–199.

Taillé, J. (1975). *Courbes et surfaces*, 4th ed, Que sais-je No. 564. Paris: Presses Universitaires de France.

Talib, A., and Poddar, P. (1980). *User's manual for the program system ARCHIMEDES 76*, translated form the original of Poulsen, ESS Report No. 36, Technical University of Hannover.

The New Encyclopedia Britannica (1989). Vol. 18. Chicago: Encyclopedia Britannica.

Thuillier, P., Belloc, J.C., and de Villèle, A. (1991). *Mathématiques – géométrie différentielle*, 2nd ed. Paris: Masson.

Tuite, A.J., and Renilson, M.R. (1997). The effect of principal parameters on broaching-to of a fishing vessel in following seas. *RINA Transactions Spring Meetings Journal*, Report AME CRC C 98/12.

Tuohy, S., Latorre, R., and Munchmeyer, F. (1996). Developments in surface fairing procedures. *International Shipbuilding Progress*, **43**, No. 436, 281–313.

Umeda, N., and Peters, A. (2002). Recent research progress on intact stability in following/quartering waves, in *Proceedings of 6th International Ship Stability Workshop*, Webb Institute.

Ünsalan, D. (2006). Contribution to "An investigation into the loss of the steel bark Admiral Karpfanger ex L'Avenir. *Marine Technology*, **43**, No. 1, 40–41.

van Harpen, N.T. (1971). Eisen te stellen aan de stabiliteit en het reserve-drijfvermogen van bovenwaterscheppen der Koninklijke Marine en het Loodswezen. *S. en W.*, **38**, No. 4, 1972.

Wagner, P.H., Luo, X., and Stelson, K.A. (1995). Smoothing curvature and torsion with spring splines. *Computer-Aided Design*, **27**, No. 8, Aug., 615–626.

Watson, D.G. (1998). *Practical Ship Design*. Amsterdam: Elsevier.

Wawrzyńczyk, K.A. (1996). *Geometría de curvas y superficies*, Barcelona: Anthropos, México: Universidad Autónoma Metropolitan-Iztapalapa.

Wegner, U. (1965). Untersuchungen und Überlegungen zur Hebelarmbilanz. *Hansa*, **102**, No. 22, 2085–2096.

Wendel, K. (1958). Sicherheit gegen Kentern. *VDI-Zeitschrift*, **100**, Vol. 32, 1523–1533.

Wendel, K. (1960a). Die Wahrscheinlichkeit des Überstehens von Verletzungen. *Schiffstechnik*, **7**, No. 36, 47–61.

Wendel, K. (1960b). Safety from capsizing, in *Fishing Boats of the World: 2* (Traung, J.O, ed). London: Fishing News (Books), pp. 496–504.

Wendel, K. (1965). Bemessung und Überwachung der Stabilität. *Jahrbuch der Schiffbautechnischen Gesellschaft*, **59**, 609–627.

Wendel, K. (1970). Unterteilung von Schiffen, in *Handbuch der Werften*, Vol. X, pp. 17–37.

Wendel, K. (1977). Die Bewertung von Unterteilungen, in *Zeitschrift der Technischen Universität Hannover*, Volume published at 25 years of existence of the Dept. of Ship Technique, pp. 5–23.

Zigelman, D., and Ganoni, I. (1985). *Frigate seakeeping – A comparison between results obtained with two computer programs*. Haifa: Technion – Department of Computer Sciences and Faculty of Mechanical Engineering.

Ziha, K. (2002). Displacement of a deflected hull. *Marine Technology*, **39**, No. 1, Jan., 54–61.

Zucker, S. (2000). *Theoretical analysis for parametric roll resonance in trimaran*, MSc work, University College of London.

Index in English

Page numbers in italics refer to tables and figures.

Index in French

A

Aire
   de la surface de la flottaison, *18*
   du couple milieu, *18*
Angle de bande, de gîte, *18*
Arrière, *5*
Assiette, *5*

B

Bâbord, *5*
Bande, *5*
Bouge, *4*
Bras de levier, *124*

C

Caréne, 9
Carènes isoclines, 106
Cavalement, *302*
Centre
   de carène, *35, 124*
   de gravité, *124*
   de gravité de la flottaison, 44
Cloisons étanches, 262
Coefficient
   de bloc, *18*
   de remplissage, *18*
     au maître couple, *18*
     de la flottaison, *18*
     vertical, *18*
     prismatique, *18*
Couple, *5*
Courbe
   de stabilité, *124*

des longueurs envahissables, 285
Creux, *4*

D

Déplacement, *18*
Développée métacentrique, 49
Distance du centre de gravité à la
   ligne d'eau zero, *124*

E

Embardée, *302*
Expérience de stabilité, 179–83

F

Flottaison normale, *4*
Franc-bord, *5*

G

Gîte, *5*

H

Hors membres, *5*

I

Isocarènes, 42, 117

L

Lacet, *302*
Largueur, *4*
Ligne
   de base, *4*
   d'eau, *5*
   de flottaison en charge, *5*

de surimmersion, 262
Longueur
   à la flottaison, *5*
   entre perpendiculaires, *5*
   envahissable, 262
   hors tout, *5*
   hors tout immeregé, *5*

M

Méthode des carènes perdues,
   267–8
Méthode par addition de poids,
   268–71

P

Pantocarènes, *124*
Perméabilité, 264
Perpendiculaire, *4*
Pilonnement, *302*
Plan
   des formes, *5*
   longitudinal de symétrie, *4*
Point le plus bas de la carène, *124*
Pont de cloisonnement, 262
Poupe, *5*
Profondeur de carène, *4*
Proue, *4*

R

Roteur, 319–20
roulis, *302*
   sur houle, 302–5

Index in German

Index in Italian

Index in Spanish

Printed in the United States
By Bookmasters